Testing of Agriculture Technological Processes

Testing of Agriculture Technological Processes

A Systems Approach

S.V. Kardashevskii
L.V. Pogorelyi
G.M. Fudiman
P.I. Lobko
and
V.V. Brei

United Book Prints

First reprint in India, 2017

Published by:
United Book Prints
4806/24, Ansari Road,
Daryaganj,
New Delhi - 110 002
Tel.: 011-41511055
E-mail: unitedbookprints@gmail.com

Print: 2017

ISBN: 978-93-83692-74-3

Printed in India

UDC 631.17.001.4

The authors of the present book have examined organizational and technological principles and scientific methods of testing agricultural technological processes. They discuss in detail modern methods of agricultural technological evaluation, power rating, operational and technological evaluation and economic estimation of agricultural machines and complexes. They also describe methods for estimating the reliability of agricultural machines and the working conditions for machine operators.

The authors have introduced the fundamental principles of systems analysis and methods for forecasting the effectiveness of an agricultural technological process in the testing stage.

The book will be useful for engineers and technicians involved in the research and development, and testing of agricultural technological processes.

Foreword

Accelerated development, the application of new agricultural technological processes and the transition of agricultural production to an industrial base are ensured to a marked degree only because of the improvements and raising of the technical level of testing. This is characterized by (1) the development of methods and techniques of experimentation, (2) the application of mathematical models, (3) the use of modern instruments and equipment by the testing centers and machine-testing units of the State Committee on Agricultural Technology, as well as by extensive use of computer techniques. The technical skill of specialists has considerably increased in recent years, with training for professional testers of agricultural technological processes having begun in higher educational institutions of the country. All-Union technical conferences on testing methods for agricultural technological processes are conducted regularly. Collections of articles under the title 'New Developments in Testing Methods' (*Novoe v metodakh ispytanii*) are continuously being published, with the number of publications concerning this subject steadily increasing.

The present technical level of testing of agricultural machines is determined considerably by the extent of development of and improvement in the generally accepted methods and scientific principles of investigations. The progress in modern methodology and the practice of theoretical and experimental investigations in the field of agricultural mechanics and agricultural production processes are the result of the efforts of a large number of scientists, beginning with Academician V.P. Goryachkin, the founder of agricultural mechanics. The creation of a scientific basis under the auspices of the State Committee for Agricultural Technology has, in the last decade, clearly been reflected in industry standards which are already implemented or are being evolved. Such a scientific basis has become a major step in perfecting testing techniques.

Methods of investigations and testing have been integrated, on the basis of a modern systems approach, into an independent discipline and have, therefore, acquired a wider interdisciplinary character.

In view of the aforementioned, it becomes necessary to standardize the experience accumulated in the field of perfecting testing techniques. There is also a need to systematize scientific achievements and to determine trends in the development of a theory of testing. Such a theory plays an important role in controlling machine quality. The works of V. Koshak, G.M. Kukta, A.A. Vysotskii, V.G. Georgievskii and others have provided a definite basis for the comprehensive examination of testing problems on the whole. At the time of preparation of the present book, it became clear that within the framework of a limited volume, it would only be possible to deal with the most general principles of testing, proposed on the basis of the so-called types of estimations. For this reason, it became necessary to avoid a description of particular testing methods of individual machines, and systems of instruments, and to briefly discuss testing for reliability.

This book is written by a group of authors who have, over many years, collaborated in examining and developing testing methods of agricultural technological processes.

The authors have tried to include information of a fundamental nature in the book. Within the framework of this volume, they have used only those sources which, in their opinion, permit the description of the concepts, methods and results which characterize the contemporary technical level of testing of agricultural technological processes. Specialists in the State Committee on Agricultural Technology of the USSR, namely, L.A. Korbut, N.A. Stolbushkin, V.K. Fribus and others have provided invaluable assistance and support to the authors.

Contents

Introduction

An all-round progress of complex mechanization and automation of agricultural production is determined by the development and application of new tractor designs, agricultural machines and by the introduction of technical means for mechanization of processes in stock-breeding.

Efforts directed toward the general speeding up of technical progress in agricultural production have become the main theme of the Communist Party, which has been adopted throughout the years of the existence of the Soviet government as a Soviet State policy. Cutting down the period for developing new technology and reduction in the costs of introducing the achievements of science into the national economy appear as the basic elements in hastening technical progress.

An enormous potential for accelerating the development of economic production is inherent to technological growth. It is estimated that as a result of technical progress, production increased by about 2% per year. If it would be possible to reduce the time lapse between the stage at which the concept gets formulated and its implementation for production purposes from 5-8 or 8-12 years to 1-4 or 4-8 years, i.e. by 4-6 years, then this would provide an increase in productivity by 4% to 8% per year. Such enormous reserves are made available by speeding up the formulation, development and implementation of scientific and technical ideas in agricultural production, with testing holding a significant role for these processes.

The development of a technical base for agriculture does not only lead to the quantitative growth of the means of production, but also proposes deep-rooted qualitative as well as quantitative changes in the technology and in the very character of the production process.

The development of systems of machines, which ensures comprehensive mechanization of production processes in the chief branches of agriculture, forms an integral part of the type of qualitative improvements mentioned here; these include (1) increase in power, increase in working speeds and (2) in larger power capacity of tractors and machines; (3) increase in the productivity of harvesters; (4) construction of wide-span and multiserial harvest combines; (5) construction of machines intended for new technological processes (rotary soil-processing implements, combined aggregates and others); (6) application of machines and equipment without destroying the nutritious value of fodder even on their preservation; (7) accelerated drying, briquetting and granulating; (8) development of a completed complex of machines for the production of stock-breeding products on an industrial basis; (9) a wider use of electricity and the means of automation of different processes, etc.

It is of utmost importance to accelerate transition from mechanization involving individual processes in agriculture and stock-breeding to complex mechanization and automation.

Complex mechanization includes two mutually complementary aspects: firstly, development of diverse technical methods enabling the transition from manual to mechanized labour in all technical operations and, secondly, achievement of a level of production technology such that it permits the creation of an optimal machine stock in collective and state farms within a short span of time and, in practice, to accomplish the complex mechanization of agriculture, stock-breeding and other branches of production.

An enhanced tempo in the designing and adoption of new agricultural technological processes is a fundamental requirement to solving the problems of complex mechanization.

The difficulty in achieving the proposed target does not lie only in the huge volume of design work. The chief difficulty lies in the fact that more and heavier operations (such as harvesting of vegetables, fruits, grapes; automated fodder distribution, milking, etc.) considered difficult from the point of view of replacement by mechanized technology, come under the purview of unmechanized labor. This calls for the necessity to develop new avenues in the scientific investigations in the area of mechanization, the organization of collaboration among engineers, agronomists, veterinarians and specialists in selective breeding with a view to working out new technological decisions, the creation of new varieties of cultivated crops largely meeting the requirements of mechanized processes and a wider use of the latest findings in physics, chemistry and applied sciences for designing in agricultural technology.

A wide system of industrial, departmental and state testing, which is required to provide timely completion, overall control and selection of the most prospective designs of machines and complexes for production, plays a very important role in the design and application of new agricultural technological processes [32-34, 42, 59].

Over 1500 technical processes for mechanization of plant-breeding, more than 400 machines and implements for mechanization of stock-breeding, above 500 machines for mechanization of the work on soil improvement, about 200 machines and fixtures for silviculture are included in the system of machines for the comprehensive mechanization of agricultural production for the years 1976 to 1985. All these machines must undergo preliminary (industrial) and acceptance (state) tests.

If it is considered that most of the machines used in production need modernization with subsequent complete testing, then the development and improvement of new test methods as well as the working out of progressive and, at the same time, sufficient universal means for the operational and reliable estimation of machine efficiency appears to be of prime importance.

According to the current norms, a machine is brought to the working condition and technically required quality level by means of successive finishing operations during industrial and departmental testing. In doing this, it is also considered worthwhile to bring out the prototypes of machines for state testing. The objective of state testing is multifaceted control of the effectiveness of new machine designs, their correlation with agrotechnical requirements and with the requirements of machine systems in various working conditions as compared to the imported samples and analogs produced in batches. Prototypes as well as machines from the batch production are, therefore, subjected to state tests.

Testing of prototypes of machines is carried out at machine-testing stations (MTS) with the participation of specialists in the new technology from regional and provincial institutions under the State Committee on Agricultural Technological Processes. For instance, in such tests the adaptation of machines toward working in various soil conditions is revealed. Generalized results of these tests, carried out by the group of MTSs and also by the provincial institutions of the State Committee on Agricultural Technological Processes, serve as a basis for taking decisions about the introduction of these machines and complexes into production. Such decisions are taken by the Scientific Council of the State Committee on Agricultural Technological Processes as well as by the concerned ministries and departments. An indispensable condition for the introduction of machines into production appears to be their correspondence with the highest quality requirements.

Once the production of a new machine is finalized control (periodical) testing is carried out during the entire period of production of the machine. Estimations of quality and technical level of the machines are made in these testings.

Results of state testings of agricultural technological processes are used by the leading central and zonal administrative bodies for taking decisions on the planning of new production and batch production of machines as well as of spare parts.

Based on the activities of machine-testing stations and research institutes, the main impact in accelerating scientific and technical progress in agriculture is achieved on account of the following:

1. Operative testing, the timely determination of design defects and

demerits of technology (this gives an opportunity to the design bureaus to modify and rectify the design of the developed machines, on the one hand, and to the technological organization to improve technology of field-breeding and stock-testing, on the other);

2. higher reliability of test results, making it possible to predict more accurately the effect of the use of the new machines besides eliminating erroneous estimations of the working capacity of the machines;

3. realization of the in-depth research on the quality of execution of technological processes in conditions very close to the production conditions, making it possible to take new useful decisions;

4. a study of production technology of agricultural products in different regions, the basis of rational machine complexes and the optimization of processes for the cultivation of crops; and

5. advertisement and active application of new techniques and technologies of agricultural production.

The machine-testing stations and institutes are fully equipped to solve these problems.

In order to raise the scientific and technical level and the organization of testing, machine-testing stations and institutes, solve a series of new problems alongside testing, which include the following:

1. development of research works aimed toward the improvement of methods and technical means of testing and modification in designs of machines and the optimization of mechanized technology;

2. improvement in the systems of collection and processing of information accumulated during testing, wider application of modern methods of planning of experiments and mathematical modeling in the practice of testing; and

3. improvement in the system of scientific control of testing based on the use of computers and other modern means of automatic control systems.

During the last decade, as a result of accomplished research work and the generalization of testing experience, tasks on improving methods of testing of agricultural machines at this stage were completed by the efforts of machine-testing stations and research institutes. A system of industry standards on the general conditions of testing as well as on the types of estimations such as technical expertise, power evaluation, the estimation of electric drives, working conditions, the reliability of tractors and agricultural machines; economic estimates and operational and technological estimates have been evolved. Besides, standards are evolved on individual types of machines, namely, tractors, soil processing machines and implements, machines for sowing and plantation, the application of fertilizers to protect plants, for harvesting, for the processing of harvested produce, for storage and for the preparation of fodder, for reclamation and silviculture, for mechanization of stock-breeding, loading and unloading processes, etc.

Soviet standards on methods of testing of agricultural technological processes have served as the basis for international standards in countries which are council members of the Council of Mutual Economic Aid. The preparation of these standards was done with the participation of specialists and testers from the USSR, Bulgaria, Hungary, Poland, Romania, Czechoslovakia and Mongolia.

Results of official testing are used for decision-making about production of machines and for organization of requirements in national and foreign markets. Such organizations as the DLG (Federal Republic of Germany), National Centre for Research and Experimentation in Agricultural Processes in France, the National Institute of Agricultural Technological Processes in Sweden, National Center of Agricultural Mechanics in Torino (Italy), the Portuguese Center for Agricultural Mechanics, the Austrian Federal Machine-testing Center, the Institute of Agricultural Technological Processes of Norway, etc., as a rule, inspect their work in commercial operating conditions , besides testing the machine.

These organizations lay special emphasis on the estimation of working conditions, determination of reliability factors on special testing units and methods of rapid tests. Published materials are used by consumers and manufacturers of machines. The development of a unique system for machine-testing of agricultural technological processes on the international scale is governed by its specific peculiarities (the huge variety of production technology of the products, the great number of types of machines and conditions of their exploitation, the necessity of constant adaptation of new technology against changing circumstances of environments and plants).

PART ONE

Scientific Principles of Testing

CHAPTER 1

Basic Concepts of Testing of Agricultural Technological Processes

1. MODEL AND TESTING

One of the basic problems in acquiring an understanding of natural phenomena and in studying man-made objects lies in the extraction of information necessary for an adequate description regarding the object under study.

Having obtained information about the object, the researcher further has to deal with its information model, which is then subjected to various operations depending on the aims of the investigation and the conclusions obtained regarding the control of the modeled object. The information model of the object should fulfill specific requirements in order to get accurate results based on the properties of the object. These requirements will be further examined below:

1. The model should describe the object with optimum adequacy. By using the concept of concurrence used in the theory of modeling, i.e. conformity of the model to the original, the values of concurrence can be compared on a scale from 0-1. Value 0 corresponds to the case when the model does not have anything in common with the original and value 1—to the case of complete conformity of the model to the original, i.e. use of the modeled object itself in the design of the model. To the concurrence value 1 corresponds the hypothetical case of unjustified complex model, which takes into consideration an infinitely great number of properties of the model under study. Optimum adequacy, if measured on the above-mentioned scale, lies within the limits 0 to 1. Optimum is achieved by rational correlation between the adequacy of representation of the original and the economic viability of the achieved adequacy, considering the investigation and significance of expected results.

2. The model of the object should possess adaptability, i.e. it should permit the possibility of improvement according to the scope of the problem or in case of a change of the object. It is necessary to indicate characteristics of information in the description of agricultural machines, especially

complex objects which rapidly change their properties on interaction with the surroundings. Problems of investigation of agricultural technological processes bear a concentric nature, i.e. they use repeated reference to the model of the object when problems of investigation become complicated. To a certain degree, adaptability can be ensured by the use of not only numerical characteristics but also functions of time and other external factors as units of information. In this, while crossing to other spheres of factor-space, the adequacy is ensured by functional relationships without a change in the structure of the model.

3. The gathering of information, i.e. the establishment of values of parameters of the model along with the relationships between them, should be organized in conformity with recommendations which ensure maximum informativeness of the model. Informativeness implies the bulk of information in a specified number of indexes which are expressed as numbers, functions, functional and other mathematical formulations with the use of a set of symbols and relationships between them comprising the language of modeling. One way to increase informativeness is to select independent output factors of the model as well as the use of effective methods of planning of experiments.

From the aforementioned requirements, the problems of a scientific basis for the testing of agricultural machines directly follow as problems of optimum processes for the extraction of information and also its processing and usage as initial material (base) for taking decisions of economic, organizational and a technical nature.

Having defined the content of testing of agricultural machines as an information process, it should be noted that in this case, the concept of a 'dark box' as an entirely unstudied system cannot be used. Tests assume the use of some prior information—even of a qualitative nature. It is always necessary to know the list of output parameters and the expected levels of their variation.

One of the methods to achieve information about the object under study in testing consists in the availability of inputs to which an object gives response in the form of a complex of output parameters. Examples of inputs, which should be organized in a definite form, can be the application of loads in mechanical testing of samples, application of loads on the hook in traction testing of tractors, start of machines in service tests, etc. Definite inputs are needed to achieve information even in the process of technical examination when weighted, linear and other parameters of objects, which constitute its technical characteristics, have to be determined. Here, as in any other information process, the principle of Brillouin remains valid, according to which information can be achieved only as a result of power expenditure.

Fulfillment of requirements for the information model of the objects should be ensured by the organization of the test process, which is based on the systems approach, a correct selection of the model, the use of mathematical apparatus, identification of specific types of estimations and the provision of required apparatus.

Contents comprising the parameters of the test process are developed in the respective parts of the book. A complex of methods and the means used in the testing of agricultural technological processes, which ensure optimization of the information process, form the contents of testing as a scientific discipline.

The presence of scientific language of testing, a set of axioms and facts and sufficiently developed normative base, which includes standards of various categories and also other recommendable normative documents and materials, appear as yet another index for testing as a scientific discipline.

The basic directions of scientific and systematic provisions for testings will be briefly considered.

Methodology and the theory of testing of agricultural technological processes are still in the stage of formation into an independent scientific discipline. Characteristic features of this process are the erasure of the boundary between testings and investigations and the integration and wide mutual interaction of methods and concepts from various scientific disciplines and branches of technology. A rapid development of systems analysis and the material accumulated by a number of scientists already represent a sufficient basis to formulate and to identify fundamental features of this new applied scientific approach, which is assuming an interdisciplinary character.

2. SCIENTIFIC TESTING AND THE ANALYTICAL APPROACH TO THE PROCESS OF DEVELOPMENT

The development of theory and the practice of machine testing is undoubtedly called forth by speeding up the formulation of new agricultural technological processes and the achievements of related sciences, such as agricultural mechanics, agrozoological techniques, the theory of probability, mathematical statistics, the queuing theory and reliability of systems and, lastly, modern methods of analysis and synthesis of technical objects.

First of all, mentioning the link and the succession of test methods with the agricultural mechanics created by V.P. Goryachkin and his pupils are worth mentioning. Basic approaches of testing were first formulated in solving problems of agricultural mechanics. In the works of V.P. Goryachkin, *Obshchie printsipy ispytaniya sel'skokhozyaistvennykh mashin i orudii* (General Principles of Testing of Agricultural Machines and Implements) and *Otsenka sel'skokhozyaistvennykh mashin i orudii* (Estimation of Agricultural Machines and Implements), basic viewpoints are clearly defined with the use of which an estimation of agricultural machines, implements and mediums (agronomical, mechanical, technical, economic and working estimation) should be conducted.

Significant problems in the dynamics of mobile aggregates using the concepts defined by V.P. Goryachkin and I.I. Artobolevskii were later investigated by academician P.M. Vasilenko and also by academicians V.N. Boltinskii, S.A. Alferov, and in the studies based on the probability viewpoint by A.B. Lur'e, V.D. Shepovalov, V.Ya. Anilovich, V.P. Roslyakov, A.P. Terekhov and others [15, 25, 46, 64, 72, 88]. Works of V.N. Boltinskii, G.V. Vedenyapin, V.S. Kramarov, A.S. Iofinov, V.V. Katsygin, I.I. Trepenenkov, M.S. Runchev, F.S. Zavalishin, V.V. Gus'kov, M.P. Sergeev, N.M. Orlov, G.M. Shatunovskii, R.Sh. Khabatov, N.K. Didenko and others are dedicated to the problems of improving the efficiency of machine aggregates as well as to the justification for and estimation of their parameters and to the development of speedy processes of production. The system of indexes for a complex estimation of mobile aggregates was founded by D.N. Saakyan.

The above-mentioned works dealing with a large number of agricultural machines and processes and their methodological approach appear as an example of the fruitful and effective use of the modern theory of the operation of agricultural machines and mathematical apparatus. However, works on statistical analysis of agricultural technological processes are to date not so numerous, and the majority of authors limit the use of statistical methods to the processing of experimental data alone. P.M. Vasilenko, A.B. Lur'e, F.S. Zavalishin, A.P. Terekhov and others are among those scientists who have developed the concept of the probability approach to investigations in the sphere of agricultural technological processes.

Along with the solutions to fundamental problems, many effective methods are developed to study the dynamic characteristics and to design objects which are related to the numerous concepts of cybernetics [7, 18, 19, 46, 67].

At present, when the basic problem of technical progress lies in increasing productivity and the technological effectiveness of machines, a further development of theory and problems connected with productivity acquires special importance. As stated by I.I. Artobolevskii, the aim of the theory of productivity of working machines and automatic systems, lies in the establishment of the relationships between the technical parameters of machines or systems and the level of efficiency of the productive system and the specific expenditure of labor of the production system. Many researchers relate optimization problems with the efficiency level of a productive system.

The principles for agricultural technological process testing cannot develop without referring to modern progressive scientific disciplines. In the preface to the book of Kh. Shenk [89], N.P. Buslenko has stated that an inherent element of testings is a mathematical model without which it is not possible to correctly plan the experiment. But to formulate a mathematical model of the test process, it is necessary to have a very clear idea about its structure, the behavior of its individual parameters and the interaction among them, the effect of various factors as well as its reaction to changing test conditions.

A mathematical base to conduct effective testing is composed of the modern theory of experiment, which is based on the solid foundation of mathematical statistics. The theory of experiment permits testers of agricultural technological processes to achieve effective tools for increasing the reliability of data and to perform rapid testing [2].

Plans of dispersion analysis have been specially created for comparison testing [48].

In modern understanding, the theoretical basis for methods of estimation of agricultural machines and tractors and also other technical systems with requisite properties belongs to the fundamental scientific problem of analysis and synthesis. The basis for this is provided by modern methods of experiments, the solution to variational problems, the theory of random processes, queuing theory, statistical solutions, etc. as well as the specific theory of operational processes and the mechanics and dynamics of machines.

The problem of analysis and optimum synthesis occupies a central place in modern science and technology. This is confirmed by the abundance of sources in this field and an unending stream of investigation along various directions. A very detailed analysis of several modern problems and mathematical methods in the field of agricultural mechanics can be found in the works of P.M. Vasilenko, V.A. Zheligovskii, M.V. Sablikov and others [15, 25].

General theoretical problems of analysis and optimum synthesis are based on fundamental achievements in mathematics and especially in the theory of random processes (A.B. Kolmogorov, N. Wiener, A.Ya. Khinchin, V.N. Pugachev, L.S. Pontryagin, R. Bellman, B.N. Petrov, A.M. Letov, A.A. Fel'dbaum, N.I. Andreev and others) [17, 44, 80]. Analysis of accuracy, synthesis of structure and determination of optimum parameters of dynamic systems appear as the basic problems comprising the subject of theory and applied solutions. Methods of synthesis in the form of the theory of optimum filtration are used for the designing of a system, which realizes the proposed problems with a given quality and satisfies the restrictions imposed on them.

The objectives of analysis and synthesis as a complex problem, applicable to specific conditions of agriculture, have been more clearly formulated by I.I. Artobolevskii, P.M. Vasilenko, V.S. Sergeev, Yu.K. Kirtbaya, V.V. Katsygin, S.A. Iofinov, L.V. Pogorelyi and others.

I.I. Artobolevskii, B.S. Svirshchevskii and Yu.K. Kirtbaya have formulated the general scientific approach to the estimation of operational parameters of the machine to the aggregates and, in particular, to productivity. V.V. Katsygin considered the patterns of deformation in soil and developed the theory for selection and estimation of optimum parameters of mobile machines and implements. He also pointed out the general scientific aspects on which further developments—power, dynamics and statistical principles—should be based. Such an approach to the solution of the problem of analysis already contains almost all necessary elements of systems investigations.

So far, most methods of analysis and synthesis, developed especially for the estimation and optimization of parameters of machines and production processes, can be effectively applied only with the use of mathematical modeling [13, 72]. In this field, various methods of technological modeling, production and other processes were already developed and have been projected. Modeling is now widely used in testing (P.M. Vasilenko, V.A. Kubyshev, S.V. Kardashevskii, A.P. Terekhov, R.Sh. Khabatov, V D. Shepovalov and others).

Along with the theory of analysis and the synthesis of dynamic systems, during the postwar period a new scientific discipline appeared. It included the use of agricultural terminology, 'the methods for analysis of working systems' and was created in conjunction with the necessity to arrive at optimum decisions on the organization of working of increasingly complicated objects that could be grouped into 'large systems'. This discipline has also evolved into a fundamental approach, based on modern mathematical apparatus (theories of random processes, the queuing theory, reestablishment and also methods of operational science, the mathematical theory of reliability, etc.).

A rapid development of the systems methods, using modern concepts for investigations is brought about by posing new problems relating to the increase in production efficiency and the decrease in design time as well as testing and the introduction of sophisticated machine complexes into operation.

As a result of the foregoing, in testing of agricultural technological processes, there is an increasing tendency to study the totality of objects not as a conglomerate, but as a unified organic whole. Objective reasons favouring this treatment appeared to be the general scientific progress and spreading of modern scientific concepts.

A systems approach of the aims for testing of agricultural technological processes is revealed in the actual existence of the complicated processes of interactions and interrelationships between complexes of machines or tools of production (such as production equipment) and the external environment (soil, plants, animals and collectives of workers, thereby implying the purpose and the bearer or labor). This also manifested in the necessity of determining a large number of various operational factors not only of individual machines, but of entire complexes in various external and regional conditions and in the establishment of objective interrelations between these factors, the parameters of machines, the methods of organization of production processes, etc., while keeping the development of methods for control of these processes in mind.

Even a casual glance at the basic approaches of the development of scientific methods for the cognition of functioning principles of complicated samples confirms that there already exists a sufficiently vast repertory of scientific and applied methods for solving problems of analysis and synthesis in the testing of agricultural technological processes. To the known measures, these approaches are interwoven, in a few cases, they have developed in isolation, but even then the testing experience gives one a basis to hope for the possibility to unite them into a single methodological base under the general concept of systems analysis.

CHAPTER 2

Systems Methods for Estimation of Agricultural Technological Processes in Testing

3. FUNDAMENTALS OF A SYSTEM CONFIGURATION

The aim of analysis during testing in a systems context lies in the reliable determination of the necessary and sufficient characteristics of systems and subsystems under specific and variable conditions of work, the estimation of basic parameters of machine aggregates, improvements and interlocking of working machine parts and also in the determination of possible variates for the construction of their structure and ways for improvements.

The modern methodology of systems analysis arose on the basis of a dialectic understanding of the interrelationships and interdependences of phenomena in real production and on the progress in cybernetics [4, 13, 17, 19, 34, 81, 91].

The totality of techniques and methods used in the study of various processes carried out by agricultural machines, machine aggregates and complexes should be included under systems analysis in testing of agricultural technological processes. In the simplest case, the problem of analysis lies in the determination of characteristics of elements and of the system as a whole, i.e. the criteria of their efficiency under given working conditions.

From the systems viewpoint, it is necessary to distinguish agrotechnical, physical, production-technological and economic aspects of the problems of analysis.

A system is usually represented by an array of objects, their properties, relationships and relevant functions. In applied investigations, a system is generally broken up into subsystems that are studied separately and then followed by the study of the system in its entirety.

It is naturally difficult to strictly draw boundaries between systems considered in testing. A basic requirement in this regard can be the consideration of such objects and phenomena in which a certain hierarchical structure can be observed, and specific patterns of interrelations at various levels of systems object found. For this purpose, the definitions of systems based on the works of L. von Bertalanffy, W. Ashby, M. Mesarovic, such as the *Complex of Elements Involved in Interaction, Reflection of Inputs and Conditions of the Object on Output of the Object*, etc. may be used [7, 18, 49].

All material objects which take part in the complex of agricultural operations in testing (working machines, power and transportation means, auxiliary and loading equipment and also the object of operation and external environment) could be united into a single system for the purpose of testing; this system is generally complicated.

According to the analysis of different complexes in soil-breeding and stock-breeding, the following features should be considered the basic sign and characteristic peculiarities of complicated systems:

1. the availability of an hierarchical structure, i.e. the possibility of breaking up the system into numerous subsystems and elements interacting among themselves, while performing various technological, organizational and production functions;
2. the stochastic nature of processes in functioning subsystems and elements, which consists in their interaction with constant and randomly varying factors of external conditions and internal controlled or uncontrolled disturbances;
3. the availability of a purposeful problem and competitive aims, common to the system of the functioning of subsystems on imposing restrictions of technical, technological and economic nature; and
4. the systematic exposure of the system to control by means of a guided action of servicing personnel for the purpose of achieving certain indicators of efficiency.

The majority of systems in agricultural production appear to be open because in their working, there is exchange with other systems as well as the exchange of materials, power and information (sowing material, fuel, fertilizer, crop, information about the course of work, etc.). For example, a primary member of the system, which provides the fulfillment of industrial functions, could be the simplest ergodic link, which includes the agricultural worker, having a certain qualification, and the machine itself as implements of production. This member enters as an element in the upper system of the second order. The upper system combines several primary members, which carry out interconnected and often different processes such as soil-operation, sowing, harvesting, feeding, looking after animals, milking, cleaning of dung, etc. Several members comprise a system of the third order, namely, the soil-breeding brigade; the stock-breeding complex; several brigades form an industrial sector, i.e. a system of the fourth order and so on, up to the system of the highest order, which is the agriculture of the soil-climatic region or zone.

Any agrozoological-technical or mechanical system can be represented in the same way.

The structure of a system generally refers to its organization from certain elements with the help of forces, power, material or organizational interrelationships. These interrelationships are determined by functions which are fulfilled by the system. The structure of a system usually depends on the material nature of technological and industrial processes taking place in it and also on the processes of interaction and control.

The hierarchical structure of any system can be represented in a form of a certain diagram or the so-called structural flow chart (Fig. 1). The apex of the flow chart represents the main executive and control elements of the system, while the rows represent various types of interconnections and interrelationships between the elements of the system. In such a system, the links of subjection (continuous lines), subordination (dotted lines) and interactions (lines with stars) can be differentiated [71].

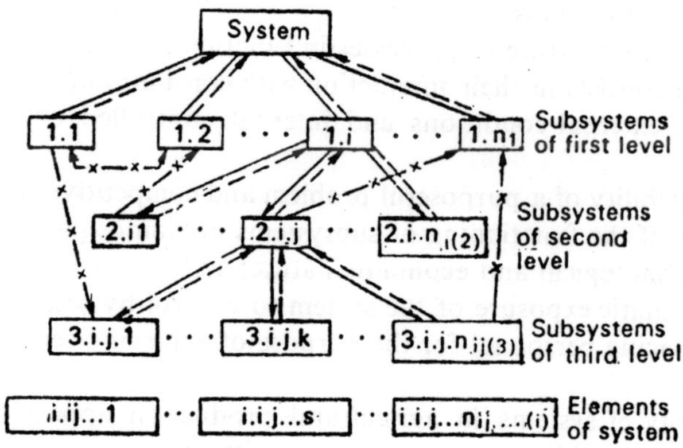

Fig. 1. Structural flow chart of a complicated system.

The relationships between the elements of a subjection and subordination type system exist in the form of information links, for example, information about the condition of working organs, the availability of seed, fodders, completion of refueling, movement of the aggregate, etc. This information describes the states of the subjected element or object with interaction between the elements of agricultural machines generally expressed in terms of power or material exchange.

On the structural diagram (Fig. 1), the elements of the first level (1.1, 1.2, . . . , 1.*i*, 1.n_1) and the second level (2.*i*1, . . . , 2.*i*.*j*, . . . , 2.*in*$_{i2}$), etc. are marked out. Thus the functioning process of the system consists in continuous interrelated changes in the conditions of the elements and subsystems which are expressed in terms of physical parameters.

The structural and functional aspect of the systems approach proposes that there are at least three levels of consideration of the system: first—setting of the structure on the level of abstract objects, second—investigation of specific links, and the third—investigation of operations.

Simplification of the nature of relationships between the elements of the given system, construction and investigation of mathematical models of functioning of its elements and the system on the whole and, lastly, a practical interpretation of results constitute the essence of the systems approach.

Generally, the values of various indexes at certain given levels of independent factors (parameters of machines, setting up speed of motions, kinematic regimes of working organs, depth of operation, etc.) are estimated in testing.

Nevertheless, the establishment of general principles for the functioning of complexes and systems of machines (with the behavior of the subsystem being looked upon as a special property) is of great interest. In this, according to the peculiarities of the cybernetic approach, it is necessary to realize the control of external and internal characteristics of the working of the system and to take the possible control on these processes into consideration.

Problems of the next class of complexity (a study of the mechanism of phenomena during interaction between the working organs and the material under treatment) cannot be so far predicted by the methods of testing. However, the knowledge of mechanisms of phenomena is useful not only from the technological and engineering points of view, but also because it considerably eases the development of methods of testing and warns against serious mistakes.

In testing, it is important to work out a reliable estimation of basic parameters of new designs of machines and their complexes (for example, the width of the hook, set power of the motor and other technical and service parameters) and also the general economic effectiveness as a function of basic parameters and specific conditions of service [15, 33, 38, 58, 60].

The last problem is formally solved in accordance with acting industrial standards; however, the results achieved do not relate to all but only to particular, randomly laid-out conditions of testing. At the same time, an objective economic estimation of the basic parameters of machines in systems approach should be more widely realized for various conditions of soil-climatic zones. This is made possible by the use of mathematical modeling of indicators of machine effectiveness.

Thus the classification and establishment of structure (structurization) of systems under consideration, the selection of a limited number of criteria and determination of connections between factors of effectiveness, natural-industrial conditions of exploitation, mechanical-technical properties of the material under process, working states and parameters (parametrization) of machines for testing should appear as basic stages of systems methods for testing. This can be achieved only by constructing mathematical models.

Various methods of mathematical statistics, statistical dynamics, the queuing theory, the theory of games, methods of agricultural mechanics, dynamic programing, etc. can be used as mathematical methods of investigation of systems and elements in relation to their physical nature.

At present, voluminous literature is available on the construction of mathematical models of various processes and, in particular, on the planning of experiments and also on methods of optimization [1, 2, 5, 40, 48, 50, 53, 54, 72, 76, 78]. However, a practical use of the accumulated knowledge is connected with serious difficulties, which are brought about by the specific peculiarities of agricultural processes (the qualitative and quantitative dependence of indicators of effectiveness on highly changeable, randomly varying external natural-industrial conditions varying in space and time; time-dependent parameters of natural-industrial conditions; the unending trend of average values and uncontrolled changes in the structure of external conditions; a high level of interference including uncontrolled disturbances, etc.).

The aforementioned characteristics of external conditions and the indexes of the working of machines, the multidimensionality and well-marked probabilistic nature, the variety of acting factors and a wide range of their variation result in the systems, under consideration, appearing to be highly diffused.

In these conditions, the role of the complex approach to the solution of testing problems considerably increases. This approach is based on the study of natural-industrial conditions, the physiotechnological properties of the material under processing and the resulting mechanical-technological phenomena as well as on the construction of mathematical models of processes with use of known postulated mechanisms of phenomena, which widen the possibilities of studies of phenomena by modeling methods.

For the solution of assigned problems, it is necessary to accumulate exhaustive information about actual industrial physiotechnological properties of the objects under processing and external working conditions. It is also essential to have a repertory of theoretical and experimental methods for the development and investigations of models for processes with the use of the methods of experiment planning, mathematical processing and interpretation of results as well as to have a standard procedure for mathematical modeling.

It is, therefore, necessary that researchers must find perfect methods for determining the characteristics of various processes over sufficiently long periods of time that are commensurate with the duration of the change of labor, agricultural periods or even the service life of machinery. This is required because the aggregate of random variables and random functions (being characteristics of external conditions and the results of machine operation) are objectively evaluated from selected data and from comparison between predicted performance and experimental findings. These results have been obtained only after testing over an extensive territory of soil and climatic zones and are formulated after experimenting over a fairly large time interval (from a few years to several decades), which is presumed to be the 'life-time' of the farming system.

4. CONCEPT OF THE STATE OF THE PROCESS AND ITS APPLICATION IN SYSTEMS ANALYSIS

Averaging is necessary in the estimation of indexes of the working of machines under test and in construction of models of various processes. In this connection, attention should be drawn to the correct determination of the sets of characteristics of external conditions and working conditions in relation to which the averaging of indexes is carried out.

The most important point in the construction of models of external conditions and functioning of systems appears to be the classical concept of the state of the process.

In order to construct methods, plans for experiments and, finally, the mathematical model of the processes under study, it is necessary to correctly define the concept of the state of process of external conditions and the functioning of technical implements under test.

The state of a system in the modern theory of systems is defined as a variable which, at any moment of time, fully determines its behavior along with the output value.

In statistical physics, the problems of the theoretical and probabilistic description of the states of systems comprising many particles have been successfully solved. In doing this, complicated systems are divided into relatively smaller subsystems (microsystems) and the behavior of the entire system is described by statistical distribution functions of the states of the set of microsystems. If an ensemble of identical systems is considered, instead of a single mechanical system, and the probability distribution of the changes in characteristics of systems in some abstract space is observed, then it is possible to describe different dynamic systems and processes.

A machine aggregate in the process of functioning can be characterized by several types of states, which describe the temporal changes in the working conditions by agrotechnical, technological, kinematic, working, power and other indexes.

Generally, all the variables of the state of a process cannot be directly measured. On the whole, estimation of the states of the object is closely related to the problems of identification.

If the machine under test is observed for a sufficiently long time in the steady-state condition, then in stabilized external conditions, the process of changes of its states can be considered to be in statistical equilibrium.

The physical state of the soil, biological masses and other objects of external environment are determined in the scales of the soil-climatic zone by meteorologic and climatic conditions and by complicated energy processes which take place on earth and in the atmosphere along with numerous acting factors. Since the probability that the majority of factors acting in the same direction are small, whereas large deviations in the factors are less probable; it may, therefore, be assumed that the space-time distribution of states of characteristics of external conditions, and accordingly, of the results of the functioning of machines over a sufficiently large interval of time can in statistical terms be considered in equilibrium. In practice, this is confirmed by well-known stability over a long duration and the finiteness of the averaged meteorological, rheological, agrotechnical and other cycles, although at certain moments of time or in different regions of the zone, there exist considerable finite deflections from averaged characteristics.

From the point of view of applied investigations, it is very important to have exhaustive statistical data about physical parameters, which characterize the state and the temporal change of characteristics of soil and plants over large areas during agricultural periods of conducting different operations or even during the vegetation period.

Hence, in testing, it is necessary to look for a reflection of the space of random states in the space of technological operational and other indexes of machine functioning.

Without giving the mathematical details, it needs to be mentioned that in such conditions, the determination of the so-called probability measure on the sets of all allowable input and output values is required, and the transformation of the input signal into the output signal by the machine should also preserve the measure, which is convenient synonym for probability.

If some process is observed at a large number of points of the field or zone, then the change of process in time will represent itself as a set of functions and trajectories. The concept of measure can be easily introduced in the space as some analog volume or area. While solving the problem for the description of probability of occurrence of a process, following the method of M. Kats, in the interval α_1, β_1 at the moment of time t_1, in the interval α_2, β_2 at the moment of time t_2, in the interval α_n, β_n at the moment of time t_n (when future behavior is independent of the past), it is necessary to find the probability (Fig. 2):

$$P(x_0/x_1, t) = \int_{\alpha_1}^{\beta_1} \cdots \int_{\alpha_n}^{\beta_n} P(x_0/x_1, t_1) \times P(x_1/x_2, t_2 - t_1)$$

$$\ldots P(x_{n-1}/x_n, t_n - t_{n-1})\, dx_1\, dx_2\, dx_3 \ldots dx_n.$$

The probable measure, given by the foregoing formula, is also related to subsets of the above-mentioned processes.

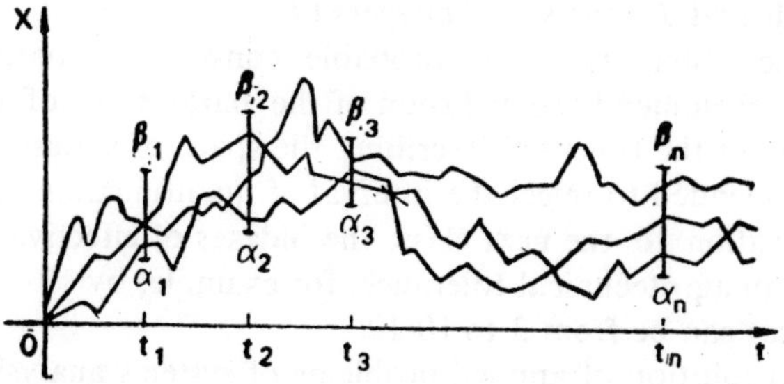

Fig. 2. Trajectories of a random process and their profile.

The concept of measure is necessary so as to compute the distribution of values of certain functions which qualitatively and quantitatively characterize the process of the working of a machine. Such a distribution function will represent itself as a measure of the set of those indexes, for which the function is given by

$$\sigma(\alpha, t) = \mu \left\{ \int_0^t v[x(\tau)]\, d\tau < \alpha \right\},$$

where μ is the measure.

Furthermore, it is possible to find an average value of the investigated characteristics of the processes and the corresponding dispersion.

The number of states of external stochastic environment and, therefore, also of output factors of machines is practically infinite. However, considering the finiteness of the magnitudes of agrotechnical, technological, power and other factors, for practical purposes, it is worthwhile to limit the range for parameters of external situations to which the machine is exposed, and, therefore, also to limit every parameter of input and output by a certain finite number of possible values. For this, it is necessary to obtain a suitable quantification of the state of the aggregate, and the sets of discrete input-output values will represent its state.

Then the set of values for characteristics of the aggregate will be represented by the vectors of the states in a multidimensional space, which are united for each factor in the so-called hypersphere. This hypersphere can be filled by a finite number of hypercubes in accordance with the established idea of quantification. Such a system, with a finite number of states, can be described by using the analogy of the theory of finite automatic systems as the product of spaces of three factors by a pair of reflections f and q related to the aggregate input and output states in the time interval from t to $t+1$:

$$f: \sum \times X \times T \rightarrow \sum;$$
$$q: \sum \times X \times T \rightarrow Y,$$

where $\sum$ is the space of the state of the system, X—the input signals, Y—the output signals and T—the set of integers ($T = \ldots, -1, 0, 1, \ldots$).

In case the system possesses probable constructive parameters, some authors also recommend the inclusion of the parameters of the object for determination of the function describing the space of states of the object.

It is recommended to select the interval of quantification in such a way that on transition to the next class, the indexes of effectiveness change by much less than agrotechnical tolerance, for example, by 10-15%. A number of such classes can be from 3 to 10-12.

A practical solution of applied problems of systems analysis demands the establishment of specific relationships (analytical and empirical) between parameters, the working regimes of machine aggregates, the space of their states (states of output values) and the external environment characteristics as well as other actual industrial conditions of the given region or zone.

Hence, the difficulties in the use of the systems method for testing and for developing the models of processes are connected, first of all, with the determination of characteristics of the set in relation to which averaging is done, i.e. a hypothetical generalized set (sets of all possible states of the external environment). It is necessary to have in mind that the selection of schemes for experimentation is determined by the set and the structure of the expected states of the subject and should be governed by requirements for ensuring necessary and sufficient information about these states.

A study of characteristics of external environment and physiotechnological properties of agricultural materials should serve as the initial material for the solution of the problems put forward.

5. METHODS OF FORMULATING TESTING PROBLEMS

After putting forward the problems, the next stage in systems analysis is structurization [18]. Agricultural machines and implements, their working parts, power means, external environment and also the industrial collective of workers and means of control are basic components of these systems with which investigators and testers are dealing. Components of the system in every individual case can be further separated into elements and their interaction with the material under processing or the external environment can be considered in the form of elementary interactions.

Identification of such elementary interactions and their consideration with respect to time as simple processes, which are characterized by some factors, are generally called as parameterization. In this, it is necessary to distinguish fundamental terms: working process, operation and industrial process.

By the structure of industrial processes, one implies the totality of the functional elements of machines, aggregates, auxiliary links, etc. and also the working processes which are related in a particular manner.

The most important condition for solving applied problems of testing and investigations is the formalization of various processes. Formalization comprises, in particular, an indispensable operation of numerical modeling of working processes. So as to reduce the infinite set of possible models of processes, it is worthwhile to classify them. For example, in soil-breeding, it is logical to extract three basic groups of processes (technologies): soil processing, distribution and proportioning of materials and primary processing of materials and products. Each of the above-mentioned groups of processes differ by their specific peculiarities, which are physiomechanical properties of the objects under processing and also by the nature of the action and parameters of the working organs. Approximate classification of working processes and operations is given in Fig. 3.

Working processes can be easily arranged from standard operations, for example, harvesting of beets can be written [28] by using the derived classified indexes as follows:

$$I_{2A}+I_1B_{bx}II_{1B}+II_{2B}+VI_3+I_1B_{b0}+II_{2A}$$

$$+I_1B_{b0}+VIII_4 \begin{cases} \rightarrow +I1B_B+I_{1A} \\ \rightarrow I_{1A}+I1B_{bx}+I_{1A} \end{cases}$$

where I_{2A} is the movement of the combine along the rows, I_{1B}—the control of the combine on the rows, II_{2B}—the copying of the microrelief of the field, II_{1B}—the digging of roots, VI_3—the working of pulling (digging) devices, II_{2A}—the leveling of roots, VIII_4—the cutting of beet leaves, I_{1A}—the transportation of beet leaves in a container, $I_{1B_{bx}}$—the transportation of roots to the collector, I_{1A}—the unloading of collectors, etc.

Industrial processes can be classified in line (industrial) production, noncontinuous line or periodical production (loading, milking, etc.). A diagram of a typical complicated industrial process of harvesting beets is shown in Fig. 4.

In applied systems investigations, traditional methods of the theory of agricultural machines, the theory of mechanisms and machines and the dynamics of machines, economics of agriculture, etc. should also be used along with general systems principles. But their use requires frequent introduction of corresponding interpretations.

Formulation of problems for analysis and a synthesis of various processes can be carried out by mechanical and mathematical or by the probability description of the functioning of machines and the operations associated with it.

In the practice of machine testing for various types of estimations, firstly, one has to deal with the estimation and analysis of various operations and working processes. In this, depending on the problem, either their static characteristics are established and the corresponding statistical models are developed, which reflect relations between parameters, the criteria of effectiveness and important variable values of the system in equilibrium states or

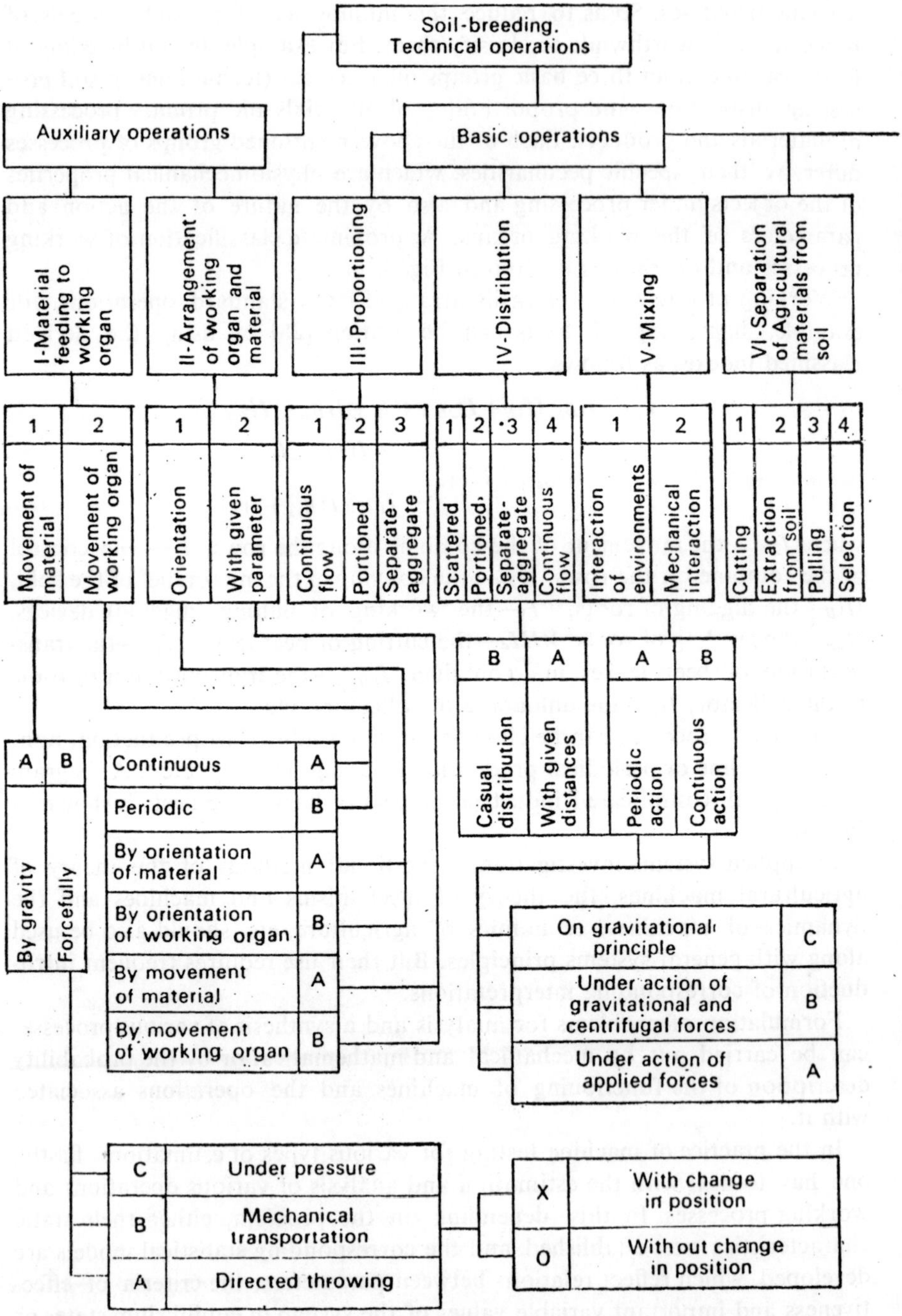
Soil-breeding. Technical operations
Auxiliary operations
Basic operations
I-Material feeding to working organ
II-Arrangement of working organ and material
III-Proportioning
IV-Distribution
V-Mixing
VI-Separation of Agricultural materials from soil
1 Movement of material
2 Movement of working organ
1 Orientation
2 With given parameter
1 Continuous flow
2 Portioned
3 Separate-aggregate
1 Scattered
2 Portioned
3 Separate-aggregate
4 Continuous flow
1 Interaction of environments
2 Mechanical interaction
1 Cutting
2 Extraction from soil
3 Pulling
4 Selection
B Casual distribution
A With given distances
A Periodic action
B Continuous action
A By gravity
B Forcefully
Continuous A
Periodic B
By orientation of material A
By orientation of working organ B
By movement of material A
By movement of working organ B
On gravitational principle C
Under action of gravitational and centrifugal forces B
Under action of applied forces A
C Under pressure
B Mechanical transportation
A Directed throwing
X With change in position
O Without change in position

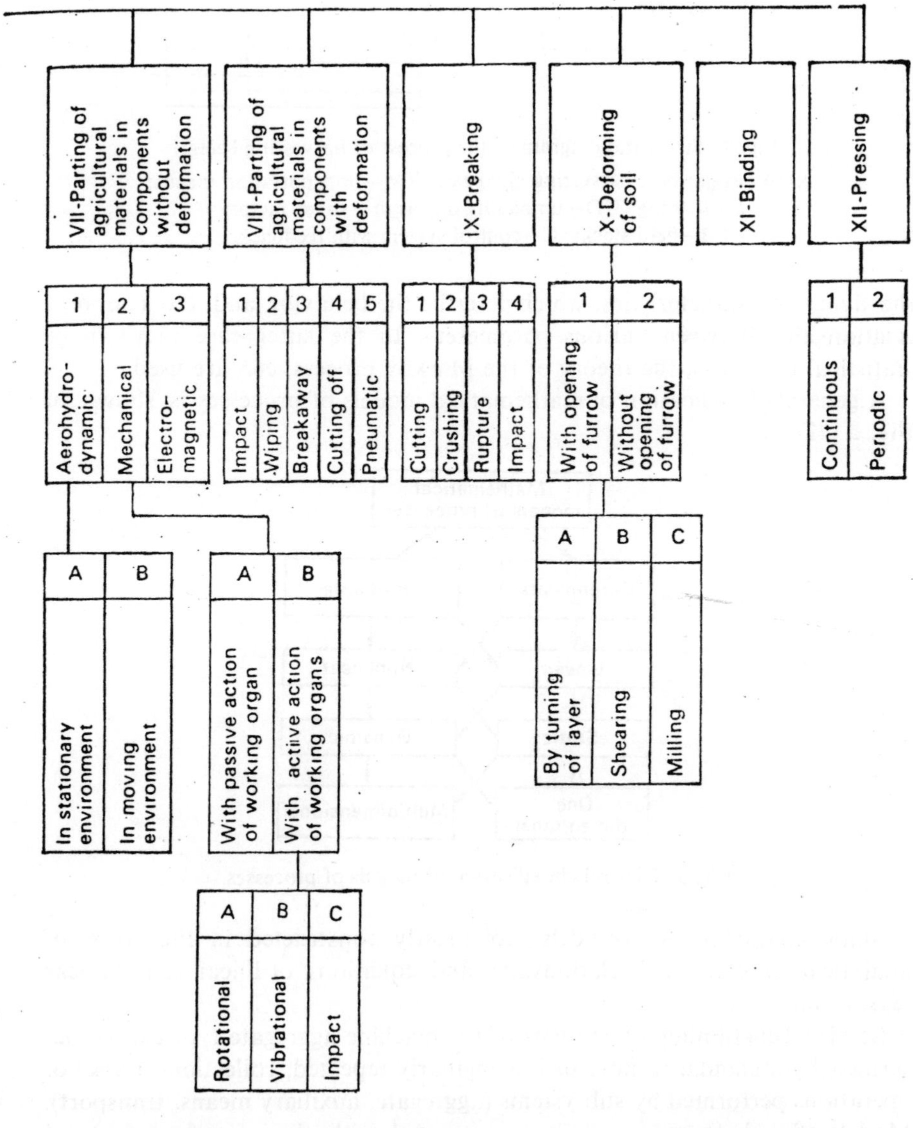

Fig. 3. Classification diagram of technological operations and methods of their fulfillment.

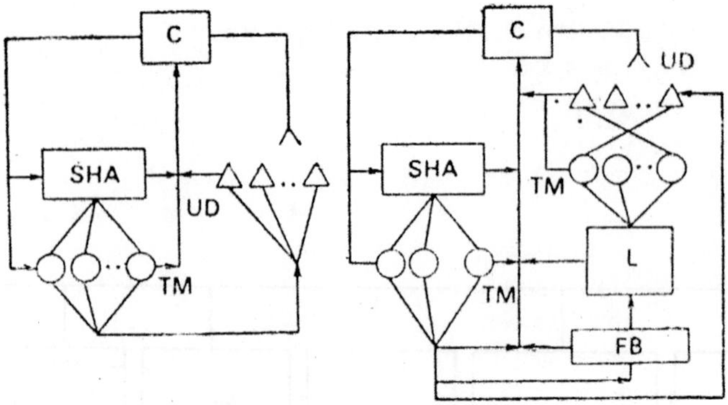

Fig. 4. Structural diagram of the process of harvesting beets:
SHA—sugar beet harvesting aggregate; TM—transportation means;
L—loading; UD—unloading means in a sugar factory;
FB—field stacks; C—control system with feedback.

the dynamic characteristics, which describe the behavior and the temporal relationships between various parameters. In the latter case, methods of statistical dynamics, the theory of the Markov process, etc. are used.

A general classification of mathematical models of processes is shown in Fig. 5 [72].

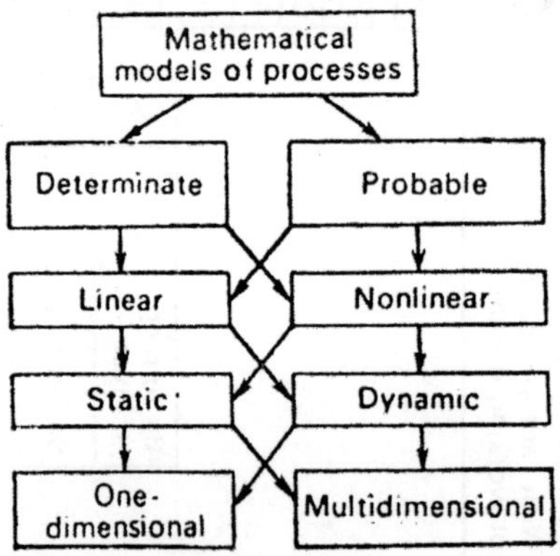

Fig. 5. General classification of models of processes.

Static characteristics (models) are mostly constructed in the form of analytical or empirical relationships and equations of linear or nonlinear regression.

Service functioning of systems of the machine aggregate type can be described by a standard, more or less regularly repeated, collection of a set of operations performed by subsystems (aggregate, auxiliary means, transport). Models of such processes can be constructed with the use of operational research [54, 61, 70, 72].

Introduction of modern methods of formalization of various technological and industrial processes and on this basis, the construction of corresponding methodical guidance in a systematic way would permit the problems of the applied use of the methods of systems approach to be solved. In the construction of the model of objects under test, with the aim of describing and investigating efficiency indicators and the functioning of machines, it is necessary to take all the basic properties and primarily the structure of the system, the values of parameters of its elements, external and internal random factors, etc. into consideration.

The basic difficulty in the use of the systems methods in testing and the estimation of objects in agriculture lies not only in the large dimensions of the problems of analysis, determined by the large number of subsystems, elements and independent indexes and, consequently, partial efficiency indicators, but also in the necessity to achieve reliable statistical estimates of these indicators. They reflect the results of functioning of the machine on numerous external conditions over a sufficiently large span of time under a direct effect of the system of random action of the environment and also controlling and organizing actions from servicing and managing personnel.

It is noteworthy that while testing by the existing methods, a few arbitrary distribution function variables under study are generally measured. These functions correspond to the massives of characteristics or the realization of the random processes $x(t, \rho)$, which appear to be the functional of some random event (or events) ρ, apart from being a random function of time. The origin of the events depends, apart from other reasons, on the set of characteristics of external conditions. The absolute characteristic efficiency indexes can be obtained by averaging the results of machine operation for all possible values of external conditions indexes. A random process can be characterized exhaustively by a complete set of realizations $x_\Omega = \{x_i\}$, which can be naturally expressed as a distribution functional for random functions $x_i(\Omega, \rho)$ in the set of external conditions.

The process of functioning of agricultural machines or aggregates with parameters Π_k, formally represents the transformation of input variables $\{x_i\ (t)\}$ (random characteristics of external conditions and the material under processing) of some system of operators L_{Q^i} in output Q^i_Ω, which are considered as criteria of effectiveness in the set of possible values of working conditions.

The construction of a mathematical model of an object under testing, generally speaking, is equivalent to finding the operator LQ_Ω.

The basic criteria of complex technological and working estimations of aggregates such as systems can be the following: specific technological and agrotechnical indexes $A_t(x_\Omega)$, productivity $W(x_\Omega)$, indexes of functioning, specific mass and power characteristics and also the prime cost of works $I_c(x_\Omega)$ operating expenses $I_0(x_\Omega)$ and expenditure of labor $Z_t(x_\Omega)$ and overheads $I_p(x_\Omega)$, etc.

In order to reduce the subjectiveness in the availability of absolute characteristics to a minimum the most powerful criterion of effectiveness in the form of the probability maximum is used:

$$P_\Omega = \sum_{k=1}^{n} P_k = \sum (P_1 \geqslant a_{1t}, P_2 \geqslant a_{2t}, \ldots, P_k \geqslant a_{kt}).$$

It uniquely characterizes the degree of correspondence of machines to the requirements by all partial criteria a_k.

Therefore, as a methodical base of the systems approach for estimating the efficiency indexes in the testing of the machine aggregate or the complex of machines with common parameters and parameters of subsystems $\Pi_k (k = 1, 2, \ldots, n)$, the determination of these factors in the set of characteristics of external conditions is carried out [58]:

$$Q^i_{x_\Omega}(\Pi_1, \Pi_2, \ldots, \Pi_m) = L_{Q^i}(x_\Omega, \Pi_k),$$

where L_{Q^i} is the certain operator of the machine (generally nonlinear), $x_\Omega = \{x_i\}$—the set of values of the abovesaid independent variables of the characteristic of external factors (this set is determined as a direct multiplication of its values).

The average values of effectiveness criteria, determined from the average values of characteristics, differ considerably from the average values determined from the full range of functioning of aggregates by 5-25% [58]. This is because of the nonlinearity of operators $L_{Q^i_\Omega}$, (in the general case) and also because of consideration of the fact that the distribution laws of many working factors and the external environment are different from the normal distribution and that they clearly possess an asymmetrical nature.

Hence, the expansion of experimental conditions not only makes it possible to achieve material which is necessary for analysis and modeling, but which can also prevent serious mistakes.

Necessary absolute characteristics of effectiveness criteria (mathematical expectations $MQ^i_{x_\Omega}$ and dispersion $DQ^i_{x_\Omega}$) can be achieved by the averaging of working indexes of the machine for all values of x_Ω occurring in the given zone. This is done by using combined methods of experimental investigations and mathematical modeling.

So as to avoid unresolvable difficulties in analysis, modeling and decision-making, it is recommended to estimate the necessary degree of detailing of the process under study in relation to the number of indexes taken into consideration by the functional $N = 1 - e^{-n}$. This functional indicates the usefulness of introducing new indexes into consideration, starting from the certain value of n.

In constructing a statistical mathematical model, three essentially different approaches can be used: methods of multidimensional mathematical statistics developed by Fisher; methods based on a postulated description of development of processes with respect to time (in particular, the Markov processes); various phenomenological methods, including the well-known method of polynomial description of the effectiveness criterion.

If only independent characteristics of external environment and work material are taken into consideration, then mathematical expectations and the dispersion of the effectiveness criterion can be determined from the following expression:

$$MQ^i_{x_\Omega} = \int\int_{-\infty}^{\infty} \ldots \int L^i_Q(x_1, x_2, \ldots, x_n) \prod_{j=1}^{n} f_j(x_j)\, dx_1\, dx_2 \ldots dx_n;$$

$$DQ^i_{x_\Omega} = \int\int_{-\infty}^{\infty} \ldots \int [L^i_Q(x_1, x_2, \cdots, x_n) - m_{Q^i}]^2 \times \prod_{j=1}^{n} f_i(x_j)\, dx_1\, dx_2 \cdots dx_n,$$

where $f_j(x_j)$ is the density of the probability distribution of the jth characteristic of the external conditions determined by observation on the values of x_j.

However, the method of multidimensional functions, which demands consideration of the complex combinations of random values with different laws of distribution, does not lead to clear analytical solutions in practical applications and is quite complicated.

In most of the cases of large complicated modern machines and the insufficient physical study of operations and working processes, approximate mathematical models can be successfully developed with the use of methods of experimental planning without any reference to the physical significance of processes in the form of a polynomial expansion:

$$L_Q = a_0 + \sum_{i=1}^{k} a_i x_i + \sum_{i \neq j=1}^{k} a_{ij} x_i x_j + \sum_{i=1}^{k} a_{ii} x_i^2,$$

where a_0, a_i, a_{ij}, a_{ii} are the selective estimates of regression coefficients of a generalized set.

Such types of models remain the most useful for the study of various technological operations and processes in a localized sphere of conditions. The following chain of models can be constructed on this basis from known or derived relationships, models of individual elements and subsystems are first constructed on the level of a mathematical (preferably stochastic) description of technological operations and processes, even of an operative character, then complex models for the modeling of interrelated operations and processes are created and, finally, basic parameters of machine aggregates and characteristics of the structure of industrial processes are introduced for consideration; lastly a system of corresponding algorithms is created. The laws belonging to the complex object under test can be established on the realization of the algorithm on the computer, while estimating the dynamic properties (identifications) of working organs and machine aggregates as a whole; the dynamic properties of the process on the system can be achieved by observing (computing) the trajectory of the system in the space of states. For a multidimensional system, this will be a trajectory of an n-dimensional vector of states.

In order to describe the system in terms of its states, a set of differential equations of the first order, which connect the rate of change of the states of the system with an input of $x_i(t)$ and an output $y_i(t)$ of values, can be used:

$$\sum_{j=0}^{n} a_j(t) \frac{d^j y(t)}{dt^j} = \sum_{i=0}^{m} b_i(t) \frac{d^i x(t)}{dt^i}.$$

In general, in accordance with developed principles in the theory of dynamic systems, during identification of complex objects, it is recommended to determine the degree and form of the relationship between the input and output variables; the steadiness of random functions under investigation; the quantitative estimation of the degree of identification (isomorphic value, adequacy) of the model to the original object; the estimation of the degree of nonlinearity and the possibility of achieving the operator of the object in a class of linear operators, etc.

After achieving necessary statistical characteristics of input and output values, it is also possible to determine the conditional density of distribution of the output variable $y(t)$ with respect to the input $x(t)$ as a general characteristic of the machine in the form:

$$f(y_t/x_t) = \frac{f(y_t, x_t)}{f(x_t)},$$

where $f(y_t, x_t)$ is the combined probability density of $y(t)$ and $x(t)$.

Hence, the conditional probability density can be considered as some operator of the object:

$$f(y_t) = \int_{-\infty}^{\infty} f(y_t/x_t)\, f(x_t)\, dx_t.$$

In particular, a known relationship between correlation functions

$$K_{yy}K_{xx} - K_{xy}^2 > 0,$$

can serve as the criterion, verifying the linearity of the operator of the object.

If the characteristics of input values vary within wide limits (which is characteristic for agricultural aggregates) and the relationship of effectiveness criteria with a variation of input values is substantially nonlinear, then such an operator can be used only within the limits of variation of the values, which do not exceed one or two discrete states when linearization is still permitted.

Along with the aforementioned, the technique for achieving density of distribution of quality indexes in relation to the external conditions and parameters of machines within a wide range of conditions is highly cumbersome. The final model of the process, in terms of states [61], is more convenient for achieving the general absolute characteristics of objects under investigation, when the dynamics of the system and the final statistical characteristics can be achieved with methods of the Markov processes involving a definite number of states.

6. GENERAL STATISTICAL DESCRIPTION OF PROCESSES OF MACHINE OPERATION

Specific peculiarities of working conditions of agricultural machines lead to the fact that with the passage of time, on account of the action of meteorological and other external factors, the state of processes in the working of machines varies randomly and the processes observed during testing acquire a unique nature of random fluctuations in the space of states. It is natural that observed processes in the areas of large dimensions or in a large interval of time, when the conditions of uniformity are not satisfied, bring to light the properties of unsteadiness with respect to mathematical expectation and, similarly, dispersion.

Here one inevitably comes across difficulties which are caused by the so-called space-time scale of the characteristics of external conditions. While considering limited realizations, differences in scales of external factors lead to the observation of the known phenomenon of trend, i.e. unsteadiness even in generally stable weather and other conditions.

The concept of the systems approach allows one to overcome these difficulties. According to this concept, any system (process) can be considered as a totality of localized subsystems (subprocesses), which possess certain hierarchical properties. This signifies that the contribution of individual components in the overall result decreases with the decrease of their scale, while components with sharply differing scales can be considered practically independent.

It is first of all necessary to establish the nature of scales (linear or temporal) contained in the processes under consideration. This will permit splitting up the processes into elementary components and selecting the rational form for representation of the characteristics of external conditions of indexes for the working of machines under different scales of time and space.

It can be assumed that at any moment of time, in a large soil massif, it is possible to find a certain number of plots (which are limited in dimensions) with statistical, isotropical, agrotechnical and physical characteristics that ensure certain steadiness and uniformity for different indexes of the working of machines. In analogy with the ideas developed in statistical physics, linear and time scales of indicated plots (time of existence of the steady state of physical parameters or factors of working) are those scales which characterize the steady microstructure of the jth process having time-independent statistical parameters. Then an unstable process, as a whole, during a long interval of time, can be interpreted as some random change of one state (event) by another or as a flux, consisting of 'pieces' and possessing an array of characteristics of stable structure that randomly replace one another.

Hence, logically one arrives at the possibility of using a model for quasi-steady processes, represented by the sequence of a finite collection of stable pieces (processes) corresponding to the above-introduced discreteness of states which replace one another smoothly or in jumps. For every scale of process, these pieces will have a different length (lesser for small-scale, than that for large-scale).

A number of mathematical models of steady and quasi-steady processes are known. The most simple and accessible model is that in which the space of states is formed by an additive set of structures of different scales.

Steady microstructure in this case appears as a nonuniformity or disturbance of the smallest scale (micro-irregularities and nonuniformity of soil, material under processing, etc.) and, naturally, should have parameters practically independent of time. Such processes can be observed on small, even and uniform sections of the field; while, in the presence of large-scale components in the process, they can be obtained in the form of increments with respect to the mathematical expectations which are calculated for the sliding interval of time:

$$\Delta x = x(t) - \frac{1}{T} \int\limits_{t-T/2}^{t+T/2} x\ t)\, dt.$$

Increment $\Delta(x)$, in many cases, possesses highly useful properties: slow changes in level and constant components have little influence on its values, whereas an unsteady process for most agricultural processes, at fairly non-rigid assumption, can be interpreted as a process with steady increments [66].

A steady microstructure of processes, i.e. the collection of microprocesses, is highly important for investigation of minute agrotechnological and dynamic properties of working organs and aggregates (distribution of seeds with respect to depth, crushing of materials, analysis of different structural parameters, springs, seats on tractors, etc.). This gives basis to establish a scale for the microstructure of processes on a natural time scale—of the order of characteristic sizes of seeds, particles, etc. and for dynamic processes—of the order of the period of natural vibrations of structural elements, in most cases, 0.1 or 2-5 s or 0.01-10 m.

An unsteady microstructure exists on the intervals of time of the order of some microstructures and is conditioned by disturbances of external scales (large-scale nonuniformity of the soil massif, microrelief or distribution of crop yield), external meteoclimatic factors and also the slowly varying mass of the aggregate due to the storage or decrease of produce in the bunker, etc.

For a limited realization of such a process, statistical characteristics will already be depending on time, the nature of the action of the large-scale component becomes determinate; it can be measured and estimated by developing cause-effect relationships.

Using the method of Zade and Ragazin, the given structure can be considered as one consisting of a nonrandom function of time (or the sum of function $p(t)$, which is represented by polynomials of the jth degree) and a steady random function $m(t)$ with a known correlation function:

$$S(t) = m(t) + p(t).$$

Parameters of a nonsteady microstructure are required basically for investigation of technological and power characteristics, the processes of the working of an engine and also for solving problems of automatic control on movements of aggregates.

For the most common problems, mainly of an operational-technological nature, it is necessary to know not only localized characteristics, but the entire totality of processes during the working of aggregates at various regimes and different conditions; besides, it is also necessary to gain averaged parameters of processes on all possible micro- and macrostructures, i.e. their megastructure. In this, the totality of micro- and macroprocesses will be represented by some multidimensional quasi-steady process. A steady megastructure exists on sufficiently large time intervals or areas, when the process averaged on all microstructures neither depends on scales of components nor on time.

Corresponding scales of processes are determined as an integral of the correlation function of process in the direction of the corresponding co-ordinate axis:

$$l_i = \int_0^T R_{xi}(\tau)\, d\tau.$$

The approximate time scales for components of the process can be estimated from the order of the ratio of the characteristic dimension of nonuniformity to the average speed of the process.

Scales of described structures must be selected in such a way that the correlation functions between parameters of structures are practically absent, i.e. $l_1 < l_2 < l_3 \ldots$

In conformity with the above-mentioned aspects of the problem, for investigation of micro- and macroprocesses, the quantification of the time interval or elementary area can be different and can stretch from several seconds, centimeters or meters to days, ten-day periods, months or years and kilometers.

The necessary degree of the detailing of the system and its description should be established from specific considerations and peculiarities of the working of the machine.

The description of macroprocesses for estimation of the working and technological properties of machines can be based on ignoring the fluctuations of instantaneous values of characteristics in relation to the sliding averages; this originates from the assumption that these fluctuations do not visibly affect macroprocesses of this type, for example, the speed of movement, productivity, etc.

An acceptable mathematical description for processes under consideration during the working of the machine can be taken as: the manifestation of their structure, the establishment of the pattern of variation of states with respect to time and space and also as the establishment of laws of distribution (multidimensional in general) and determination of the first two central moments for basic factors of external environment and indexes of the functioning of machines. A qualitative description of the final model of the process can be mathematically expressed as follows [11].

Let n represent steady and steadily linked microprocesses $\xi_i(t)$ with known mathematical expectations, second moments and correlation functions (Fig. 6).

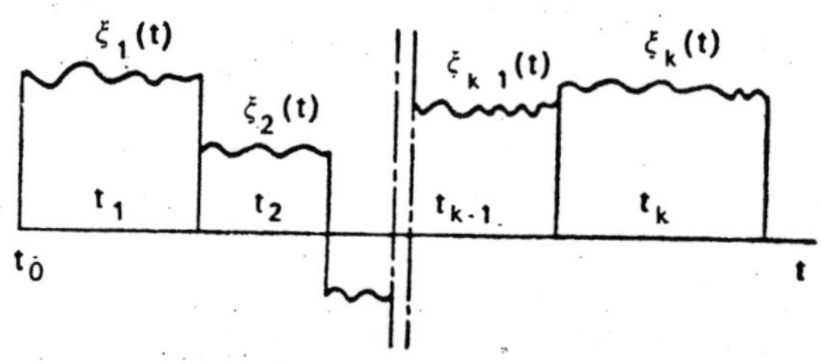

Fig. 6. Diagram of a quasi-steady process.

Micro- and macroprocesses are formed by random selection of realizations which coincide with one of the discrete states by some law.

A sufficiently large number of such 'pieces', are selected randomly from the universe of the process under observation. This process characterizes external conditions (input value) $X^i_\Omega(t)$ or effectiveness criteria (output value) $Q^i_\Omega(t)$.

It follows from the definition for each moment of time that the process $x_i(t) = \xi_k(t)$ with probability P_k. Thus from the formula of total probability, the following can be found:

$$P\{x_i(t) < u\} = \sum P_k P\{\xi_k(t) < u\}; \quad \sum_{k=1}^{n} P_k = 1,$$

where P_k is the probability of appearance of the kth realization of the process $\xi_k(t)$ with characteristics m_k and σ_k; u—the certain value of the index under investigation, for example, the agrotolerance or technical requirements.

If one-dimensional densities of distribution are known, then the density of probability distribution $f(u)_x$ of the process does not depend on time and is equivalent to

$$mx_i = MX_\Omega^i(t) = \sum_{k=1}^{n} P_k M\xi_k(t) = \sum_{k=1}^{n} P_k m_k;$$

$$DX_\Omega^i(t) = M[x_i(t) - mx_i]^2 = Mx_i^2(t) - m^2x_i$$

$$= \sum_{k=1}^{n} P_k M\xi_k^2(t) - m^2x_i = \sum_{k=1}^{n} P_k\sigma_k^2 + \sum_{k=1}^{n} P_k m_k^2 - m^2x_i.$$

The law, which is controlling the change of structures (states), can be assumed to be steady. As for example, the simplest Poisson law or the law of variable (random) intensity $\lambda_n > 0$, which is often confirmed by observations, has its distribution function in the following form:

$$p(t) = \lambda e^{-\lambda t}.$$

Considering that most of the real processes have an auto-correlation function of the standard type, $r_{cc}(\tau) = \sigma_x^2 \exp(-\alpha_k(\tau)) \cos \beta_k\tau$, the following expression for the correlation function and spectral density of process $x_i(t)$ on the basis of numerous experimental data is obtained:

$$RX_\Omega^i(t, S) = \sum_{k=1}^{n} P_k(1 - P_k)\sigma_k^2 e^{-(\lambda_n+\alpha_k)\,|t-S|} \cos \beta_k\tau$$

$$+ \sum_{k=1}^{n} P_k^2\sigma_k^2 e^{-\alpha_k\,|t-S|} \cos \beta_k\tau + e^{-\lambda_n\,|t-S|} \sum_{k=1}^{n} P_k m_k^2 - m^2x_i;$$

$$SX_\Omega^i(\omega) = \frac{1}{\pi} \sum_{k=1}^{n} P_k\sigma_k^2(1 - P_k)(\lambda_n - \alpha_k)$$

$$\times \frac{\omega^2 + (\lambda_n + \alpha_k)^2 + \beta_k^2}{[\omega^2 - (\lambda_n + \alpha_k)^2 - \beta_k^2] + 4(\lambda_n + \alpha_k)^2\omega^2}$$

$$+ \frac{1}{\pi} \sum_{k=1}^{n} P_k^2\sigma_k^2\alpha_k \frac{\omega^2 + \alpha_k^2 + \beta_k^2}{[\omega^2 - \alpha_k^2 - \beta_k^2]^2 + 4\alpha_k^2\omega^2}$$

$$+ \frac{1}{\pi} \frac{\lambda_n}{\lambda_n^2 + \omega^2} \sum_{k=1}^{n} P_k m_k^2 - m^2x_i,$$

where λ_n is the parameter of the exponential distribution of the time interval in which the abovesaid realizations are observed and which are variable in time in the general case, α_k, β_k—the coefficients of correlation functions of the observed realizations.

Values of P_k and λ_n depend on the properties of the system under consideration and on the degree of detailing of the processes, i.e. the values for the interval of the quantification of their states to be established practically required distinction between the values of investigated indexes, while crossing from one discrete level to another.

The type of model presented here, to a known extent, appears universal for describing processes during the working of real aggregates. For example, it can describe characteristics of external conditions as well as the change in the speed of movement and also technological, power, working and other parameters.

The density of the distribution of parameters of processes with respect to a steady megastructure can be obtained in terms of the corresponding densities of microprocesses and their weight functions, which can be determined from the distribution laws for micro- and macrostructures:

$$P(u_{\Sigma}) = \sum_{i=1}^{N} P_i f(\xi_i).$$

If the experimental information about the particular interval to which the microstructures may belong is known, then it is easy to calculate the probability of their existence in time and the frequency of repetition.

The characteristics of a megastructure for the conditions of the given region or zone are necessary for solving important problems of analysis and synthesis: estimation and selection of the basic parameters of aggregates and states of their working from the technological, techno-operational and economic criteria with the use of modeling methods.

Over a period of 10 years, such an essential meteorological factor as precipitation was believed to follow, in a number of cases, the Poisson temporal law of distribution (Table 1). It is possible to attribute the process and

Table 1

Indexes	Law of distribution	Mathematical expectation (parameter of distribution)	Root mean square value (standard deviation)	Criterion of correspondence, κ^2
Number of changes of states:				
in a month (September-October)	Poisson	9.4	3.10	10.52
in that number, because of fall of precipitation	”	3.15	1.76	22.24
because of soil desiccation	”	6.25	2.38	19.20

changes in the parameters associated with it, describing the microstructures of humidity, soil hardness and biological masses over large areas to random fluctuations. It also confirms the above-mentioned hypothesis regarding the possibility of using the exponential distribution of time intervals (paths) of existence of microstructures.

It is also necessary to mention the physical manifestation of the law of large numbers in the formation of external conditions and indexes of working machines for the stretch of large time intervals. The law of large numbers leads mainly to normal space-time distributions of most characteristics. In fact, an experimental check-up confirms that such generally accepted characteristics of the physical state of soil as humidity, compaction and density and also the crop-yield of different cultivation with a sufficiently high accuracy (which increases with the increase in the area of observations within the assigned limitations) correspond to the normal law of distribution. Mostly, different technological, power and operational characteristics, of the working of aggregates under fixed conditions of work also follow the normal law [55].

Shennon has proposed a hyperbolic type of distribution for large sets. He believes that in this case, the density of distribution obeys the inverse logarithmic law of distribution $P(x)=[x \ln (x_2/x_1)]^{-1}$.

Similar judgements about the possibility of the existence of an inversely proportional relationship of characteristics of objects and the probability of their existence on a limited section of space have been stated by other researchers as well.

The mathematical description of processes given here can be generalized for the multidimensional case, when the indexes of effectiveness are determined by the set of factors X_Ω. In this case, if each index accepts a value from the set $\{x_{kj}\}$, $k=\overline{(1, n)}$, $j=\overline{(1, m_k)}$ with probability $\{P_{kj}\}n$, then the criterion of effectiveness should be determined for $\prod\limits_{k=1}^{n} m_k$ states of conditions

$$MQ^i_{X_\Omega}=L_{Q^i}\sum_{S=1}^{\prod\limits_{k=1}^{n} m_k}\left(\prod_{k=1}^{n} P_k l_{kS}\right) Q^i(x_{1l_{1S}}, x_{2l_{2S}}, \ldots, x_{nl_{nS}}),$$

where

$$x_{kl_{kS}}\in\{x_{kj}\};\; P_k l_{kS}\in\{P_{kj}\};\; l_{kS}=\overline{(1, m_k)};\; j=\overline{(1, m_k)};\; k=\overline{(1, n)}.$$

At present, such characteristics can be achieved only partially because of the tedious labor required for the experiment. Real prospects in this field obviously lie in the development of methods of mathematical modeling and the storage of data regarding external conditions and the principles of the functioning of machines.

CHAPTER 3

Formulation of Decision-making in Testing of Agricultural Technological Processes

7. FORMULATION OF COMPLEX CRITERIA WITH A PHYSICAL CONTENT

Formulation of decision-making from the results of testing of agricultural technological processes is based on the comparison between a multidimensional vector and a certain etalon (standard) vector. The components of the multidimensional vector could be the experimental data, while the etalon vector could be the predicted data possessing the characteristic of the etalon specimen of their combination.

The chief peculiarities of the procedure for decision-making are given below.

1. Data on the formalized comparison can serve only as the source material for decision-making, because it disregards considerations of a techno-economic nature (the availability of industrial capacities, the expected changes of agricultural technology, market, potential, etc.).

2. Vectors to be compared can have different dimensions (units) and can also have sets of components which do not coincide even for identical dimensions of vector spaces.

3. In each individual case of comparison, the measure of discrepancy can be formulated differently.

4. Results of comparison have different meanings for different regions and even farms.

The peculiarities enumerated above as also other factors of formalized decision-making, or rather for formalized data analysis, help in picking out such pertinent directions for solving the problem as the formation of complex criteria, the introduction of concepts about zones for application of machines, the formation of information massives, which represent the complex characteristics of zones for applications, and the formulation of modeling methods of processes on electronic computers.

Decision-making on individual effectiveness criteria is realized by comparing the indexes obtained with normative data. These indexes characterize the effectiveness of machine application from the viewpoint of the quality of fulfillment of the technological process, the energy consumed, the expenditure of labor and direct expenses, working conditions of servicing personnel, etc. The abovesaid decision-making is also realized through the evaluation of quantitative values of these indexes, which correspond to standard recommendations or to the indexes of the etalon machine. Such a comparison is conducted with the use of statistical methods, and differences in the conditions of the experiment are taken into account by corresponding methods of its planning and the subsequent processing of information.

In the general case, a complex index can be evolved by those methods which are also used for the criterion of optimization. This is because of the fact that the process of decision-making, which represents a selection of the optimum alternative among a number of possible alternatives, is basically an optimization problem.

In conformity with these methods for decision-making, either a complex index with physical content (preferably economic) or a composite index, which is achieved as a result of some transformation of weighted and standardized (in relation to the selected basis) individual indexes, can be set up [3, 32, 47, 86, 93].

The setting up of criteria with a physical content must follow the determination and quantitative estimation of relationships between complex (integrated) indexes (for example, profit or loss from the application of estimated technology or the expenditure of labor, means, etc.) and the quality of the working of the machine as well as between power ergonomical indexes and those regimes and conditions of work. This is ridden with difficulties.

A promising approach for the solution of such a problem lies in changing the scale of organization of the problem by introducing the object under investigation as a subsystem into a more complex system and in estimating it in terms of the main effectiveness criterion for the entire system.

From the viewpoint of enhancing the effectiveness of production of agricultural produce, for which technological means are being tested, the introduction of new machines would be useful only with the expanded volume of production (higher yield) and deduced production costs.

Analysis of the structure of expenditure on the fulfillment of technological operations for the manufacture of some product shows that for the indexes of operation, power and reliability estimations, it is not difficult to achieve a complex criterion of optimization in the form of direct cost of production.

It is worthwhile to consider yield, achieved by the use of a given machine or technology, as a complex index for the quality of fulfillment of the technological process. Relationships which are necessary for its quantitative estimation can be determined by experimental methods in combination with modeling methods.

When it is not possible to achieve such relationships, factors which determine the productivity of the machine should be used as a complex index for the quality of fulfillment of the technological process by the machine. Those values of productivity for which the quality of fulfillment of the technological process satisfies established limitations are used for the estimation.

A more complicated problem appears in setting up a complex index, which reflects the expenditure or expansion of the bulk of the manufactured product in relation to the indexes of ergonomical estimation.

At first glance, more favorable working conditions of service personnel promote the increase of productivity, which in an ideal case should serve as a complex index for ergonomic criteria. However, experience in testing shows that determining the quantitative expression for the extent of increase of productivity due to ergonomic factors in the process of testing is extremely complicated.

Hence, an integrated complex index appears to be the most acceptable for an ergonomic estimation of machines. However, recent tendencies of higher labor payment, for the conditions of factors unfavorable for health, create grounds for the use of cost as the complex criterion for ergonomical factors.

Using the preceding considerations from the viewpoint of estimation, the particular criteria of effectiveness can be reduced to two more general criteria namely, crop-yield (or the quantity of the processed product of a given quality) and the expenditure expressed in terms of cost. The transformation from these to the complex criterion with economic content is fairly simple. Besides crop-yield and the expenditure of means expressed in terms of cost, for decision-making, it is also necessary to take into account the live labor involved in the use of the machine under testing or technology. Hence, for the formalization of decision-making in regard to the usefulness of the application of new technology, while considering these three indexes which basically determine the effectiveness of industrial process, it is advantageous to use the more general criteria developed by A.N. Golubentsev. These criteria characterize the change in entropy of the economic system (industrial process) and determine its effectiveness, while taking into account capital investments on acquisition and the introduction of new technology, the degree of lowering expenditure on means and labor and the increase in quantity of the manufactured product [22].

The estimation of effectiveness on the introduction of new technology is not permissible in terms of the quantity of economized concrete labor in man-hours or man-shifts. Such an estimation should be realized by considering the entire economy of the systém. The industrial process being analyzed or the industrial sector of that type of product, for which the machine (being tested) and the technology are intended, comprise the said economy of the system.

In this, the totality of the sectors, which are connected to each other by the flow of production from one sector to the other, usually comprises an economic system. For such a system, the economic efficiency of labor in the realization of steps, directed toward its intensification, is determined by the degree of lowering of the entropy of the system.

In view of the fact that allowing for certain assumptions, the independent economy with developed stock-breeding, soil cultivation and other sectors of the system for obtaining agricultural production can be considered as a system belonging to the type mentioned here. Such an approach for the estimation of the effectiveness of investments, earmarked for application of new technology, can be used for estimating the expected effectiveness due to the application of new agricultural technology.

The following two situations are possible: the introduction of the subject of testing in production leads to an increased bulk of the manufactured product, i.e. to the increase of crop-yield with the same number of workers; the introduction of the subject of testing has no connection, whatsoever with the increase of crop-yield, i.e. with the expansion of production, but it saves live labor. The introduction of the subject of testing is connected with the change of production (crop-yield) and the volume of live labor.

The first situation is characteristic of the machines and technologies whose introduction does not affect the number of workers and duration of the technological cycle, but only the bulk of manufactured production (production of chickens in incubators, cocoons of silkworms, etc. without changing the number of workers). The advantage of introducing such machines and technologies is determined by the inequality:

$$(1+A)\ln\frac{p_2}{p_1}+\ln\frac{z_2^*}{z_1^*}<0, \tag{1}$$

where $A = z_k^*/z^*$ is the coefficient of capital intensity, which characterizes the relation between expenditure on increase of the unit production capacity and the cost of this production unit, P_1, P_2—the bulk of manufactured products before and after the introduction of the subject of testing, z_1^*, z_2^*—the cost of production, measured in the total working time.

In the second situation, which is characterized by the introduction of soil-processing, sowing, harvesting and other machines with a constant volume of production (for example, treated area) in a specific farm or zone. The machine or technology is effective only in that case when the following condition is fulfilled:

$$(1+A)\ln\frac{T_2^*}{T_1^*}-A\ln\frac{z_2^*}{z_1^*}<0, \tag{2}$$

where T_1^*, T_2^* is the hypothetical number out of the total number of workers, corresponding to the body of socially obligatory.

For the remaining situation, which is perhaps most widespread in the estimation of new agricultural technology, one can easily obtain the inequality which determines the effectiveness of introducing the subject under test:

$$\ln\frac{T_2^*}{T_1^*}\left(\frac{(1+A)\ln\dfrac{P_2}{P_1}+\ln\dfrac{z_2^*}{z_1^*}}{\ln\dfrac{P_2}{P_1}+\ln\dfrac{z_2^*}{z_1^*}}\right)<0. \tag{3}$$

In expressions (1) to (3), parameters of the economic system in two of its states are expressed in terms of the expenditure of socially obligatory labor. The following relationships are used for transition into the actual number of laborers being employed:

$$\frac{z_2^*}{z_1^*}=\frac{(x_2)\,\|x\|_2(\lambda^*)_2}{(x_2)\,\|x\|_2(\lambda^*)_1},\quad T_{1,\,2}^*=\frac{T_{1,\,2}}{\|x\|_{1,\,2}},$$

where $\|x\|_1$ and $\|x\|_2$ are the norms of factors, individual members of which, x_{1i}^* and x_{2i}^*, are obtained by solving the following system of equations:

$$\begin{cases}(1-\lambda_1)(x_1)_1+a_{12}x_2+\cdots+a_{1n}x_n=0;\\ x_{21}(x_1)_1+(1-\lambda_1)\,x_2+\cdots+a_{2n}x_n=0;\\ \cdots\cdots\cdots\cdots\cdots\cdots\cdots\cdots\\ a_{n-1}(x_1)_1+a_{n2}x_2+\cdots+(1-\lambda_1)\,x_n=0;\end{cases}$$

here $(\lambda^*)_1$ and $(\lambda^*)_2$ are the characteristic numbers for reducing expenditure of actual labor to the hypothetical and are obtained from the condition det $\bar{A}=0$, $\bar{A}$ being the transposed matrix of the means of production, and the coefficients a_{ij} reflect the flows of production between different sectors of the system, $T_{1,\,2}$—the number of working personnel engaged in production of the product in the sector, under analysis, before and after the introduction of new technology, $x_2, x_3, \ldots, x_n$—the actual labor consumption in the related sectors, which are determined as the ratio of the number of workers (or number of man-hours, man-shifts, etc.) to the volume of manufactured production over a given time interval.

The above-mentioned assumptions for setting up a complex criterion of effectiveness with physical content bear a very general nature, but that does not imply the impossibility of developing them in methodical detail for practical applications.

In setting up composite complex criteria, it becomes necessary to solve the following problems: standardization or reduction of indexes to dimensionless values and determination of the degree of importance (weightage) of each index setting up generalized complex criteria.

Depending on the methods of solving these problems, it is possible to create a set of methods of complex estimation. These methods, because of certain constraints on the fundamental index, have a limited scope.

In a number of works [30 et al.], a conditional index, which represents the ratio between the generalized quality index k_0, having power α, and specific reduced expenditures which are normalized relative to their basic value $k_{es} = k_0^{\alpha}(S/S^{basic})^{-1}$, is recommended for use as a complex index of effectiveness.

For calculation of the generalized quality index k_0, indexes, under estimation foreseen by the system of standards are classified on levels of hierarchical totality; by methods of expert interrogation, their number is reduced, and the weightage of each index on the corresponding level of hierarchy is determined; the transformation from absolute values of indexes to relative, is realized by formulas $y_i = P_j/P^{basic}$ or $y_j = P^{basic}/P_j$, and the complex index k_0 is calculated by the formula

$$k_0 = \sum_{j=1}^{n} \alpha_j \sum_{i=1}^{k} \alpha_{ij} y_{ij}; \quad \sum_{j=1}^{n} \alpha_j = 1; \quad \sum_{i=1}^{k} \alpha_i = 1.$$

Individual indexes, which are also determined by methods of expert interrogation, appear as the veto* index. In the presence of a wide deflection of the indexes from the established (by experts) values of veto, $y_i = 0$.

The procedure for calculation of the main index of effectiveness is sufficiently simple in the realization (if the organizational difficulties of expert interpretation are not to be considered), but contains a series of deficiencies which limit the effectiveness of the estimation of machines on the recommended criterion. Unjudicious use of this method can lead to undesirable results.

For practical realization of methods for setting up composite criteria as the base criteria, it is recommended [30] that optimum values of agrotechnical, power and other indexes are selected for the zone, depending on the type of estimation, i.e. $P_{basic} = P_{opt}$. Then, so as to determine P_{opt}, it is necessary to solve the problem of optimization, i.e. to select such a value of P at which some criterion of effectiveness will accept the extremal value under imposed limitations. A similar approach is recommended for a scientifically based selection of standard values of indexes contained in the agrotechnological requirements for the machine.

However, if P_{opt} can be successfully determined, then in that case, the necessity of setting up a composite criterion may be eliminated. This is because of the fact that on the basis of formalized relationships, while solving optimization problems, it is possible to determine complex criteria with physical content on the results of testing. Therefore, it is worthwhile to take the values of indexes of the etalon machine for reference.

It should also be mentioned that in standardizing, relative indexes y_j (in the case of preferential increase of absolute values) are connected with P_j linearly $y_j = P_j/P^{basic}$, and indexes with preferential decrease are connected

*Refer Chapter 2.

nonlinearly $y_j = P^{\text{basic}}/P_j$ and, therefore, calculation of

$$k_0 = \sum_{j=1}^{k} \alpha_j y_j,$$

as the sum of linear terms in the variables α_j, y_j cannot be accepted as sufficiently well-founded; for a nonlinear relationship when $P_j \to 0, y_j \to 0$; and if $P_j < 1$, then y_j increases to the extent when k_0 becomes insensitive to the remaining values of y_j, i.e. because of certain merits in the machine, a series of other, more important, demerits get accounted for.

So as to avoid these demerits to a certain degree, it is worthwhile to use the following expression in the calculation of y_j:

$$y_j^* = \frac{P^j_{\text{test}} - P^j_{\text{etalon}}}{P^j_0 - P^j_{\text{etalon}}},$$

where P^j_0 is the value of index P under the condition when the process is being ideally realized by the machine [59].

For new machines, which are better than etalon machines, y_j^* is positive but does not exceed $+1$; and for machines, which are inferior to the etalon, this index is negative with no lower limit (in this lies its demerit).

Interpretation of y_j^* from the positions of systems analysis consists in the following: if $y_j^* = 0$, the subsystem, which is characterized by the index P^j_{test}, does not in any way affect the main index on replacement of the standard machine by the machine under test; if $y_j^* > 0$, this effect is positive and at $y_j^* = 1$, the subsystem operates ideally; if $y_j^* < 0$, the abovesaid effect is negative.

In some works, other methods of setting up generalized criteria of quality are proposed.

In order to reduce absolute values of estimated indexes to dimensionless values, it is recommended to give the value of unity to the most desirable level of the index and value zero or somewhat higher (0.05)—to the least desirable. A value in the interval between 0 and $+1$ is given to levels of indexes in the interval of $P_{\min}$ and $P_{\max}$. The attempt to limit relative indexes within the range 0 to $+1$ and to evolve a procedure for standardization, while taking into account the significance of change of a concrete estimated index within the limits $P_{\min}$–$P_{\max}$, is an achievement of this method of treatment. However, the process of converting relative to absolute values with the use of qualitative considerations is not formalized. This hinders the use of the present approach for practical applications.

The value 0.5 is recommended to be given as the average value of the index from the data of the last ten years P_{av}, and to the center of the interval ($P_{\max}$–P_{av}), the value 0.8; to the limiting permissible value—0.2. This approach is also subjective and, besides, it is not clear which values of functions of correspondence are to be taken in intervals $P_{\max}$–P_{av}–$P_{\min}$, because it is not possible to plot, at any rate, even one curve with the desired properties through three points (1, 0.5 and 0.2).

The essence of the method for determining weighted coefficients in terms of the generalized index, in particular, from the agricultural yield due to the use of the machine under estimation, consists in setting up the function of effectiveness, which changes along with the variation of the index under consideration, and in finding out its weightage in accordance with the proportion of contribution of each index to the function of effectiveness. As already mentioned, if there is a possibility to calculate the function of effectiveness with physical and still better, with the economic content, then the setting up of partial functions of correspondence and weightage coefficients for the complex estimation of agricultural technology does not make sense. Therefore, for composite (conditional) criteria for the estimation of effectiveness, apparently, the method of expert sampling remains the basic method to obtain weightage coefficients.

Besides the generalized index $k_0^* = \sqrt[n]{\prod_{i=1}^{n} y_i^{\alpha_i}}$, which is the weighted geometric mean (unlike the arithmetic mean), it is recommended to use the sum of weighted squares of the differences between the absolute of standard indexes of the machine under test and of the etalon (or agriculture requirements) as the complex criterion:

$$r_0^2 = \sum_{i=1}^{n} \alpha_i^2 (y_i^{\text{test}} - y_i^{\text{etalon}})^2;$$

$$r_0^2 = \sum_{i=1}^{n} \alpha_i^2 (y_{in}^{\text{test}} - 1)^2.$$

But for the case when part l of the indexes of the machine under test is better and part $(n - l)$ is worse than the etalon machine, two estimation indexes are recommended as follows:

$$r_1^2 = \sum_{i=1}^{n-1} \alpha_i^2 (y_i - y_i^{\text{etalon}})^2;$$

$$r_2^2 = \sum_{i=1}^{l} \alpha_i^2 (y_i - y_i^{\text{etalon}})^2.$$

Decision-making on two factors is not formalized, and this makes the use of the given method for solving the applied problems difficult.

The criteria 'minimum of some of losses' and 'minimax of losses' are most widely used.

Solution $| y^* |$ is found from the first criterion mentioned here. This solution is an element of the region ω in which the total loss of optimality, while taking into account the weightage coefficients r_i in terms of all partial criteria K, should be minimum:

$$\theta(y^* r_i) = \min \left[\sum_{j=1}^{k} r_i \Phi(y) \right]. \quad (4)$$

The essence of the second criterion (minimax of losses) consists in the following: it is necessary to find the solution $| y^* |$, which is an element of the region ω where, the maximum losses of optimality in terms of partial criteria (while accounting for weightage coefficients) should be the minimum possible. Its mathematical model is:

$$\theta(y^*, r_i) \max [r_i \Phi(y^*)] = \min \max [r_i \Phi(y^*)]. \quad (5)$$

On the whole, in spite of all the imperfections in the methods for setting up conditional criteria of effectiveness deprived of physical content to a certain degree, and because of the limited value of the results achieved by their use, further work in this more effective (from the point of view of practical applications) direction appears to be necessary and important. This is so because any formalization for decision-making guarantees logic in the analysis and decrease the degree of subjectivity in the estimation of effectiveness of agricultural technology.

8. FORMULATION OF ZONAL PRINCIPLE IN DECISION-MAKING

The problem of comparison of subjects on a given criterion can lead to the independent fulfillment of a series of successive logical operations of the origin of the criterion, which characterizes the quality and technical level of the subject for testing. In this, the effect of the zone, in which the testing of machines was performed and over which the results of the tests are spread, is substantial.

The experience accumulated so far permits one to make certain generalizations and to propose general principles for the approach to statistical consideration of the zone of testing in problems of comparison.

The main initial hypothesis of the statistical analysis of the zone are:

I. The concept of a zone which disintegrates somewhat on modifications and cannot be used without more accurate definition.

A. The zone of activity of the MTS (machine-testing station). This concept is widely used in titles, not always reflecting the region attached to the MTS: Central, Altaic, Siberian, Northwest. Clear boundaries to the zone of activity of the MTS are not correctly defined, and, probably, there is no necessity to make efforts in this direction.

B. The zone of possible use or zone of application of the machine. This viewpoint is more comfortable although it also does not guarantee the unique identification of the zone. Thus the zone of use of machines for the pilling of hay is not correctly determined territorially (Latvia, part of Lithuania, part of Estonia), but also changes in time with the distribution of other more progressive technologies.

II. An important viewpoint in the determination of the zone is the type of machine. Without coordination of the type of machine, the concept of zone does not bear any meaning. A note in recommendation 'in the zone of activity of the MTS' should be considered equivalent to saying 'in the zone of application of the machine, investigated by the given MTS'. Analysis of applicability of the machine, based on mathematical modeling of its interaction with the external conditions of the zone, should necessarily precede the citation on the zone of application.

During testing on one and the same MTS, different machines, as it follows from the above, can have different zones of application. The zone of application of incubators envelops the entire span of the Soviet Union; the zone of application of combines, tested on an MTS near Moscow for harvesting of green peas, is Yaroslavsk and, probably, Orlovsk districts only.

The specialization of machine testing stations is organized in such a way that they territorially enter or are close to zones of applicability of the corresponding machine, but this does not give a basis to identify the concepts of the 'zone of MTS activity', which is purely geographic, and the 'zone of application of the machine (specific for the given MTS, studied and considered for the said MTS)'.

III. For the realization of the zonal pinciple of testing, it is necessary to recognize the existence of two viewpoints regarding the achieved efficiency because of the application of the machine.

A. The viewpoint of the consumer. The machine should guarantee increased efficiency compared with the existing or with unmechanized technology:

(1) in the given farm; and

(2) overall (without exception) farms in the region, district.

From the viewpoint of farms, where this condition is not fulfilled, it is worthwhile to retain old machines and traditional technology for the purpose of production.

B. The viewpoint of the producer (supplier). The machine should provide statewide efficiency by replacing the entire fleet of machines. In individual farms this machine can turn on to be ineffective or even unacceptable due to technological conditions (rarely applied schemes of cultivation, narrow directions of activity, peculiarities of soil-climatic conditions, locations, etc.).

Decision-making on the results of testing should be based on indexes of effectiveness, which are calculated for the zone while considering the foregoing viewpoints.

IV. One of the important initial hypotheses for the zonal principle of the testing of machines is the random nature of the data, accumulated while testing specimens.

In view of the fact that conditions of testing represent a subset of some general totality, the indexes of effectiveness (for example, reduced expenditures) represent a mean value determined with a certain confidence.

Comparison of similar values should be carried out with the use of the 'student' test if values can imply the estimation of mathematical expectations.

In a number of cases, it is possible to make assumptions regarding the laws of distribution of output values—for example, about the distribution of the following normal law. Knowing the distribution of the radom values of the indexes of effectiveness, it is always possible to determine the entropy of prediction and also the probability of the dismissal of output values beyond certain limits.

The law of distribution of output values can be established from experimental data or indirectly through the laws of the distribution of indexes, which are contained as arguments in the functions of effectiveness. In a number of cases, it is worthwhile to be limited to numerical characteristics of output parameters (OP). The following formulas are useful for their calculation:

$$M[f(x)]M = \int_{\infty}^{\infty} \varphi(x) f(x)\, dx, \tag{6}$$

where $f(x)$ is the density of the distribution of the factors.

Similarly, for two indexes

$$M[f(x_1, x_2)] = \int_{\infty}^{\infty}\int \varphi(x_1, x_2) f(x_1, x_2)\, dx_1\, dx_2; \tag{7}$$

$$D[f(x)] = \int_{\infty}^{\infty} [\varphi(x) - mf]^2 f(x)\, dx, \tag{8}$$

where $mf = M[f(x)]$ is the mathematical expectation of the function $f(x)$ and $f(x)$—the density of distribution of factors.

Similarly, dispersion of the function of two arguments is expressed as

$$D[f(x_1 x_2)] = \int_{\infty}^{\infty}\int [\varphi(x_1, x_2) - mf]^2 f(x_1, x_2)\, dx_1 dx_2. \tag{9}$$

If the linear relationships are true, then there is no need to even find the laws of distribution of factors. Therefore, it is worthwhile to linearize regression models in all cases, wherever possible.

V. The last of the hypotheses for developing a method of modeling for zonal conditions, which is worthwhile quoting here, is the selection of basic forms of quantitative analysis of relationships.

1. In experiments with quantitative factors:

(a) the use of models based on the physical content of the process, i.e. models of rational relationships. Here the problem of planning the experiment turns out to be in rendering the parameters of the model more precise; and

(b) developing models on a formal approach, selection of the relationship on the criterion of adequacy by increasing the order of the polynomial.

2. In experiments with qualitative factors:

(a) multifaceted classifications, complete and incomplete; and

(b) hierarchical classifications.

After selecting the type of machine under test as the main viewpoint for extraction of the zone, two cases will be considered:

(a) the machine performs that technological process for which it is intended, but its effectiveness at different points of the zone is low in comparison with the effectiveness of traditional technology. By the points of zone, one understands a farm, and by traditional technology—technology which is used on the given farm. By effectiveness the value of some factors/index of quality estimation of the machine (in a particular case, it can be reduced expenses) is understood. In such a way a zone, which can be called the zone of use, is marked out. In determining the zone of use, it is assumed that the machine replaces various traditional technologies. Thus the potato-harvesting combines, used in the Northwest zone of the USSR, are designed to replace the throwing type machines, potato-lifters and others. In short, the zone of use is determined only from the principal possibility of the use of the machine; and

(b) that region, in which the machine provides a guaranteed effect in comparison to the existing technologies, principally differs from the abovesaid zone and can be called the region (zone) of applicability. In the abovementioned example, this zone corresponds to the totality of elements of area, for which the values of x satisfy the condition $OP \geqslant OP_0$. And, in the second case, the concept of the zone of applicability is introduced (OP_0—the output parameter for original technology).

When there are several factors, the relationship of output indexes with them permits one to mark out a multidimensional region on the condition $OP \geqslant OP_0\ (x_{1c} < x_1 < x_{1d};\ x_{2c} < x_2 < x_{2d};\ x_{ic} < x < x_{id} \ldots x_{nc} < x_n < x_{nd})$. This region corresponds to the area inside the zone of use, where use of the machine under consideration is undoubtedly effective.

By returning to the one-dimensional case, the relationship $OP = f(x)$ which is linear in the region of change of the index from x_{min} to x_{max} will be found out.

For example, let the qualitative factor be the crop-yield of grass and OP—the losses which are selected as a criterion. Relationships of x can be put in conformity with the distribution of the values of the factor; for example, close to the normal, it is $f(x)$. Then calculation of the combined effect, from the machine in the zone of its application, should be carried out with a correction on the law of the distribution of factors. Without taking into account the distribution, the effect (cumulative) from the application of the machine will be expressed by the following value:

$$E_m = \int_{x_{min}}^{x_{max}} \lambda' f(x)\, dx = \lambda' \int_{x_{min}}^{x_{max}} f(x)\, dx, \tag{10}$$

where λ' is the constant coefficient which takes into account the load as a factor, for example, the portion of the area of the zone which comes on the unit factor.

Let the common area of the zone be 100000 hectares, factor x—the crop-yield, which varies within the limits from $x_{min} = 10$ quintals per hectare to $x_{max} = 40$ quintals per hectare; then λ'—the load on one quintal of crop-yield—comes to:

$$\lambda' = \frac{100000}{40 - 10} = \frac{100000}{30} = 3330 \text{ ha/quintal.}$$

In such a way, the coefficient λ' gives a possibility to 'lay' the machine in the zone.

On dividing λ' by the total area, one gets the normalized value of the coefficient:

$$\lambda = \frac{\lambda'}{S},$$

where S is the area of the zone of application.

In this case, expression (10) takes the following form:

$$E_m = S\lambda \int_{x_{min}}^{x_{max}} f(x)\, dx. \tag{11}$$

If the density of distribution of the area of the zone on the factors is not constant, which happens in reality, λ becomes a variable quantity $\lambda(x)$. This variable is standardized on using the condition $\lambda(x)\, dx = 1$, and the calculation of the effect of application of the machine in the zone should be carried out by the formula

$$E_m = S \int_{x_{min}}^{x_{max}} \lambda(x) f(x)\, dx. \tag{12}$$

So, for determination of the effectiveness of the machine in the zone where no machine finds its application, it is necessary to know the relationship $\text{OP} = f(x)$ and distribution $\lambda(x)$.

The first relationship can be determined by testing with the use of methods of planning of experiments and the second—by the results of investigation of the zone.

For simplicity, the case shall be considered when the relationships of interest are continuous. In reality, factors can be substantially discrete or qualitatively variable. This will be reflected in the technique of calculations and will not affect the statement of the problem.

The role of distribution of $\lambda(x)$ shall be examined in more detail. For the case of normal distribution of the index and linear relationship $OP=f(x)$, it is sufficient to calculate numerical characteristics of the quantity OP. And only then, on the basis of the known properties of distribution, the probability of appearance of undesirable values of output index should be predicted.

Continuity of the factor should not be identified with any of the topological properties of the zone. So territorially the zones of application and the applicability of any two machines can be schematically represented by sections of the surface which are not interconnected.

Hence, as a supplement to the testing of agricultural technological processes and in conformity with organizational and structural principles which are laid down as the basis of the selection of machine-testing stations, the zones should be differentiated on the following basis: activities of the MTS and the possible and effective application of the machine. So as to estimate the effectiveness of the use of technological process at all points in the country where agricultural crops are produced, it is necessary that the zone of activities of the MTS (S_{MTS}) should be limited to distinct territorial borders in the administrative order, and care should be taken that the entire territory engaged in agricultural production is covered. The zone of the possible use of the machine S_p is determined by the zone of cultivated culture of the agricultural operation for the mechanization of which the machine under test is meant. This zone should not necessarily be included in the zone of activity of the MTS, i.e. condition $S_{MTS} \cap S_p > 0$ can exist.

The zone of effective application M (zone of applicability) is the zone in which a new machine or technology is more effective than the previous one or it provides guaranteed effectiveness.

For the zone S_p and M, relations $M \subset S_p$ and $S_p \geqslant M$ are obvious.

The relation between the zones of effective application for two subjects M_1 and M_2 under test (comparison) can be of four types (Fig. 7):

$$\text{I}—M_1 \subset M_2; \quad \text{II}—M_2 \subset M_1; \quad \text{III}—M_1 \cap M_2 = 0;$$

$$\text{IV}—M_1 \cap M_2 < M_1 \text{ and } M_2.$$

As a criterion, on the basis of which the above mentioned relationships are established, the average values E_c of the generalized (complex) index of effectiveness can be used. This index is derived while taking its statistical

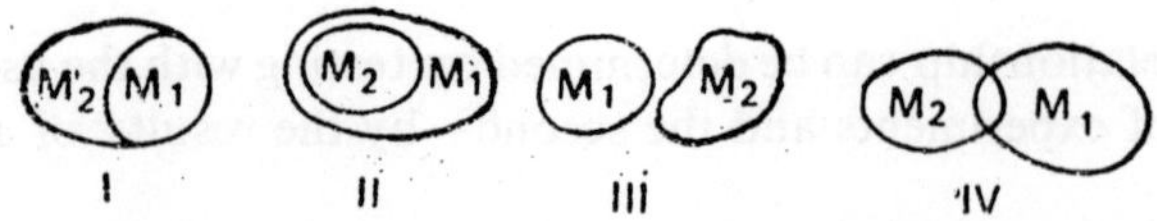

Fig. 7. Relationships between zones of effective application of subjects M_1 and M_2 under test.

nature into account:

$$E_c = \iint_S \ldots \int f[(x_1, x_2, \ldots, x_n), p(x_1, x_2, \ldots, x_n)]dx_1, dx_2, \ldots, dx_n.$$

For such a criteria as yield, expenditure and profit for the entire zone, $E = SE_c$ where $f(x_1, x_2, \ldots, x_n)$, $p(x_1, x_2, \ldots, x_n)$ are the criteria of effectiveness, E_c is described as the functions of the factors of external conditions and production situations, $p(x_1, x_2, \ldots, x_n)$—the density of the probability distribution of systems of factors, which is determined by experimental methods.

Function f is determined from the results of the factorial experiment in the entire range of action of factors.

The procedure of comparison of two subjects under test as an example of one-dimensional factorial space shall be considered.

Let $x_1 < x < x_2$ (Fig. 8) be the values of the factor, which for comparison correspond to the region of possible application of machines under test. The zone of effective application of the first machine is $x_3 < x < x_4$, of the second, it is $x_5 < x < x_2$ (Fig. 8). If the effectiveness of machines is expressed by the functions $E_1(x)$ and $E_2(x)$, respectively, then the combined effect from the replacement of machine A_1 by machine A_2 in the zone of their possible application is expressed uniquely by the function:

$$\Omega = S_1 \int_{x_3}^{x_4} E_1(x)p(x)\,dx - S_2 \int_{x_5}^{x_2} E_2(x)\,p(x)\,dx, \tag{13}$$

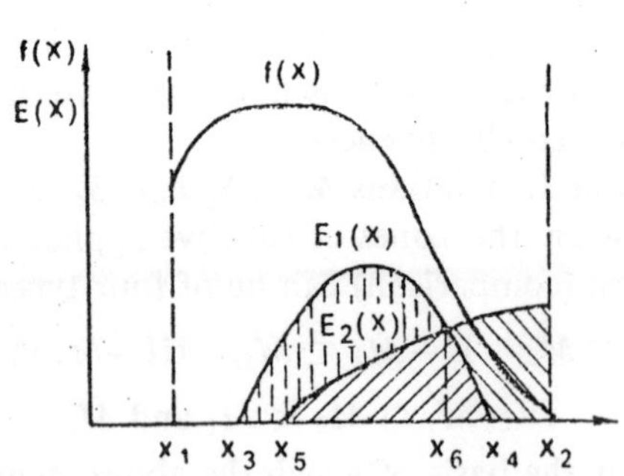

Fig. 8 Determination of the effect from the replacement of machine A_1 by machine A_2 having effectiveness corresponding to $E_1(x)$ and $E_2(x)$, respectively.

where S_1, S_2 are the areas of the zone of the applicability of machines.

The relationship of the zone of applicability of machines for the given example corresponds to relationship IV in Fig. 7.

Analyzing these relationships in combination with the values of combined effect Ω, eight typical cases of comparison of two machines can be distinguished (Fig. 9). Symbols taken in the diagram: $+A_2$—preference is given to machine A_2; $-A_1$—preference is given to the other machine; $+A_2+A_1$ —the zone of applicability is divided in two, due to the fact that both machines A_1 as well as A_2 are required; A_2+A_1—both machines can find application but for decision-making, additional information is required. Questions related to the relationship IV lie beyond the limits of rigid methods.

Conclusions of the type A_2+A_1, characteristics to the intersection of the zone of applicability, reflect transition to the production of new machines which will provide a higher combined effect in comparison to the one being replaced, besides the fact that some farms would refuse to bear the losses.

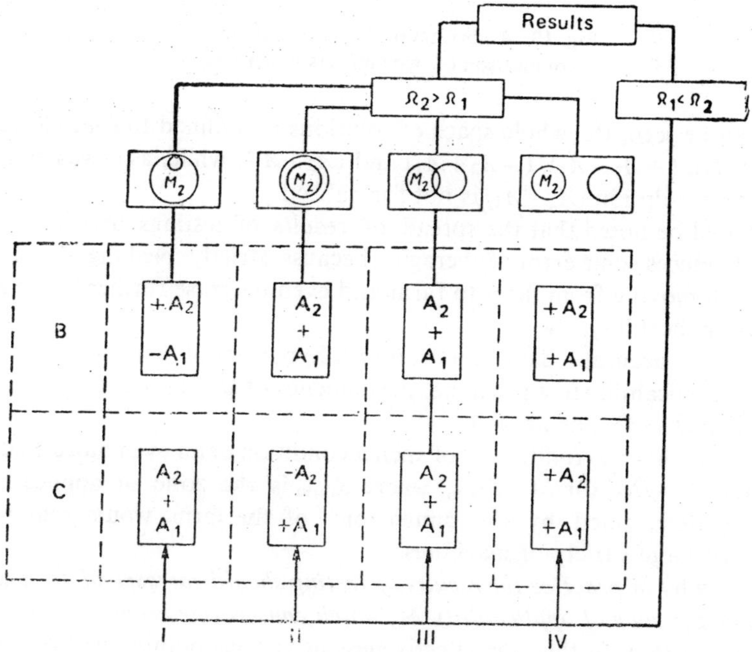

Fig. 9. Typical cases of comparison of two subjects, taking into account the zonal conditions.

The method of comparison and decision-making in these conditions deserves separate treatment, taking into account the possibilities of reconsideration of farm specialization, the production of different types of specialized machines, etc.

Combination IC, IIB, IIIC, IIIB (see Fig. 9) cannot be formalized.

The algorithm of decision-making, considering the effect of the machine in the zone of applicability in its general form is shown in Fig. 10.

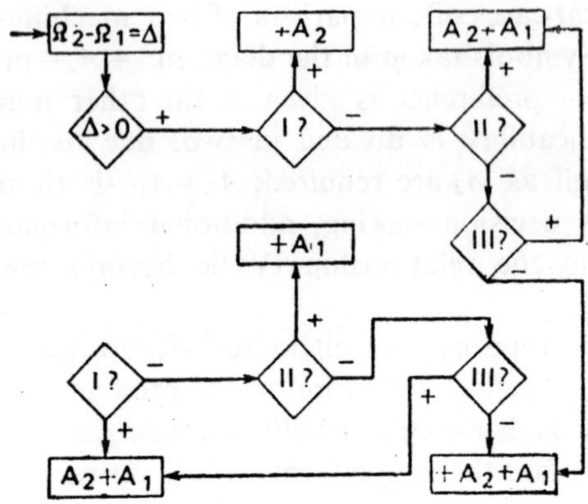

Fig. 10. Algorithm of decision-making for comparison of two subjects under test.

As can be seen, the whole space of solutions is confined to four elements: $(+A_2+A_1)$, $(+A_2-A_1)$, $(-A_2+A_1)$ and (A_2+A_1), while, as it was pointed out earlier, solution (A_2+A_1) is not formalized.

It should be noted that the spread of results of testings over a certain zone introduces some error of averaging because, strictly speaking, the factors change on moving from farm to farm and even decrease further in the zone of their applicability.

The most accurate data can lead to a hypothetical 'contraction' of the zone of applicability to a point, i.e. the tendency of Ω to approach its limiting value Ω_{lim}: $\Omega \to \Omega_{lim}$ as $M \to 0$.

It is clear that the fulfillment of such a condition or an even more realistic condition $\Omega \to \Omega_{lim}$ for $M - S_{farm}$, where S_{farm} is the zone of applicability, which is determined by the dimensions of the farm, would require an unjustified large variety of machines.

That is why *in practice the results of testing should correspond to a sufficiently larger zone of applicability M, which can provide economically viable batch production.* In this, the effectiveness at various points turns out to be different; at some points, it can be negative besides the total positive cumulative effect in the zone.

CHAPTER 4

Statistical Methods of Analysis and Synthesis

9. METHODS OF ANALYSIS

A significant number of special works are devoted to the analysis methods of massive statistical information. Academician V.P. Goryachkin recommended that the results of investigations should be processed with the help of effective mathematical techniques, which include the method of least squares, the theory of errors, the theory of approximate calculations, the technique of graphical interpretations and the comparison of empirical data with the calculated results such that the degree of accuracy and the reliability of results in all cases remain absolutely clear.

Groups of modern testing problems in which statistical methods of analysis are widely used will be considered. These problems include the primary statistical analysis of the experimental data on the basis of the use of parametric and nonparametric tests and the estimation of the significance of the difference between samples under test [52, 69], the estimation of the significance of the effect of condition and regimes of working and also of the design peculiarities of machines under test on the criterion of effectiveness; the estimation of optimum parameters and conditions of working of machines under test; the search of sources of disturbances in kinematic chains, units, mechanisms and technological processes of agricultural machines; the formalization of conditions of functioning of objects under testing; the building of generalized criteria of effectiveness and formalization of decision-making on the results of testing; and lastly the interpretation of experimental results.

Estimation of the quality of functioning of the agricultural technological process is based on the investigation of the complicated environment-machine stochastic system. One of the main methods of investigation of such systems appears to be the factorial experiment, which is oriented toward the use of plans with qualitative factors (plans of dispersion analysis) and with quantitative factors (plans of regression analysis). A rigorously worked out theory of experiment decision [1, 2, 48, 53, 65, 74, 76, 79, 89] makes it possible to use certain processes for analysis, estimation and forecasting at the testing stage for the effectiveness of agricultural technological processes. However, specific questions of the theory of testing related to the statement of applied problems, the analysis of justification and the selection of factors, procedures of experiment and lastly the interpretation of the results of experimental investigations are not sufficiently worked out. In the majority of specific problems on testing, their main content lies in the objective achievement and comparison of certain complexes of output factors, which can be represented in the form of a multidimensional vector $y_1 = | y_1, y_2, y_3, \ldots, y_n |$, with a set of standardized data $y = | y_1, y_2, \ldots, y, \ldots, y_n |$.

At present, there is no complete scientific justification of the methods of achievement of multidimensional characteristics and their comparison during testing of agricultural technological processes. One of the possible ways of solving this problem appears to be the use of methods, which were considered in the previous chapter, and development of concepts pertaining to the peculiarities of experiments in the agricultural field.

Generally, experimentally achieved information is used in different ways, depending on the problem of testing.

Very often achieved data are compared with standardized agrozootechnical requirements, which correspond to a given technical level or with data on a competitive object. The comparison results give a basis for the solution of questions about the effectiveness of a new machine. Such a problem is characteristic to the information model of testing of experiment samples.

A significant part of the problems is concerned with the control for maintaining the complex of factors which characterize the machine on a given level in mass production. Experimental information in this case represents itself as knowledge about the degree of correspondence between the complex of the standardized (dictated by technical and other conditions) factors of the working of the machine and the actually obtained data.

One of the important problems of statistical analysis is the search for cause and effect relationships between operational and technological processes.

From the point of view of the use of mathematical methods, the most general problem appears to be the estimation of the effectiveness (technological, technical and other characteristics) of working parts or the machine as a whole [59].

For the solution of these problems, it is necessary to take into account the states and properties of external conditions which correspond to the actual factors in nature. In the opinion of Prof. V.V. Nalimov, these factors control the experiment, i.e. they determine the general pattern of functioning of the machine and the processes.

In order to understand them and to qualitatively analyze them in the real processes observed in nature and in the working of agricultural machines, it is necessary to distinguish three components (three-point logic): to determine the part which can be exactly accounted for and even provisionally measured; to determine the probable part manifested during prolonged observations and arising in the process of the working of the machine or the interaction of external factors; to determine the purely random part which cannot be predicted (error in the operation of working organs, error of instruments which cannot be measured or controlled, etc.).

It is known that the study of the determinate part is possible through inductive methods for a unique exposure of cause and effect whereas, the random component can be determined only through deductive methods by using a statistical approach rather than a microscopic approach. These situations should be accounted for not only in developing methods for testing but in the complexes of instruments for the multidimensional analysis of processes.

For the estimation of machine aggregates and complexes, it is necessary to determine the techno-economic effect of the new process relating to the zone of the application of the machine. An increase in the reliability of data and the justification of recommendations requires consideration of the probable error which always accompany the process of decision-making.

There are two such errors in the general case: errors of the first kind namely, the rejection of the machine under test if it is more advanced in comparison to the standard one; and an error of the second kind, i.e. the acceptance of the subject under test for production even if it does not provide substantial improvements.

Observance of scientific principles of reliability can be achieved by a wider introduction of the methods of experiment planning and also by methods of mathematical modeling of processes, environment and farm situations.

Presently, the estimation of a machine is carried out from the point of view of its correspondence with specific technological and operational requirements. However, incomplete analysis of effectiveness taking zonal conditions into account gives rise to a number of demerits—the most important of which is that the data obtained on the MTS and the indexes, which characterize the machine in operation under widely varying farm conditions, do not correspond, i.e. there is an apparent unsatisfactory reliability of prediction of machine efficiency in the zonal content. Reasons for the abovesaid divergence may be of two types: noncorrespondence of the method for obtaining the factors, idealization of the parameters and conditions in laboratory tests, use of nonrepresentative samples, errors in measurements and processing of data; and also the nonrepresentability of backgrounds, the conditions inw hich testing is performed without an understanding of their characteristics and the probability of their belonging to one or another state. It is difficult to eliminate reasons for the second group. Unfortunately, this is so because that would require a scientific study of the zone and the organization of experiments on the elimination of factors and on the study of their effect on the output parameters of effectiveness. It will also receive the working out and the realization of procedures of the estimation of the effectiveness of imposition of the subject under test on the given region or the zone of its application and decision-making on the basis of formalized rules.

In view of the foregoing, great significance is attached to the prediction of effect from the mass application of the machine in the zone; optimization of the regimes of use of the machine and organization of production processes; more precise technical characteristics of the object under test, agro- and zootechnical requirements, etc.

The aforementioned problems of testing pose specific requirements toward the organization of methods for obtaining experimental information and also toward the technique of processing.

Solution of these problems is possible only with the application of complex mathematical methods; probability and statistical methods for the verification of hypothesis [52, 57, 63, 69]; dispersion and regression analysis [56, 57, 77, 82, 90]; methods of experiment planning, *a priori* modeling and qualimetry [3] and also methods of sequential statistical analysis and mathematical modeling of processes (digital and analog) [13, 51, 59, 72].

As mentioned earlier, the criteria for the effectiveness of agricultural technological processes possess an objective randomness; that is why decision-making about the degree of perfection of the test subject is based on the comparison between obtained values of the criteria of effectiveness and the base (etalon) values, which necessarily require probability and statistical methods for verifying the hypothesis about the inherent difference between the characteristics compared.

Results of comparison can be represented in the form of a matrix of preferences $\| y_{ij} \|$, where y_i can be equal to zero or one. Elements of the matrix correspond to conclusions of the type 'better-worse' for each of the criteria of effectiveness y_α. If the values y_{ij} are normalized and are represented by continuous values, then it is possible to obtain a quantitative measure of preference.

The principal basis for comparison of the subject of testings appears to be the equalization of conditions of their functioning (which can be realized by various methods) in relation to the nature of the components of external conditions which determine the working process of the machine. If it is necessary to equalize conditions on determined components, which can be taken into account, machines are then checked for selected parts of backgrounds. For the equalization of conditions on purely random, unmeasurable, hard-to-consider components, the experiment is arranged while considering the random nature of the tests. In cases with partial control or uncontrolled conditions, which can be accounted for and measured in terms of characteristics, one can use methods of correlation and regression analysis for the equalisation of conditions of testing and the estimation of the machine to identical values of external factors. In this case, for the machines to be compared, factorial experiments are conducted at several levels of factors: the dependence of the criterion of effectiveness on the qualitative characteristic of factors is described by the regression equation, whereas the comparative estimation is realized from values of criteria of effectiveness calculated from the regression equation for similar values of factors.

Dispersion analysis is a fundamental mathematical method of the interpretation of results of testing for dealing with qualitative factors. This method can be used to determine which of the selected factors affect the value being examined, what is the extent of their effect, what is the numerical equivalent to the effect of the factors and their interactions under randomized and stochastic working conditions.

Considering the variety of factors (Fig. 11), which act on the technological processes under test and their effect on output values, the authors came to the conclusion that the method of analysis should take into account specific aims of testing, conditions of experiments, the nature of factors, peculiarities and the degree of their interaction with the subject for testing.

The practice of experimental investigation shows that by any of the most advanced methods of experimental data analysis, it is not possible to obtain reliable results and conclusions in those cases when the experiment is planned and realized without taking into account peculiarities of factors, objectives and aims of the experiment. That is why experiment planning methods should find wide application in the testing of agricultural technological processes [51].

The mathematical tools for design of the experiment include the totality of methods for selection of the number and sequence of observations (which are optimal to some extent) for recording procedures of observed processes and also for the statistical mehtods of processing and interpretation of results.

In the following chapters, an attempt has been made to present mathematical methods for analysis of the effectiveness of agricultural technological processes and to cite examples of their application in solving the problems of estimation of subjects for testing.

10. PRELIMINARY STATISTICAL ANALYSIS OF EXPERIMENTAL RESULTS OF TESTING

The effectiveness of the preliminary statistical analysis of experimental data from the testing of agricultural technological processes is mainly determined by the correlation between the actual situation and the scheme of performing the experiment with the selected method of analysis and is also with the degree of fulfillment of the initial assumptions underlying the development of the method.

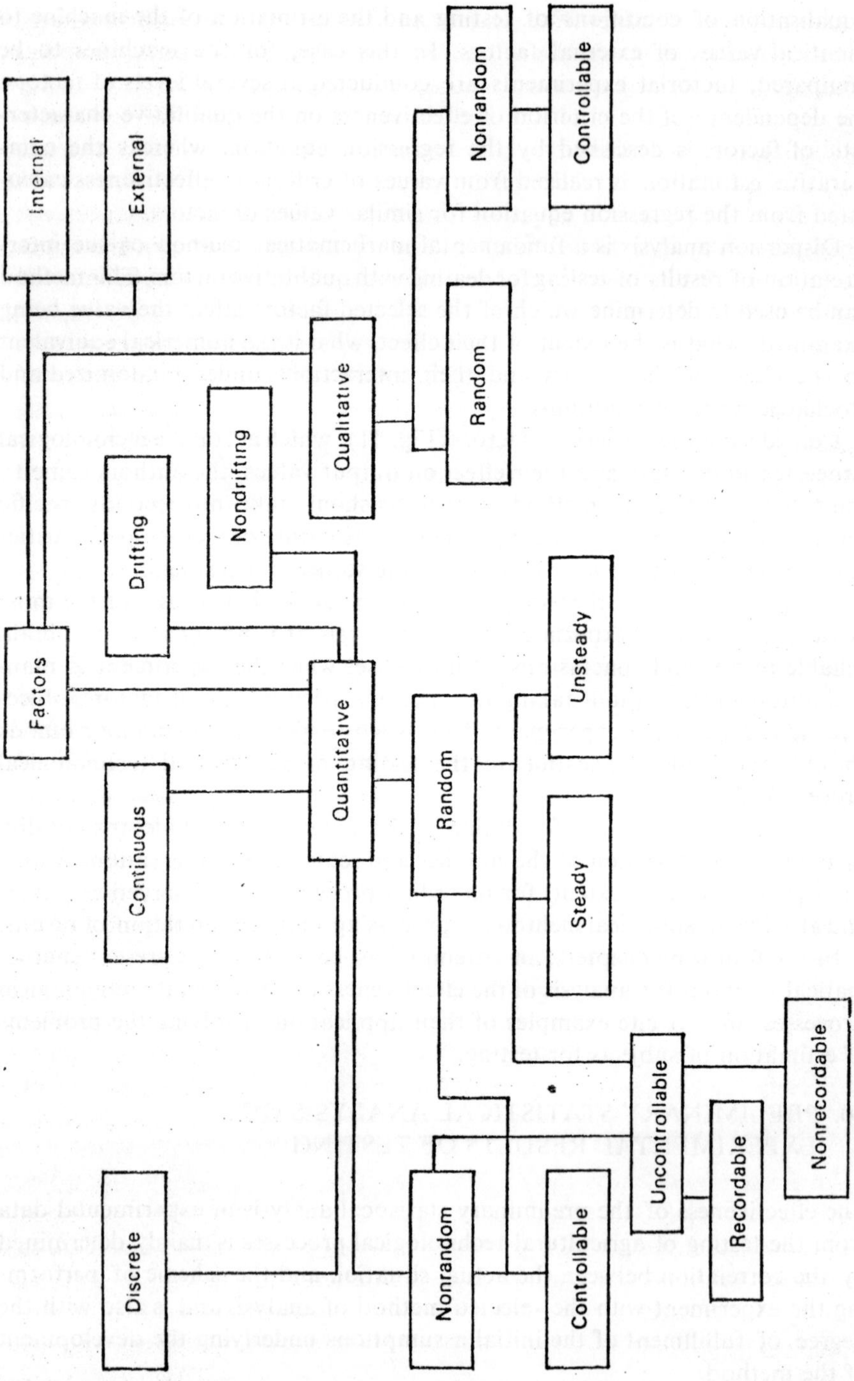

Fig. 11. Classification of factors acting on the subject under test.

As a rule, in the stage of the initial processing of experimental results, the following problems can be solved: checking of whether the two samples belong to one general totality; analysis with the view to discard sharply differing data; calculation of numerical characteristics along with a mandatory estimation of reliability on the basis of the concept of confidence intervals; comparison between mean dispersions and coefficients of variation in the respective criteria; analysis of distribution laws of values under study; estimation of parameters of theoretical distribution; testing of the hypothesis about the adequate description of empirical distribution as inferred from theoretical distribution.

A large number of works [8, 16, 52, 63, 69] are devoted to methods of statistical analysis in studying experimental results. Principles of the method of estimation of statistical characteristics and the testing of statistical hypothesis are included in a number of industry standards on agricultural technological process testing. However, despite their apparent simplicity, they have still not obtained sufficient application in the practice of testing. In particular, it is not always possible to use the criteria for verifying that the two selections belong to the same general set, not the methods for rejecting coarse errors in measurements. However, experience shows that disregard for such checking of experimental data can lead to incorrect conclusions.

Comparison of statistical samples and rejection of coarse errors in measurements. The problem of comparison between two samples or two series of independent measurements arises while estimating the uniformity of conditions of testing of machines and states of working, the time structure of industrial processes and the factors of effectiveness of standard machines and of machines under test, etc.

The given problem can be solved by using the Wilcoxon test based on calculation of the so-called inversion for checking of the null hypothesis H_0 [69]. In this, the values of two samples are ranked, and the number of values of the second sample, prior to each value of the first, are calculated.

Suppose that two ranked samples, combined into one sequence in numerical values, were laid down as follows: y_1, x_1, x_2, x_3, y_2, y_3, y_4, x_4, y_5, x_5, y_6, x_6, x_7, y_7, y_8, x_8, y_9, x_9, x_{10}, x_{11}, y_{10}, x_{12}, y_{11}, y_{12}. Then, the calculation of inversion for x_1 gives one, for x_2 and x_3, it also gives unity, for x_4—4, for x_5—5, etc. Hence, the total number of inversions $z=1+1+1+4+5+6+6+8+9+9+9+10=69$.

The essence of checking the hypothesis with the use of the above test lies in the fact that the hypothesis is rejected if the calculated number z exceeds the limit selected in accordance with the level of significance. This limit is determined from the calculation that for volumes $n>10$ and $m>10$ of samples, the number of inversions z is distributed approximately normally with respect to the center $Mz=(mn)/2$ and the dispersion [69] given by:

$$Dz=\frac{mn}{12}(m+n+1).$$

For the given example

$$Mz = \frac{12 \cdot 12}{2} = 72; \qquad Dz = \frac{12 \cdot 12}{2}\,(12+12+1) = 300.$$

If the level of confidence is fixed at 5%, for which $t = 1.96$, then the lower limit of the critical region $z_l \leqslant 72 - 1.96\,\sqrt{300} \approx 38$, whereas the upper limit is $z_u \geqslant 72 + 1.96\,\sqrt{300} \approx 106$.

As the obtained number of inversions $z = 69$ does not lie in the critical region $106 < z_{cr} < 38$, hypothesis H_0 is accepted, and there is no basis to consider that the samples being studied belong to two different population groups.

The random nature of the estimated values with sufficiently complicated and, as a rule, unknown distributions, the limitation of samples and the impossibility to repeat experiments, make it difficult to use the known [68, 69] criteria of the exclusion of coarse errors of measurements. It is considered that when there are sufficient grounds for rejection in the conditions of the experiment [69], the most reliable method of rejection of coarse errors is to exclude doubtful results.

That is why, before carrying out experiments for the estimation of technological processes under test in extremely nonuniform and variable conditions of agricultural production, it is necessary to measure (or fix at a level of qualitative description) a sufficient number of characteristics of external conditions which accompany the experiment and the state of the subject for testing working materials so as to make it possible to exclude coarse errors.

In those cases when it is not possible to do this for some reason (the experiment being carried out in uncontrollable conditions, errors in measurements of the characteristics of conditions of the experiment, the state of the subject or the work material exceeding the errors of measurements of the analyzed values of the criteria of effectiveness, the absence of a qualitative description, the content analysis or any other *a priori* information about the behavior of the subject for testing in the experiment conditions, etc.) the formalized rules of exclusion of coarse errors in observations, which are based on the methods of statistical testing of a hypothesis, are used. In this, the null hypothesis about the value under analysis x_{max} belonging to the totality to which other results also belong and the alternative to the null hypothesis, that x_{max} is a result of action at a moment (or at the point) of measurement of any factor which strongly affects the result and which is not accounted for (the erroneous recording of the measurements by the laboratory assistants, the transverse furrows while measuring the depth of the plow or plowshare) are checked.

The null hypothesis may be discarded if the value to be checked exceeds a certain critical value that corresponds to a given level of confidence. Because of the fact that event $x_{max} < x_k$ is equivalent to saying that all n

measurements will be less than x_k, and assuming that values which are obtained as a result of measurements are independent then, on the basis of the theorem regarding the multiplicity of independent events one can write

$$P(x_{max} < x_k) = (1 - q) = [\Phi(x < x_k)]^n,$$

where Φ is the distribution function of random values.

From here, it is clear that for the determination of critical value x_{max}, it is necessary to find the quantile of distribution of the measured random value which corresponds to the probability $(1-q)^{1/n}$ [17]:

$$x_k = \arg \Phi(1-q)^{1/n}.$$

For sufficiently strict rejection of widely differing results of measurements with the help of formalized rules, it is necessary to know the law of distribution of the measured random value. Unfortunately, in the practice of experimentation, such a law is not known in the majority of cases, and it is replaced by some hypothetical law (as a rule by the normal distribution) with parameters which are also determined on the results of observations. This argument is inherently present in the fact that formalized methods of rejection of widely differing results of measurements can be recommended for use only in extreme cases because the reliability of these methods depends on how close the selected hypothetical distribution and value of its parameters are to the real ones.

From special literature, the methods of rejection of coarse errors of measurements are known for the cases of the normal law of distribution, either if the dispersion is known before the experiment or if the average value and dispersion are to be evaluated from the results of experiments. For these cases, the quantile of distribution of the centered and standardized random value are tabulated in correspondence with the level of confidence and number of measurements, while the methods of checking are distinguished only by the tables [68, 69] which are used.

For normally distributed results of measurements, other tests are also applied, which are distinguished by the completeness of the use of values of the sample under analysis and by the random values of the quantiles which are tabulated, etc. In using, for example, the Romanovskii test, the average values and dispersion are evaluated uniquely; in using the Irwin test, the quantile for a given level of probability and set of measurements are tabulated with respect to $\frac{x_n - x_{n-1}}{\sigma}$. Sometimes, the so-called rule of 3σ is used when values, which do not lie in the interval $(\bar{x} = 3\sigma, \bar{x} + 3\sigma)$, are rejected. This test may be applicable only in the conditions of a sufficient number of measurements and when the distribution of the parameter is very close to normal distribution.

Example: In chronometric observations for setting up an aggregate for working, the following values of time for joining the agricultural machines and the tractor were recorded: 6, 6, 4, 3, 4, 5, 8, 1, 6, 5, 4, 4, [min]. As a result of calculations, the following estimates were obtained for the average value $\bar{x} = 4.66$ and the mean-square value $\sigma = 1.7$.

It is required to check whether values 1 and 8 are coarse errors of measurements. For a 5% level of confidence with $n = 12$, on the assumption that for the time of linking the machine with the tractor is normally distributed, critical random values are obtained $(x - \bar{x})/\sigma : \beta_2 = 2.28$. Corresponding to the value 1, $\beta = (1 - 4.66)/1.70 = 2.15$; and for the value 8, $\beta = (8 - 4.66)/1.70 = 1.97$.

Comparing these values (2.15, 1.97, and 2.28), it is seen that there is no basis for the rejection of values 1 and 8 ($1.97 < 2.15 < 2.28$). If the Romanovskii test is used, then with probability 0.95, it can be confirmed that $x = 1$ and $x = 8$ do not belong to the given population, and those can be therefore discarded.

Thus it is seen that checking on those two tests leads to contradictory conclusions, which once again points out the necessity of thoroughly weighing the possibility of using the hypothesis of normal distribution.

In the above example, the distribution of the time for connecting the machine with the tractor and also the time for fulfillment of other production and technological processes bear a distinct asymmetric character. This distribution is limited by the minimum essential time of fulfillment of the operation being analyzed and can be very roughly described by the normal law of distribution. That is why the given criteria of the rejection of widely differing values find a very limited application in the testing of agricultural technological processes. If for the given example, a detailed analysis is carried out for the process of linking the machine with the tractor, then it turns out that this operation cannot be physically realized in one minute; whereas eight minutes can be spent on the fulfillment of this operation only due to unmatched links or insufficiently processed design solutions. From these considerations, the value unity should be discarded while there is no basis for rejection of the value eight.

Testing of the hypothesis regarding the equality of statistical characteristics. The majority of testings are confined to comparisons of machines and technologies regarding individual statistical characteristics of effective working conditions. Very often the average values, dispersions and coefficients of the variation of the parameter under study are compared.

Suppose, as a result of testing, the average values and dispersions of productivity of two machines x and y, being calculated from the results of observations n_x and n_y, are equal to m_x, σ_x^2, and m_y, σ_y^2 respectively.

Average values (assuming a normal distribution) are compared with the help of t (Student's test). To do this, at first dispersions are compared with the help of the F criterion. This ratio between the higher dispersion and the lower one is calculated and if the ratio $F = s_1^2/s_2^2$ for $s_1^2 > s_2^2$ or $F = s_2^2/s_1^2$ for

$s_1^2 < s_2^2$ is higher than the tabulated value for a given level of confidence α and the number of degrees of freedom $k_1 = n - 1$, $k_2 = n_2 - 1$, then the hypothesis regarding the inequality between two general dispersions is applicable. In case, $F \leqslant F_{1-\alpha/2}$ then the hypothesis of equality of dispersions is applicable.

For testing of the hypothesis regarding equality of two average values m_x and m_y in the case when $\sigma_1^2 - \sigma_2^2$, the total dispersion is calculated by the following formulas

$$s^2 = \frac{(n_2 - 1)\, s_1^2 + (n_2 - 1)\, s_2^2}{n_1 + n_2 - 2};$$

$$t = \frac{m_x - m_y}{s} \sqrt{\frac{n_1 n_2}{n_1 + n_2}}.$$

The obtained value of t is compared with the tabulated value $t_{\alpha,\,k}$, which is found for the level of confidence α and $k = (n_1 + n_2)$ degrees of freedom. If $t < t_{(\alpha,\,k)}$ then the hypothesis about equality $m_x = m_y$ holds good. In these cases, when $\sigma_1^2 \neq \sigma_2^2$, then the equality of two average values is checked with the help of the approximate t criterion by the formula

$$t = \frac{m_x - m_y}{\sqrt{s_1^2 n_2^{-1} + s_2^2 n_2^{-1}}},$$

for the following number of degree of freedom

$$\frac{1}{k} = \frac{C^2}{n_1 - 1} + \frac{(1 - C)^2}{n_2 - 1},$$

where

$$C = \frac{s_1^2}{n_1} \left(\frac{s_1^2}{n_1} + \frac{s_2^2}{n_2} \right)^{-1}.$$

For $|\, t \,| \leqslant t_{\alpha,\,k}$, $m_x = m_y$. Otherwise, $m_x \neq m_y$.

There are certain simplified, though less accurate, methods of testing the statistical hypothesis regarding the absence of nontrivial differences between the average values.

If the number of measurements for each sample exceeds 25, then as an approximate estimation of the nontrivial difference, the inequality $(\bar{x}_1 - \bar{x}_2)/\mu_p > 3$ is used, where μ_p is the so-called error of difference which is calculated for independent values by formula $\mu_p = \sqrt{\mu_1^2 + \mu_2^2}$. Values μ_1 and μ_2 in their turn are calculated by formula $\mu = \sigma_m = \sqrt{s/n}$, where s—the estimate of the mean-square deviation and n—the number of measurements.

For estimation of the nontrivial difference of coefficients of variations y_1 and y_2, the following test is applied [68]:

$$t_v = \frac{v_1 - v_2}{\sqrt{v_1^2 (2n_1)^{-1} + v_2^2 (2n_2)^{-1}}},$$

where n_1 and n_2 are the volumes of two samples.

For $t_v > 3$, the difference in the coefficients of variation is assumed to be trivial.

Testing of hypothesis regarding the laws of distribution. It is worthwhile to determine the laws of distribution, the characteristics of conditions and the efficiency of the machine operation measured during tests, in the following sequence [16, 68, 69]: to select a proposed type of the law of distribution and express it in the form of a formula; to estimate parameters of distribution; to test the hypothesis regarding conformity of experimental data with the selected law of distribution.

In order to determine the law of distribution, the method based on the qualitative analysis of the histogram, plotted from experimental results, is most widely used. The number of ranks in the histograms should be sufficiently large so as to bring out the nature of the curve of the probability density, and the interval of quantification should be sufficient so that at least 5-6 measurements lie in it. The recommended number of ranks is not less than eight [69].

It is recommended to combine the ranks in which an insufficient number of observations are placed. The histogram, thus obtained, is compared with the graph of densities of standard distributions, and on the basis of the qualitative considerations regarding their closeness, the most suitable is selected. Ths abovesaid method is most widely used in practice of testing due to its simplicity, but it does not provide high accuracy. The more accurate and informative method appears to be the determination of the laws of distribution on the basis of a study of the mechanism of the phenomenon, for example, the laws of distribution of particles during fragmentation, which was obtained by A.N. Kolmogorov, the laws of destruction and multiplication for systems of Borodachev [9] curves. In the present work, an example is given for the construction of the law of distribution of intervals between seeds and between beet plants on the basis of probability considerations about the mechanism of their formation. Estimation of parameters of the anticipated law of distribution are obtained by methods of the moments of maximum likelihood.

In using the method of moments for evaluation, the unknown parameters of distribution are selected in such a way that certain important numerical characteristics (moments) of the theorotical distribution describe the corresponding characteristics, determined from experimental data. If characteristics are determined with the help of the distribution of grouped data, which is widely used in the practice of testing, then dispersion is calculated while taking into account Sheppard's corrections [68]:

$$\sigma^2 = \frac{1}{n-1} \sum_{i=1}^{n} n_i(x_i - \bar{x})^2 - \frac{1}{12}n^2.$$

The method of moments, in the case of many-parameter distributions, leads to less effective results because of the large errors in estimating moments of a higher order [16, 68] in view of the high simplification.

The method of maximum likelihood permits, in the best possible way, the use of all information about unknown parameters and to forecast their determination from the condition of transformation to the maximum of the function:

$$L(\alpha_1, \alpha_2, \ldots, \alpha_l) = \prod_{j=1}^{n} f(x_j, \alpha_1, \alpha_2, \ldots, \alpha_l),$$

where $\alpha_1, \alpha_2, \ldots, \alpha_l$ are the parameters under estimation. For the use of this method, in practice it is necessary to solve the following complicated system of equations:

$$\left.\begin{aligned} \sum_{j=1}^{n} \frac{\partial \ln f(x_j, \alpha_1, \alpha_2, \ldots, \alpha_l)}{\partial \alpha_1} &= 0; \\ \sum_{j=1}^{n} \frac{\partial \ln f(x_j, \alpha_1, \alpha_2, \ldots, \alpha_l)}{\partial \alpha_2} &= 0; \\ \sum_{j=1}^{n} \frac{\partial \ln f(x_j, \alpha_1, \alpha_2, \ldots, \alpha_l)}{\partial \alpha_l} &= 0. \end{aligned}\right\}$$

Use of computers help to overcome the difficulties in the estimation of the parameters of distributions by the method of maximum likelihood.

Of the large number of different criteria for checking the agreement of the data with the accepted law of distribution, the chi-square goodness of fit-criterion, proposed by Pearson, is most widely used:

$$\chi^2 = n \sum_{i=1}^{m} \frac{(\hat{p}_i - p_i)^2}{p_i},$$

where n is the number of measurements, m—the number of ranks in which the elements of a sample are divided, $\hat{p}_i$—the frequency, determined from the experimental series of distribution, p_i—the probability that a random value belongs to the ith class calculated for the accepted law of distribution, for example, by formula

$$p_i = \int_{x_{\min}+(i-1)h}^{x_{\min}+ih} f(x)\, dx.$$

If the calculated value of chi-square is more than or equal to the tabulated value [51, Appendix of Table 9], then it should be accepted that the selected assumption about the law of distribution of the random value x poorly corresponds with the results of experiment and should be discarded. In the contrary case, the assumption about the type of law of distribution of random values does not contradict the experimental data.

In conclusion, certain practical recommendations and examples of construction of histograms of distribution will be furnished.

While breaking the series of distributions into intervals, it is necessary to keep in mind that the value and limits of intervals should be in conformity with the nature of the approximation of the varieties in the process of the measurement of the characteristics. The simplest rule of conformity lies in the following: the limits of intervals of grouping should coincide with the limits of approximation so that all the intervals of approximation are included in the interval of grouping.

If this rule is not followed, then the feequency of certain intervals of division will have to be broken up in two parts by artificial methods, distributing them among the adjoining intervals of grouping. Correct and incorrect construction of intervals of division are shown in Fig. 12, a and b.

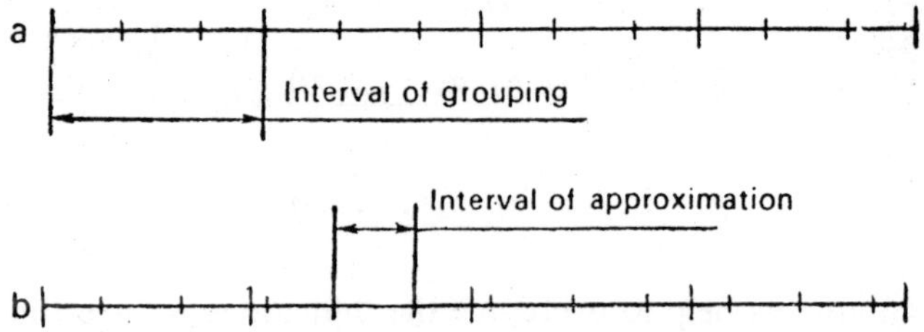

Fig. 12. Selection of the interval of approximation.

The edge intervals of approximation are broken up by the limits in two parts; their frequency should be correspondingly distributed among adjoining intervals of grouping.

Attention should also be paid to the fact that there should not be rounded numbers (1000, 1500, etc.) in the vicinity of the limits of intervals because during testing there is a tendency to round off the adjoining numbers. It would be better if such numbers lay close to the centers of the intervals.

While plotting the histograms, it is generally assumed that the density of frequency is constant with each interval and only changes in jumps at the limits of the intervals. More close to reality is the assumption about the uniform change of density from interval to interval, which leads to the representation of the interval series in the form of polygons of distribution where the ordinates denote the density of frequency rather than the frequency itself. The frequency of the interval is represented in the form of the area bounded by a curvilinear trapezium limited on the top by a section of the polygon, on the bottom by the abscissa and on the sides by two ordinates (Fig. 13).

Suppose in the measurement of the length of the beet root for the estimation of working conditions for a beet-harvesting machine, 106 observations

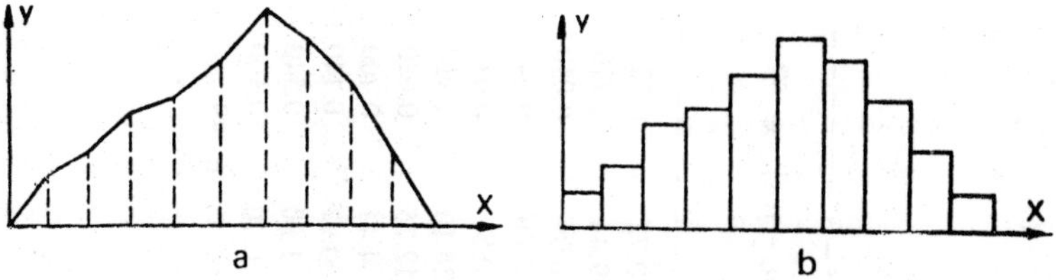

Fig. 13. (a) Polygon and (b) histogram of distribution: x—interval of grouped data; y—frequency of interval.

are obtained, which are classified in 11 classes with intervals of two centimeters (Table 2). The mean value and mean-square deviation are $m = 24.5$ cm and $\sigma = 4.56$ cm, respectively. It is necessary to test the hypothesis about the normal distribution of the length of the beet root.

The values of function $1/2\Phi(t)$ are found from the tables of normal distribution for standard values of upper and lower limits of the intervals, while values of the theoretical probability of the grouping of random values into class-intervals are determined by subtracting $1/2\Phi(t_i)$ from $1/2\Phi(t_{i+1})$. The obtained value for $\chi^2 = 6.55$, is less than the tabular value for the 5% confidence level ($\chi^2_{\alpha=0.05} = 15.51$), which gives a basis for accepting the hypothesis about the normal distribution of the length of roots in the beet plant.

Techniques of calculations are described in detail in Table 2.

The chi-square test is also used for the testing of the hypothesis regarding the closeness of two empirical distributions [69]. Such a problem is encountered in testing for the distribution comparison of beet sizes in two sections, for comparison of intervals between the plants for two seeding machines and for comparison of the time intervals of the production process for two machines, etc.

In this case, two independent populations of sets n_1 and n_2 are classified in l groups with the number of measurements in each of them being m_1', m_2', ..., m_l' and m_1'', m_2'', ..., m_l''.

In this

$$\sum_{i=1}^{l} m_i' = n_1, \quad \sum_{i=1}^{l} m_i'' = n_2.$$

It is required to test the hypothesis that the distributions of measured characteristics are identical in both the populations.

The chi-square test for this case is calculated by the formula [69]:

$$\chi^2 = n_1 n_2 \sum_{i=1}^{l} \frac{1}{m_i' + m_i''} \left(\frac{m_i'}{n_1} - \frac{m_i''}{n_2} \right)^2,$$

for degree of freedom $(l-1)$.

The chi-square test is used for the checking of the hypothesis—even in more simple cases. For example, it is known that the tomato-harvesting

Table 2

Class	$x_i - x_{i+1}$	x_i	m_i	$x_i - \bar{x}$	$t_i = \frac{x_i - \bar{x}}{\sigma}$	$1/2\ \Phi(t_i)$	$p_i = 1/2\ \Phi(t_{i+1}) - 1/2\ \Phi(t_i)$	$p_i n$	$(m_i - p_i n)$	$[m_i - p_i n]^2$	χ^2
1	14–16	14	–163	–10.5	–2.3	–0.4895	0.0210	2.22	0.78	0.72	0.324
2	16–18	16	8	–8.5	–1.86	–0.4685	0.0465	4.93	3.07	9.44	1.910
3	18–20	18	9	–6.5	–1.42	–0.4220	0.0830	8.70	0.30	0 09	0.010
4	20–22	20	13	–4.5	–0.99	–0.3390	0.1300	13.80	0.81	0.66	0.048
5	22–24	22	28	–2.5	–0.55	–0.2090	0.1650	17.10	0.90	0.81	0.047
6	24–26	24	20	–0.5	–0.11	–0.0440	0.1390	14.70	5.30	28.10	1.910
7	26–28	26	16	1.5	0.24	0.0950	0.1845	19.50	3.50	12.25	0.620
8	28–30	28	9	3.5	0.77	0.2795	0.1055	11.10	2.10	4.41	0.400
9	30–32	30	6	5.5	1.20	0.3850	0.0655	6.94	0.94	0.97	0.140
10	32–34	32	2	7.5	1.65	0.4505	0.0315	3.34	1.34	1.80	0.540
11	34–36	34	2	9.5	2.08	0.4810	0.0110	1.16	0.84	0.70	0.600
	–	36	–	10.5	2.40	0.4920	–	–	–	$\chi^2 = 6.549$	

Remarks: Degrees of freedom $k = 11 - 3 = 8$.

machine spoils 25% of the produce. In the conditions of testing a sample with 400 fruits (spoiled—90, unspoiled—310, which gives a ratio of 22.5%) is taken. It is necessary to check the hypothesis as to whether the relationship between the quantity of spoiled and unspoiled fruits is identical.

From a sample containing 400 fruits with 25% damage, there should have been 100 spoiled and 300 unspoiled pieces (theoretical relationship). The calculated values of criterion, namely

$$\chi^2 = \frac{(310-300)^2}{300} + \frac{(90-100)^2}{100} = 1.33$$

is less than the tabular value for the 5% level of confidence for one degree of freedom (for two groups $v = l - 1 = 1$), which gives the basis to accept the hypothesis about fruit damage by the tomato-harvesting machine in the ratio 1 : 3.

The chi-square test is also used for analysis of the so-called four-column tables, the essence of which can be explained in the following example (Table 3).

Table 3

Group of cows	Number of cows		Total
	with reduced milking	without reduced milking	
Milked on production equipment	a=8(10.9)	b=38(35.1)	46
Milked on experimental equipment	c=14(11.1)	d=33(35.9)	47
Total	22	71	n=93

Over a period of one month, in a herd consisting of 93 milch cows, reduction in milk has been recorded. Analysis showed that milk from 22 cows was reduced, while milk from the other cows remained at the previous level. Forty-seven cows from the herd in this period were milked on an experimental milking rig. The question arises whether the reduction in milk has been caused by the use of the new milking rig.

In the general case, the problem is put in the following way for analysis of a four-column table: distribution of subjects under study (cows) with respect to some criteria (or on two levels of one criterion, in the given example, cows with reduced milking and without reduced milking) can change under the effect of the two assumed factors or of two levels of one factor (production equipment, experimental equipment).

It is necessary to test the hypothesis about the substantiation of the effect of factors (or of levels of factors) on the distribution of volumes with respect to criteria (or with respect to two levels of the same criterion).

The theoretical number of cows, in the null hypothesis, consisting of the fact that the new milking rig has no effect on the change in milk yield, is shown in brackets near each number. These numbers are calculated on the basis of relations of the concluding line 22 : 71 : 93.

Calculation for χ^2 is shown in Table 4.

Table 4

Frequency		$(m_i - m_i')$	$[m_i - m_i']^2$	$\frac{[m_i - m_i']^2}{m_i'}$
m_i (actual)	m_i' (theoretical)			
8	10.9	−2.9	8.41	0.77
38	35.1	2.9	8.41	0.24
14	11.1	2.9	8.41	0.76
33	35.9	−2.9	8.41	0.23
				2.00

For the four-column tables, the number of degree of freedom is taken to be one; therefore, the calculated value χ^2 is less than the tabular value for the 5%-level of confidence, and there is no basis to assume that decrease in milking is caused by the experimental milking rig.

For the fourfold table, χ^2 is calculated from the formula

$$\chi^2 = \frac{(ad - bc)^2 n}{(a+b)(c+d)(a+c)(b+d)},$$

whereas for our example

$$\chi^2 = \frac{(8 \cdot 33 - 38 \cdot 14)^2 \, 93}{46 \cdot 47 \cdot 22 \cdot 71} = 2.$$

While using this formula, the necessity of the calculation of theoretical frequencies is eliminated. More accurate results can be obtained if χ^2 is calculated by the formula which defines Yates' correlation:

$$\chi^2 = \frac{[\,|ad - bc| - 0.5n]^2 \, n}{(a+b)(c+d)(a+c)(b+d)}$$

$$= \frac{[\,|8 \cdot 33 - 14 \cdot 38| - 0.5 \cdot 93]^2 \, 93}{46 \cdot 47 \cdot 22 \cdot 71} = 1.35.$$

For a small number of subjects, Yates' corrections are absolutely necessary. The value of χ^2, calculated without corrections, is always somewhat increased and indicates a greater significance of results than in reality.

It must be noticed that the results of testing of the hypothesis about the law of distribution with the help of the chi-square test are somewhat arbitrary because they depend to some extent on the fulfillment of the law of distribution of χ^2 for the given volume of the sample satisfied, and also on the division of results of measurements into sufficiently arbitrarily selected classes.

In such a division (classification), part of the information contained in the initial data is lost. Therefore, it is recommended [69] to supplement the value of the chi-square test by values of other tests.

The ω^2 (omega-square) test is based on the use of ungrouped data and is calculated by the formula:

$$\omega^2 = \frac{1}{12n^2} + \frac{1}{n}\sum_{k=1}^{n}\left[p(x_k) - \frac{2k-1}{2n}\right]^2,$$

where $p(x_k)$ is the value of the theoretical function of distribution when the measured value is equal to x_k, and n—the number of measurements.

For the values $n\omega^2$, the critical points, depending on the level of confidence, are given below [69]:

Level of confidence $q = p(n\omega^2 > z_q)$, %	5	3	2	1	0.1
Critical points, z_q	0.4614	0.5489	0.6198	0.7435	1.1679

If the calculated value of $n\omega^2$ is less than the tabular one for the given level of confidence, then the hypothesis about the law of distribution can be accepted.

Besides the chi-square and omega-square tests, a series of other tests may be used for testing the hypothesis about the law of distribution.

The Romanovskii test, which is as follows:

$$k_p = \frac{\chi^2 - \nu}{\sqrt{2\nu}},$$

where ν is the number of degree of freedom.

If k_p is less than three in absolute value, then the difference between theoretical and empirical distributions can be considered trivial.

For applying the Kolmogorov test λ, the closeness between theoretical and empirical distributions is determined by means of comparison of their integral distributions with repect to D (maximum upper limit of the absolute difference of accumulated frequencies) [16]:

$$\lambda = D/\sqrt{n},$$

where n is the number of observations.

If the value λ is less than the tabular one for a given level of confidence, then the hypothesis about random deviation of the theoretical and empirical distributions can be accepted.

In practice, the Yastremskii test can be applied for comparison between distributions:

$$J \leqslant 3\sqrt{2n+4\theta},$$

where $J = C - n$; $C = \sum \frac{(m - m')^2}{m'q}$; m are the empirical frequencies, m'—the theoretical frequencies and n—the number of groups.

When the number of groups is less than 20, $\theta = 0.6$ and $q = 1 - p$. The value of J, is less than $3\sqrt{2n + 4\theta}$, in the Yastremskii test shows an insignificant deviation between empirical and theoretical frequencies in the given distribution.

For values J, higher than $3\sqrt{2n + 4\theta}$, the deviation between theoretical and empirical distribution is significant.

Example: As a result of the comparison of empirical distribution with theoretical normal distribution, the following values are obtained: $C = 5.65$, $n = 13$, $\theta = 0.6$; therefore $(13 - 5.65) \leqslant 3\sqrt{26 + 2.4} = 3\sqrt{28.4}$; $7.35 < 3 \times 5.34$; $7.35 < 16.02$.

The results of analysis confirm a normal distribution of the population under investigation, because calculation was carried out with the use of theoretical frequencies of normal distribution.

The use of Vestergard numbers, 0.3, 0.7, 1.1, 3.0, belongs to the elementary test methods for normal distribution. For using these numbers, the basic characteristics are first determined, namely, the arithmetic mean $\bar{x}$ and the mean-square deviation σ are determined, and then transition to the scale of variance is carried out, expressed in fractions of σ.

So as to consider that the given empirical distribution follows the normal law, it is required that the following conditions of distribution of random value with respect to intervals formed by the use of the Vestergard numbers are fulfilled:

(1) a quarter of the entire aggregate lies in the interval $\pm$ 0.3;
(2) half of the entire aggregate lies in the interval ± 0.7,
(3) three-quarters of the entire aggregate lies in the interval $\pm$ 1.1; and
(4) 0.998 of the entire aggregate lies in the interval of ± 3.0.

Graphical testing method. The normality of distributions can also be checked by the graphical method. The accumulation of frequency is represented on semilogarithmic paper. The plot in this case should be a straight line.

The method discussed in this section do not fully reflect all the aspects of the preliminary statistical analysis of the results of experiments in testing of agricultural technological processes. This happens because in such an evaluation, as well as in a number of others, the real teacher is experience, and no book giving a detailed description of the methods of the statistical processing of data is in a position to replace it.

11. ACCURACY OF MEASUREMENTS IN THE TESTING OF AGRICULTURAL TECHNOLOGICAL PROCESSES

Estimates of effectiveness of agricultural technological processes under test and their correspondence to the stated requirements can be realized from the quantitative characteristics, which reflect the entire set of multifaceted properties of the subject under test. Accuracy of measurement of those characteristics determines the reliability and objectiveness of decisions made on the basis of test results. Moreover, an unjustified application of accurate, more complicated, expensive and labor consuming methods and the technical implements of measurements results in the increase in schedules and costs of testing. That is why in testing of agricultural technological processes, a choice of accuracy of methods and implements for measurements, taking into account the properties of characteristics to be measured and the required reliability of their estimation under limitations on the duration and cost of experiments, appears to be the main problem. This is because it determines the quality and effectiveness of testing. It is necessary to solve this problem simultaneously with the selection of the program and the method of carrying out the testing [27, 37, 43, 68].

Accuracy of measurements, it is well known, is characterized by absolute $\Delta x = x - \bar{x}$ or relative $\Delta x/x\ 100\%$ errors, which in their turn are represented as the sum of systematic δ and random ξ errors. Systematic errors in the measurements in testing of agricultural technological processes are the errors of instruments or methods of measurements. These errors, in the majority of cases, are known or can be determined with the help of etalons or by comparison with more accurate instruments or methods. Random errors of measurements, the occurrence of which varies from one measurement to the other and cannot be accounted for, can be provisionally divided into two groups: errors which depend on the changes in conditions of measurements and errors which occur due to the variable nature of the quantity being measured. Since the values measured in testing of agricultural machines appear to be of a random nature, and they change in the majority of cases within sufficiently wide limits, the second group of random errors appears to be dominating and determines the order of the value of the random error, and often, even of the cumulative error. In this connection, the measurement of parameters under study should be carried out so as to ensure the presentability of the random sampling.

There are two ways to enhance the accuracy of measurements: first—a decrease in systematic error, i.e. an increase in the accuracy of instruments and methods of measurements and second—a decrease in random error by selecting (stabilizing) uniform conditions of measurements (for decrease in the random error of the first group) and by increasing the number of measurements (for decrease in the random error of the second group). Which of these methods should be given preference is decided for every concrete case depending on the availability and costs of instruments, labor consumption in measurements, real comparability possibilities and the uniformity of the measurement conditions. But, at any rate, it is necessary to strive toward reduction of the error having a higher value in order to reduce the cumulative error and to increase accuracy of measurement.

There exists an incorrect viewpoint that by increasing the number of measurements, it is possible to obtain the least possible error. But, in reality, only the random error can be reduced in this way; the systematic error remains constant because that is determined mainly by the errors of instrument and the method of measurements.

For example, intervals between the seeds of beets are measured with a systematic error of 0.5 cm. The average interval is 4-6 cm. It is not possible to attain accuracy more than 10-13% despite the number of measurements which will be performed for the estimation of the average value of the interval. In the works [27, 68], the authors do not recommend the decrease of the random error by more than 10 times; whereas in practice, they recommend reducing it by 2-3 times as compared to the systematic error.

The average value is generally used as an estimate of the measurement result:

$$\bar{x} = \frac{1}{n}\sum_{i=1}^{n} x_i.$$

If the condition of limiting the theorem ($x_i \ll \sum x_i$) is satisfied and the result of each measurement is independent of the previous and subsequent results, then the distribution of the sum turns out to be approximately normal for a sufficiently large number of measurements. This important property of independent random values is used for the determination of random errors of measurement, i.e. limits of the confidence interval $t_\alpha s/\sqrt{n}$, which is determined with confidence coefficient α:

$$P\left[(x - \bar{x}) \leqslant t_\alpha \frac{s}{\sqrt{n}}\right] = \alpha,$$

where t_α is the Student test which is determined from tables for n degrees of freedom and the accepted confidence coefficient and $s^2 = \frac{\sum_{i=1}^{n} (\bar{x} - x)^2}{n-1}$ —the estimation of dispersion of measurements.

It is difficult to formalize the rule which can be used for the determination of the confidence coefficient of measured values in testing. In the majority of cases, for solving the problems of testing, the value of α should not be more than 0.95 [27]. The problem of formalizing rules for selecting a sufficient accuracy of measurements or an allowable error is also difficult. Generally, it is intended that the relative error should not exceed 5%. The allowable error should be selected on the basis of the importance of recommendations to be made from the results of the measurements as well as on the basis of availability of instruments, methods, the cost of the measurements, relationship of extra expenditure to increase the reliability of the measurements and the economy accrued out of it. Systematic errors of instruments are generally given in the form of the maximum value of the possible error without any information about their distribution. In the practice of testing, frequently, the systematic δ and random ξ errors are close to each other, and both determine the cumulative measurements to the same degree. In the estimation of accuracy of measurements, it is necessary to figure out cumulative error, $\sum$, which is approximately determined by formula [27]:

$$\sum = \delta + 2\frac{\sigma}{\sqrt{n}}. \tag{13}$$

In testing, measurements are generally carried out by linear or angular measuring instruments (by scales, measuring tape, goniometer) and other instruments, and the results are approximated to the value of the least count of the instrument (spacing between seeds, plants, thickness of swath, depth of sowing of seeds and others). In such measurements, it may be considered that the systematic errors follow the law of uniform distribution with parameters which are equal to half of the accounted division Δ. Dispersion of the systematic error in this case is determined by formula:

$$\sigma_{\bar{x}}^2 = \frac{1}{12}\Delta^2. \tag{14}$$

This dispersion of the cumulative error is given by:

$$\sigma_{\Sigma}^2 = \frac{1}{12}\Delta^2 + \frac{\sigma^2}{n}, \tag{15}$$

and the absolute error will get determined as

$$\sum = t_\alpha \sqrt{\frac{1}{12}\Delta^2 + \frac{\sigma^2}{n}}. \tag{16}$$

Many of the indexes for evaluation are determined in testing by indirect methods, i.e. after a certain modification of directly measured values. Estimation of errors in such results is equivalent to the estimation of errors of the measured value ζ, which is related to measured independent values x_1, $x_2, \ldots, x_n$ by a certain relationship $\zeta = \varphi(x_1, x_2, \ldots, x_n)$, where the function φ is known.

Each of the arguments $x_1, x_2, \ldots x_n$ is measured by different methods to different accuracies; the problem of the experiment is in obtaining the values ζ with the maximum accuracy (finding out the best estimate of $\bar{\zeta}$) and the determination of accuracy of the calculated estimation [68]. In solving this problem, it is necessary to consider two main aspects: the nonlinearity of function ζ and the intervals of variation of random values $x_1, x_2, \ldots, x_n$. If φ is a linear function then, independent of intervals of variation x_i, one has

$$M(\zeta) = \varphi(\bar{x}_1, \bar{x}_2, \ldots, \bar{x}_n);$$

$$D(\zeta) = \sum_{j=1}^{n} \left(\frac{\partial \varphi}{\partial x_j} \sigma_{x_j}\right)^2. \quad (17)$$

If function φ is nonlinear and the interval of variation x_i is small, making it possible to linearize φ with a negligible error in this interval, then $\bar{\zeta}$ and $D(\zeta)$ are calculated by approximate formulas (17). If function φ is nonlinear and intervals of variation are substantial, then for estimation of $M(\zeta)$ and $D(\zeta)$, in the one-dimensional case, it is necessary to use the following expressions [68]:

$$\bar{\zeta} = \varphi(\bar{x}) + \frac{1}{2}\left(\frac{d^2\varphi}{dx^2}\right)_{\bar{x}} \sigma^2_{x_\Sigma}; \quad (18)$$

$$D(\zeta) \approx \left(\frac{d\varphi}{dx}\right)_{\bar{x}} \sigma^2_{x\Sigma} + \left(\frac{dp}{d\bar{x}}\right)_{\bar{x}} \left(\frac{d^2\varphi}{d\bar{x}^2}\right)_{\bar{x}} \mu_3^{(x)} + \frac{1}{4}\left(\frac{d^2\varphi}{dx^2}\right)^2_{\bar{x}} \left(\mu_4^x - \sigma^4_{x_\Sigma}\right), \quad (19)$$

where μ_3, μ_4 are the third and fourth centralized moments of x, respectively.

For nonlinear functions of some variables, similar estimations are found with the help of operators describing the mathematical expectation and the dispersion to the expansion of function φ in Taylor's series, limiting the series to terms of the second order only [68].

$$\bar{\zeta} = \varphi(\bar{x}_1, \bar{x}_2, \ldots, \bar{x}_n) + \frac{1}{2}\sum_{i=1}^{n}\left(\frac{\partial^2\varphi}{dx_i^2}\right)_{\bar{x}} \sigma^2_{x_i}; \quad (20)$$

$$D(\zeta) = \sum_{i=1}^{n}\left(\frac{\partial\varphi}{\partial x_i}\right)^2_{\bar{x}} \sigma^2_{x_i} + \frac{1}{4}\sum_{i=1}^{n}\left(\frac{\partial^2\varphi}{\partial x_i^2}\right)^2_{\bar{x}} \left(\mu_4[x_i] - \sigma^4_{x_i}\right)$$

$$+ \sum_{i<j}\left(\frac{\partial^2\varphi}{\partial x_i \partial x_j}\right)^2_{\bar{x}} \sigma^2_{x_i}\sigma^2_{x_j} + \sum_{i=1}^{n}\left(\frac{\partial\varphi}{\partial x_i}\right)_{\bar{x}}\left(\frac{\partial^2\varphi}{\partial x_i^2}\right)_{\bar{x}} \mu_3[x_i]. \quad (21)$$

For random values distributed approximately according to the normal law, it may be considered that $\mu_3[x] = 0$ and $\mu_4[x] = 3\sigma_x^4$. Then, formulas (18) and (21) can be written in the following forms:

$$D(\zeta) = \left(\frac{\partial\varphi}{\partial x}\right)_{\bar{x}} \sigma^2_{x_\Sigma} + \frac{1}{2}\left(\frac{\partial^2\varphi}{\partial x^2}\right)^2_{\bar{x}} \sigma^2_{x_\Sigma}; \quad (22)$$

$$D(\zeta) = \sum_{i=1}^{n}\left(\frac{\partial\varphi}{\partial x_i}\right)^2_{\bar{x}} \sigma^2_{x_i} + \frac{1}{2}\sum_{i=1}^{n}\left(\frac{\partial^2\varphi}{\partial x_i^2}\right)^2_{\bar{x}} \sigma^4_{x_i} + \sum_{i<j}\left(\frac{\partial^2\varphi}{\partial x_i\partial x_j}\right)^2_{\bar{x}} \sigma^2_{x_i}\sigma^2_{x_j}. \quad (23)$$

Formulas (18) to (23) are applicable to the case when the arguments are independent.

In most cases, these values appear to be dependent or correlated. It is necessary to take this into account while estimating errors in values measured indirectly. In order to estimate the error in these cases, the following expressions can be used, which are given in the works [16, 68]:

the sum of related random value is

$$M\sum_{i=1}^{n} x_i = \sum_{i=1}^{n} M(x_i); \qquad D\sum_{i=1}^{n} x_i = \sum_{i=1}^{n} D(x_i) + 2\sum_{i<j} K_{ij}, \tag{24}$$

where K_{ij} is the correlation moment of values x_i, x_j; and

the product of random values is

$$M(xy) = M(x)\,M(y) + K_{xy};$$

$$D(xy) = D(x)\,D(y) + \bar{x}^2 D(y) + y^2 D(x); \tag{25}$$

(the formula for dispersion of the product is given for independent random values);

linear functions are: $\sum_{n}^{i=1} a_i x_i + b = \varphi$

$$M(\varphi) = \sum_{i=1}^{n} a_i M(x_i) + b;$$

$$D(\varphi) = \sum_{i=1}^{n} a_i^2 D(x_i) + 2\sum_{i<j} a_i a_j K_{ij}. \tag{26}$$

In these cases, when it is necessary to determine an estimate of the quotient or the product of correlated values, it is worthwhile to linearize these functions by taking logarithms of both sides; thus:

$$\ln \zeta = \ln x_i + \ln x_2. \tag{27}$$

Having determined the average dispersion and correlation moment of $\ln x_1$ and $\ln x_2$, the corresponding values of $\ln \zeta$ are evaluated by the formula (27).

Transformation to characteristics for ζ is performed by formula (21) for the function $\varphi = e^{\ln \zeta}$.

For nonlinear, but marginally differing from linear functions in the domain of variation of values of correlated random arguments, approximate formulas are recommended [16, 68]:

$$\bar{\zeta} = \varphi(\bar{x}_1, \bar{x}_2, \ldots, \bar{x}_n);$$

$$D(\zeta) = \sum_{i=1}^{n} \left(\frac{\partial \varphi}{\partial x_i}\right)_{\bar{x}}^2 \sigma_{x_i}^2 + 2\sum_{i<j} \left(\frac{\partial \varphi}{\partial x_i}\right)_{\bar{x}} \left(\frac{\partial \varphi}{\partial x_j}\right)_{\bar{x}} K_{ij}, \tag{28}$$

where K_{ij} are the correlation moments.

While estimating errors which are calculated on the results of the measurements of criteria of effectiveness in the multidimensional case, it is necessary to use not only dispersion of random errors, as it is accepted in the practice of testing, but also dispersion of the cumulative error calculated by formulas

(13) and (16). Such an estimation is necessary in order to formulate assumptions regarding reliability and dependability of obtained results, to compare them with results of other MTSs. It is also required to estimate the effect of accurate measurement of individual values on the accuracy of estimation of the criterion of effectiveness and to establish, on the basis of this measurement, which of these values and to what degree they should be increased so as to increase the accuracy of final results.

The possibility of using the given expressions for the estimation of accurate measurement of techno-operational and economic criteria of effectiveness of mobile agricultural aggregates shall be considered in greater detail. The net productivity of such an aggregate is determined as the quotient of the sum of the shift-output of the aggregate to the sum of intervals of the net working time:

$$W=\frac{\sum\limits_{i=1}^{N} Q_i}{\sum\limits_{j=1}^{N}\sum\limits_{i=1}^{n} t_{1ij}}, \tag{29}$$

where Q_i is the output of an aggregate in a shift, which is equal to the product of width B and the length of the processed area.

The accuracy of measurement of the area shall be determined, taking into account the errors of instruments used for measurement of lengths

$$Q=BL;$$

$$D(Q)=\frac{1}{12}\,\Delta_L^2\,\frac{1}{12}\,\Delta_B^2+B^2\,\frac{1}{12}\,\Delta_L^2+L^2\,\frac{1}{12}\,\Delta_B^2$$

considering that $\Delta_L^2=\Delta_B^2=\Delta^2$, $D(\sum x_i)=\sigma_x^2$.

The dispersion of systematic error $\sum Q_i$ is equal to

$$D(\sum Q_i)=\frac{\Delta^2}{12}\sum_{i=1}^{n}\left(\frac{\Delta^2}{12}+B_i^2+L_i^2\right),$$

while the dispersion of random error $D_c(\sum Q_i)=\sigma_Q^2$.

For the time of net working during a shift, the dispersion of the systematic (error of the method) is given by

$$D(t_j)=\sigma_{ij}^2+n_j\,\frac{\Delta^2}{12},$$

where σ_{ij}^2 is the dispersion of time intervals during a shift.

Dispersion of the absolute error of the net working time is given by

$$D(\sum t_{ij})=\sigma_t^2+\sum_{j=1}^{N}\left(\sigma_{ij}^2+n_j\,\frac{\Delta^2}{12}\right),$$

where σ_t is the dispersion of net working time of the machine and N—the number of shifts.

Since output $\sum Q_i$ for the time of net working of the aggregate $T = \sum t_{1ij}$ are related parameters, therefore, the estimation of dispersion of productivity is carried out by using approximate formulas.

In these formulas, instead of the correlation moment, its statistical estimate is used, having calculated that from the experimental data for values of accrued operating and net working times being described as growing sums:

$$M(W) = Q/T;$$

$$D(W) = \left(-\frac{Q}{T}\right)^2 D(T) + \left(\frac{1}{T}\right)^2 D(Q) + 2\left(-\frac{Q}{T}\right)\left(\frac{1}{T}\right) r_{QT}\,\sigma(T)\,\sigma(L).$$

Similarly, the output for unit technological shift and operational times are evaluated:

$$D(T_2) = \sigma_{t_2}^2 + \sum_{j=1}^{N}\left(\sigma_{t2j}^2 + n_j \frac{\Delta^2}{12}\right);$$

$$D(T_{\text{tech}}) = D(T_1) + \sigma_{t_2}^2 + \sum_{j=1}^{N}\left(\sigma_{t2j}^2 + n_j \frac{\Delta^2}{12}\right) + \sigma_{t_3}^2$$

$$+ \sum_{j=1}^{N}\left(\sigma_{t3j}^2 + n_j \frac{\Delta^2}{12}\right) + \sigma_{t_4}^2 + \sum_{j=1}^{N}\left(\sigma_{t4j}^2 + n_j \frac{\Delta^2}{12}\right).$$

In a similar way, dispersion of time for technological servicing and elimination of technical faults and time for idle running is determined.

For traversing from one field to another, the dispersion of the average value will be, for example, equal to

$$D(t_{1\ \text{tra}}) = \left(\frac{-L_1 B_k}{v^2 s}\right)^2 \frac{\sigma_v^2}{n_v} + \left(\frac{B_k}{vs}\right)^2 \frac{\sigma_{L_1}^2}{n_L} + \left(\frac{L_1}{vs}\right)^2 \frac{\sigma_{B_k}^2}{n B_k} + \left(\frac{-L_1 B_k}{vs^2}\right)^2 \frac{\sigma_s^2}{n_s},$$

where n_x is the number of measurements in estimating the average value of parameter x.

Using the above expressions, formulas for the estimation of accuracy of other criteria of effectiveness of technological processes under test, calculated from results of direct measurements, can be obtained.

12. DESIGNING OF EXPERIMENTS FOR TESTING

The mathematical theory of experiments makes it possible to apply mathematical methods not only in the final stage of the experiment during the processing of results, but also while setting-up the problem of formalization of the *a priori* data, i.e. selecting the optimum scheme for conducting the experiment or analyzing the obtained results of experiments and decision-making.

Recent works on the practical application of these methods for the solution of some problems of testing render experiment planning an integral part of the methodology of testing. However, for working out the methodological fundamentals for experiment planning in testing of agricultural technological processes, development of a corresponding scientific base is required.

In future descriptions, only some general principles of the design of experiments and recommendations and examples of their applications while estimating the agricultural technological process would be dealt with.

During testing of agricultural machines, specialists inevitably came across the necessity of determining a large number of the parameters of machines and processes simultaneously, viz. indicators of technological, operational, economic, technical and other types under various kinematics, dynamic and regulating states and external working conditions. In fact, any type of testing leads to carrying out experiments on the comparative investigation of multiparameter objects.

The aim of designing such experiments appears to be the search for those conditions and rules for carrying out experiments so that it is possible to obtain more information at less expense and to qualitatively estimate the reliability of this information.

I. *Experiment*
   - A. Statement of the problem.
   - B. Selection of the response or dependent variable.
   - C. Selection of indicators under variation.
   - D. Selection of levels for these indicators.
     1. Qualitative or quantitative.
     2. Fixed or random.
   - E. Selection of the combination of levels of indicators.

II. *Designing*
   - A. Necessary number of observations.
   - B. Order of conducting the experiments.
   - C. Method used for randomization, i.e. method of averaging the effect of unconsidered factors.
   - D. Mathematical model for description of the experiment.

III. *Analysis*
   - A. Collection and processing of data.
   - B. Computation of statistics for the testing of the hypothesis.
   - C. Interpretation of results for the experimenter.

Up to the present time, the designing of an experiment is the most developed aspect of the mathematical theory of experiments [48, 53, 74, 89] and depending on the objectives to be achieved for which the experiment is arranged, corresponding plans are recommended for application.

In those cases where it is necessary to establish that by a change of certain parameters, the value of the efficiency index of the machine under test varies, it is worthwhile to use methods of dispersion analysis. Factors x_i, whose effect on the efficiency index is determined, are fixed at two (or more) levels, and experiments are carried out for the selected combination of indicators. In the theory of experiment, standard plans of dispersion analysis and formalized methods for recommending these were developed. This helps in selecting the most suitable plan under specific conditions of experiment. While processing results of experiments, the portion of total dispersion, which is contributed by each of the indicators, is determined and the hypothesis about the importance of effect of each factor on the value of y is checked.

The procedure of dispersion analysis is very convenient in experiments, when investigators deal with factors which do not have a quantitative expression, i.e. with so-called qualitative factors. These can also be used when it is not required to develop a model which reflects the qualitative link of the criterion of effectiveness of y with indicators x_i affecting it. These plans have found a wide application in the testings of working organs, technologies and others, particularly while comparing various machines.

So as to clarify the behavior of the subject under test in various states and external working conditions in order to solve the problem of prediction, it is necessary to apply plans for clarification of the mechanism of phenomenon [74] and for developing the model of the subject under investigation.

The purpose of such an experiment is to describe the relationship between the criterion of effectiveness and factors which are considered in the experiment.

In connection with the fact that in testing of agricultural technological processes, as a rule, enough *a priori* information is available, the above mentioned problem can be reduced to the determination of the parameters of a certain hypothetical function or to the selection of such a function from a finite set, followed by the estimation of its parameters.

Using the methods of optimum planning of experiments, it is possible to obtain substantial economy in the number of measurements. For example, it is necessary to determine coefficients of linear regression for six variables. It is known that for this, not less than seven experiments at different values of levels of factors will be required (six coefficients for independent variables and on an independent term).

By the classical method, it is required to conduct 64 (2^6) experiments, which is substantially higher than the minimum number required.

By using the methods of fractional replications, the above problem can be solved with 8 or 16 experiments.

In the quadratic regression for solving such a problem, other plans of experiments are applied which differ in the criterion of optimality (rotating time board and D-optimal experiments). For all purposes in the practice of testing, when there are certain limitations on the domain of variation of independent variables, the quadratic regression can be used for describing the behavior of a sufficiently large number of different agricultural subjects [50].

In the testing of agricultural technological processes, the important problem appears to be the selection of the optimum conditions of the technological process or the selection of optimum parameters and states of working of machines under the influence of many factors. Experiments, related with the solution of the above type of problems, are called extremum experiments or experiments for finding optimal solutions.

When there exist a large number of factors, the solution of such problems by the classical method is practically unrealistic. Box and Wilson proposed a new method for solving such a problem, which is really a sequence (step) method of the study of the surface of response, and reminds one of the iteration method of solving problems in computer science. From the results of a short series of experiments on the gradient of linear approximation in the factorial space, the direction of movements toward the extremal region is found. On the achievement of this extremal region, a long series of experiments is performed, and the surface of response is described by the polynomial of the second and, sometimes (though rarely), even of third order. Because of such a strategy adopted for a number of experiments, the region of optimum is obtained, in which the majority of experiments is performed, i.e. that region in which the investigator is interested in more detail.

The aforementioned method is widely practiced by experimenters for investigation of stationary agricultural machines [50, 59]. Designs of experiments for the solution of such types of problems are extensively developed, and description can be found both in Soviet and in foreign literature [1].

Introduction of developed methods of experiment theory into the practice of testing faces difficulties because of the specific peculiarities of the working conditions of agricultural machines and the unfulfillment of some initial assumptions associated with it.

One of the conditions of the effective application of the methods of planning of experiments is the possibility of conducting live experiments. However, in testing of agricultural technological processes, this possibility does not always exist. For example, in investigation of the quality of work of the digging part of beet-harvesting machines, an important factor which affects the process of harvesting turns out to be the hardness of the soil. The investigator barely controls this factor, i.e. changes its importance in correspondence with the plan of the experiment.

Considering the limited agrotechnical time of working of machines under test and the variability of external conditions, it is not always possible to apply sequence methods of planning and, particularly, the method of steep ascent. It becomes necessary to immediately realize the plan of experiment and to find the optimum region on the basis of the obtained results.

While estimating an agricultural technological process, one comes across indicators, the values of which cannot be estimated and maintained at strictly given level during the span of the whole experiment. In such cases, special methods of experiment planning and the processing of obtained results are applied.

It is also necessary to mention the other specific peculiarities of agricultural experiments: the large number of complicated interconnected, interacting and often uncontrollable variables; the qualitative and quantitative dependence of results with highly varying, randomly changing external, natural and production conditions (physical state of agricultural materials, sizes of fields, their relief, organization of labor, etc.); the inconsistency of parameters of natural and production conditions with time (uninterrupted drift or the trend of average values and the change in the structure of factors of external conditions); the high level of interference, including that due to uncontrollable disturbances (background noise); the dependence of results on the qualification of operators, which is particularly sensitive in experiments for operational-economic and agrotechnical estimations; substantial consumption for experiments, the absence of necessary measuring instruments for nonstop measurements. Because of all these factors, it becomes necessary to select data which can be handled only with prolonged labor-consuming laboratory analysis.

The foregoing concepts do not cover all the aspects of the planning of experiments for testing of agricultural production; however, they do permit a general understanding of the peculiarities of the application of known methods and the facilitation of the study of special literature in this field. In connection with the fact that the methods of planning experiments are well-described in literature, especially in the works of Yu.P. Adler, E.V. Markova, G.K. Krug and V.V. Nalimov, which may be directly used by investigators of an agricultural technological process, a more narrow problem will be dealt with, i.e. planning of experiments in conditions when one of the variables is time.

13. DESIGNING OF EXPERIMENTS FOR STUDYING STOCHASTIC PROCESSES

In testing of agricultural technological processes, those experiments, in which one of the variables is time, occupy an important place. In the cases when there is a possibility to observe realization of sufficient length, then for the determination of the necessary duration of realization, the formulas given in [44] can be used.

So as to estimate the mean value of the process with the correlation function. $R_{xx}(\tau)=\sigma_x^2 e^{-\alpha|\tau|}$, the length of realization can be approximately determined from the expressions $\epsilon=\frac{\sigma_m t_\alpha}{m_x}$, $\sigma_m \approx \sigma_x \sqrt{\frac{2}{\alpha T}}$, and for the process with the correlation function from

$$R_{xx}(\tau)=\sigma_x^2 e^{-\alpha|\tau|}\times\cos\omega_0\tau$$

by the following expression:

$$\sigma_m \approx \sigma_x \sqrt{\frac{2}{\alpha T\left[1+\left(\frac{\omega_0}{\alpha}\right)^2\right]}},$$

where t_α is the quantile of distribution σ_m.

The pitch of quantization of a continuously measured process is best selected by the formula

$$\Delta T \leqslant \frac{1}{2f_0}, \qquad f_0=\frac{\omega_0}{2\pi}=(3\text{–}4)f_{\max}.$$

In those cases when the time of observation is limited, it is necessary to apply special methods of the designing of experiments, one of which is described in [21].

A problem of the estimation of the average value of the steady random process on realization, noticed only in a definite part of the interval of time during which the process is observable in general will be considered. For example, for finding the estimation indexes of the quality of the working of grain-cleaning, grain or beet-harvesting machines, machines for crushing of fodder, etc., the material processed by the machines is collected at the outlet and all components are weighed, i.e., the process is observed over a course of time. The mean value of such a process can be reliably estimated only when the value of the interval of the time of observation is sufficient. However, often in investigation of the above machines, the bulk of sample collection is limited, and the possibility of the division of the material at the outlet in one or the other components is absent, etc. That is why observation of the process is forced to be interrupted for sometime δ_0.

Mathematical calculations have shown that the quality of estimation for the given time of observation does not go down. In this case, the aim of planning of experiments consists in its building in such a way so as to obtain the best estimate of the average value of process (with minimum dispersion) for the given interval of time of observation from 0 to T).

The problem can be stated mathematically as follows: assume that the random steady process, with unknown average value m and known correlation function $r(\tau)$, can be observed in time duration t. After moment t, starting from 0, observation is interrupted, at least for time δ_0 (δ_0 given). Then again, the process can be observed in the interval, the duration of which is not more than T_0, etc. The total period of observation (interval 0, T) is given. In such a case, realization of the process takes the form shown in Fig. 14. Using the realization obtained in this way, the estimate for mathematical expectation m can be found by the method of least squares. In this, the procedure of observations should provide the achievement of estimates with the minimum dispersion; however, the analysis of various methods of solving such problems shows that even for the simplest estimates, it is not clear how best to use the possible observations of the process. But if some assumptions are introduced which do not distort the physical meaning of the problem under consideration, then for the estimate, obtained by the method of least squares, a number of conditions of the optimum use of the time of observations can be fixed.

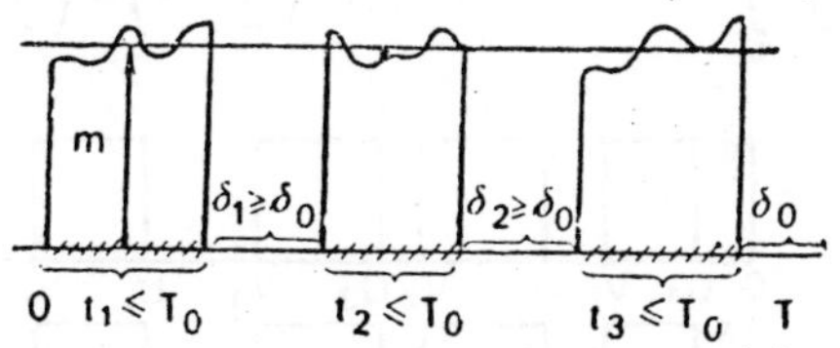

Fig. 14. Form of realization of the process in an interrupted system of observation.

Let $\varphi(t)$ denote the random process steady, in a wide sense, with mathematical expectation $M\varphi(t)=0$ and known continuous function of correlation $r(t)=M\varphi(t+\tau)\,\varphi(t)$. In this, $\varphi(t)$ accepts only real values. Assume that for some function $g(t)$, which satisfies conditions

$$0 \leqslant g \leqslant 1,\ t \in (0, T),\ \int_0^T g(t)\,dt \leqslant C,\ C \leqslant T,$$

in the interval 0, T for fixed value $T > 0$ realization of process

$$x(t)=\sqrt{g(t)}\,[m+\varphi(t)], \tag{30}$$

is observed, where m is the unknown parameter.

It is assumed that the function $g(t)$ can be integrated in the interval $(0, T)$. For the problem under consideration, it can be assumed that the function $g(t)$ is continuous over a segment of the interval. Estimation of the parameter m is obtained by the method of least squares as a result of observation of the process in the interval 0, T takes form

$$\hat{m}=\left\{\int_0^T g(t)\,dt\right\}^{-1}\int_0^T x(t)\sqrt{g(t)}\,dt, \tag{31}$$

where

$$M\hat{m}=m,\ M(\hat{m}-m)^2=\left\{\int_0^T g(t)\,dt\right\}^{-2}\int_0^T\int_0^T r(t-s)\,g(t)\,g(s)\,dt\,ds. \qquad (32)$$

In the above equations, function $g(t)$ characterizes the procedure of observations, and it should be taken such that dispersion of the estimate of the parameter is at a minimum [expression (32)].

Having solved the problem, it is established that function $g(t)$ can have form *a* or *b* (Fig. 15) for some, integer *n*, the minimum function of which has to be searched for, is expressed by the relationship

$$F(g,g)=2\int_0^{T/2}\int_0^{T/2}[r(t-s)+r(t+s)g\left(\frac{T}{2}-t\right)g\left(\frac{T}{2}-s\right)dt\,ds. \qquad (33)$$

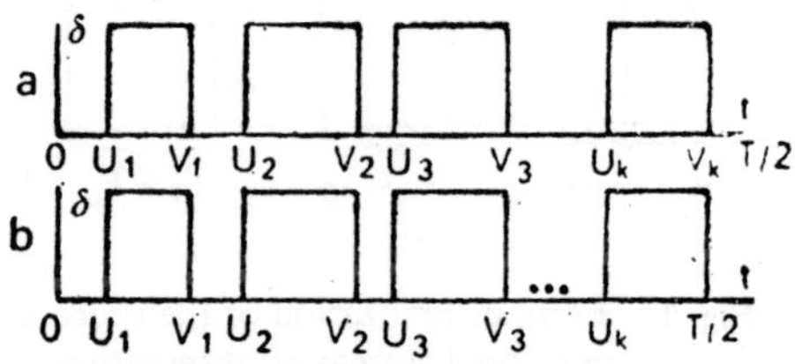

Fig. 15. Form of function, which characterizes the procedure of observation for (a) even and (b) odd numbers of intervals.

In the first case (Fig. 15, a), numbers $u_1, v_1, u_2, v_2, \ldots, u_n, v_n$ satisfy equality

$$\left.\begin{aligned}&\int_0^{T/2}[r(u_j-t)+r(u_j+t)]\,\chi\left(\frac{T}{2}-t\right)dt=\lambda;\\&\int_0^{T/2}[r(v_j-t)+r(v_j+t)]\,\chi\left(\frac{T}{2}-t\right)dt=\lambda;\\&\qquad j=1,2,\ldots,n;\\&\qquad\sum_{j=1}^{n}(v_j-u_j)=\frac{d}{2},\end{aligned}\right\} \qquad (34)$$

where λ is the intermediate conditional denotion of the function; χ—the intermediate conditional denotion of the function of the procedure of observation.

In the second case (Fig. 15, b), numbers $u_1, v_1, u_2, v_2, \ldots, u_{n-1}, u_n$ satisfy relations

$$\left.\begin{aligned}
&\int_0^{T/2} [r(u_j-t)+r(u_j+t)]\,\chi\left(\frac{T}{2}-t\right) dt=\lambda, \\
&\qquad j=1, 2, \ldots, n; \\
&\int_0^{T/2} [r(v_j-t)+r(v_j+t)]\,\chi\left(\frac{T}{2}-t\right) dt=\lambda, \\
&\qquad j=1, 2, \ldots, n-1; \\
&\sum_{j=1}^{n-1} (v_j-u_j)+\frac{T}{2}-u_n=d/2.
\end{aligned}\right\} \quad (35)$$

Then the procedure to find function $\chi(t)$, which minimizes $F(g, g)$, consists in the following: for the given correlation function, time T and integer $n>0$, functions of type a and b are determined by using the system of equation (34) or (35), and then, by changing the values of n, one of which is selected, in such a way as to render $F(g, g)$ minimum.

The above-described plan should be realized for solution of the said problem for the process using the correlation function of the type

$$r(\tau)=\sigma^2 e^{-\beta|\tau|}. \quad (36)$$

By putting in expressions (34) and (35), which are written for the whole interval $(0, T)$, the observation function is given as follows:

$$\chi(t)=\sum_{k=1}^{n} \chi_{[u_k,\, v_k]}(t), \quad (37)$$

and also the correlation function (36). By making the necessary transformations, a system of equations is obtained, whose solution helps in determining the unknown function (Fig. 16):

$$\left.\begin{aligned}
&e^{\beta\gamma}-e^{\beta\delta}=e^{\beta v_1}(1-e^{-\beta\delta}); \\
&\delta=\frac{T-d}{n-1}; \\
&2v_1=d-(n-2)\gamma,
\end{aligned}\right\} \quad (38)$$

where v_1 is the duration of the first (and last) interval of observation and γ—the duration of each of the remaining intervals of observation.

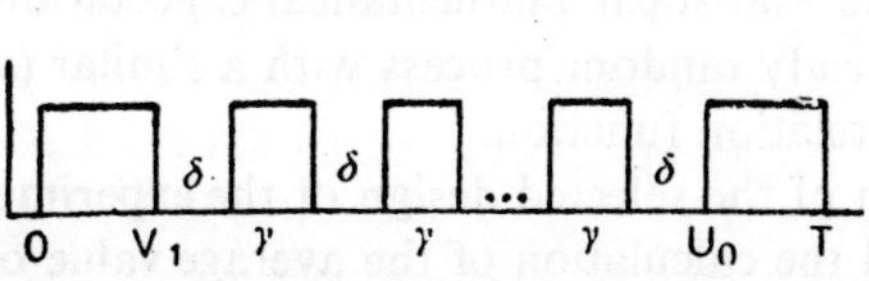

Fig. 16. Form of unknown observation function.

When $v_1 > \gamma$, an important practical conclusion can be drawn: if the experimenter can observe the process during the given interval $(0, T)$ for the time d and take n test observations, then for obtaining the most reliable estimation of the average value of the process, he should select two similar test observations. This should be done in the beginning and in the end of the interval 0, T for more duration than while taking tests in the center of the interval. The physical significance of this condition lies in the fact that the first and the last observations, lying farthest apart from one another, are least correlated among themselves, and maximum weightage is given to them as compared to others. This fact is particularly important when T and δ are small. Starting from some values of T and δ, when δ considerably exceeds the interval of correlation, values of v_1 and γ can be taken identically. Dispersion of the estimate of the average value of the process, for the proposed procedure of observations, is given by:

$$\sigma_m^2 = \frac{2\sigma^2}{d^2} \sum_{k<j} \{\beta^{-2} \exp \beta(u_k - v_j) [\exp \beta(v_k - u_k) - 1] [\exp \beta(v_j - u_j) - 1]\}$$

$$+ \frac{2\sigma^2}{d^2} \sum_{k=1}^{n} \{\beta^{-1}(v_k - u_k) + \beta^{-2} [\exp - \beta(v_k - u_k) - 1]\}, \qquad (39)$$

where $u_1 = 0$; $u_k = v_1 + (k-1)\delta + (k-2)\gamma$; $k = 2, 3, \ldots, n$; $v_k = u_k + \gamma$; $k = 2, 3, \ldots, n-1$; $v_n = T$.

The expressions given here, at fixed d, make it possible to select such values of v_1, γ and n for which dispersion of the estimate of the average value of the process will be minimum, i.e. to plan the optimum experiment. For this with the help of formulas (38), function $\chi(t)$ is determined for each (possible from the condition of the experiment) number n, and then dispersion is obtained by (39). After this, a number n is selected, the corresponding dispersion of which is minimum.

An analysis of expression (39) shows that other conditions, being the same, the dispersion σ_m^2 is inversely proportional to the square of time of observation d. It is also easily shown that if $d \to T$ and $\beta T \gg 1$, i.e. if the process is continuously observed for a sufficient duration of time interval, expression (28) can be represented in the form

$$\sigma_m^2 \approx \sigma^2 \frac{2}{\beta T}.$$

A similar expression is described in another work [44] for the estimation of accuracy of the statistical mathematical expectation for the continuous observation of a steady random process with a similar (as in the case under consideration) correlation function.

After realization of the selected design of the experiment by using the proposed method and the calculation of the average value of the process by the expression

$$\hat{m} = \frac{1}{d} \sum_{j=1}^{n} \int_{u_j}^{v_j} x(t)\, dt, \tag{40}$$

and its dispersions by formula (39), it is possible to find the confidence interval of the parameter under estimation and to check the null hypothesis for the obtained estimation and its allowable value.

For a normal process $x(t)$, due to the linearity of the function (40), the random value m is distributed normally. In such a case, the $(m-\hat{m})$ is also distributed normally with the mathematical expectation equal to zero and the dispersion σ_m^2. Then the confidence interval of the estimated average value of the process is equal to $(\hat{m}+t\sigma_m)$, $(\hat{m}-t_p\sigma_m)$, where t_p is the argument of the normalized Laplace function, satisfying the condition $p(|m-\hat{m}| < t_p\sigma_m) = 1-\alpha$ (α being the level of confidence).

Often, the so-called relative error of average $\dfrac{m-\hat{m}}{m}$ is used as an index of accuracy for estimation of the mean value. If while determining the procedure of observation, condition $(m-\hat{m})/m \leqslant \epsilon$ $(\epsilon > 0)$ holds then, from the equality $t_p\sigma_m = \epsilon_m$, one can find the mean quadratic deviation of estimation m

$$\sigma_m = \epsilon_m/t_p$$

or considering expression (39)

$$\varphi = \epsilon/t_p V, \tag{41}$$

where $V = \sigma/m$ is the coefficient of variance of the process.

By using formulas (39) and (41), it is possible to optimize the observation procedure (starting from the relative error of the mean coefficient of the variance of process) the level of confidence α as well as the limitation on the conditions of the experiment. In this, it is also necessary to know the parameter β of the correlation function.

14. DISPERSION ANALYSIS

The problem of dispersion analysis in testing, lies in the recognition and estimation of factors responsible for the variance in mean values of indicators and, the effectiveness or other parameters characterizing the quality of work or the state of the machine under test. The concept of dispersion analysis can be expressed, in mathematical statistics, as a method of separating the effect of various subsets in the set of parameters on values y under observation. Dispersion analysis consists in representing the sum of the square of deviation from the common mean value in the form of the sum of nonnegative components arising from these parameters (factors).

The mathematical apparatus for dispersion analysis is built around the assumption of mutual independence of the measurement results, the normality of their distribution, the linear effect of the factors on the parameter under study and the equivalence of dispersions. Therefore, before using this method of analysis, it is necessary to check the justification of these assumptions in regard to the performed experiment.

In dispersion analysis, the measurement of quantitative characteristics of factors is not required; however, it is necessary to have information on the combinations of levels of factors for which each measured value of the parameter under analysis is obtained so that there is a possibility to classify results of measurements with respect to these indexes.

In literature, plans of dispersion analysis are called classification [48].

If, in the realization of the experiment, values of the parameter under study are measured for all possible combinations of the levels of factors, then one obtains the complete classification; however, if the values are measured for all factors, then one obtains the incomplete classification. Classifications are also distinguished as crossed or hierarchical.

The principal difference between them can be easily seen in the following example.

Example: If machines A_1 and A_2 are used for sowing with two standard quantities of seeds per hectare and on each section, harvest is determined with three repetitions, then the results of such an experiment represent crossed classification (Table 5).

Table 5

Level of factor B	Level of factor A	
	A_1	A_2
B_1	Q_{111} Q_{112} Q_{113}	Q_{121} Q_{122} Q_{123}
B_2	Q_{211} Q_{212} Q_{213}	Q_{221} Q_{222} Q_{223}

Suppose, that two machines A_1 and A_2 are tested for three days, while each of them is served by two groups of machine operators with two persons in each group, i.e. there are four groups of machine operators in all; while the first and second groups serve machine A_1 and the third and fourth serve machine A_2 (Table 6).

In the experiment, the productivity of each machine is measured and investigated. In this case, one deals with hierarchical classification.

Table 6

Level of factor B	Level of factor A	
	A_1	A_2
B_{12}	W_{111} W_{112}	–
B_{34}	–	W_{223} W_{224}

On comparing Tables 5 and 6, one can easily observe that in hierarchical classification, factor B with four levels is grouped in two groups, which are differentiated by the levels of factor A. In this case, it is said that factor B is grouped within factor A. Examples of hierarchical classifications are often met in agriculture, where most results of measurements are divided into groups (years, economy, etc.) and then into subgroups (brand of machine, crop-yield, etc).

Methods of processing crossed classifications are described in sufficient detail in the available literature [52, 57, 69 and others].

A special case of dispersion analysis arises when there is only one observation $n=1$ in each cell, and the two factors have the forms given in Table 7.

Table 7

Source of variability	Sum of squares	Degree of freedom	Mean squares
Factor A	$s_1=q\sum_i (x_{i}-x_{00})^2$	$p-1$	$\tilde{s}_1^2=\frac{s_1}{p-1}$
Factor B	$s_2=p\sum_j (x_{0j}-x_{00})^2$	$q-1$	$\tilde{s}_2^2=\frac{s_2}{q-1}$
Relationship	$s_3=\sum_{ij} (x_{ij}-x_{i0}-x_{0j}+x_{00})^2$	$(p-1)(q-1)$	$\tilde{s}_3^2=\frac{s_2}{(p_1-1)(q-1)}$
Sum	$s=\sum_{ij} (x_{ij}-x_{00})^2$	$N-1$	–

In this case, there is no sum of squares within the cells because there exists only one observation in each cell and for the estimation of the main effects, the effect of interaction is taken equivalent to zero.

Assume that the results obtained in the testing are represented in Table 8.

Number of measurements $n=pq=2\times 4=8$

$$\text{Common mean } x=\frac{\sum_{q=1}^{p}\sum_{h=1}^{q} x_{qh}}{n}=\frac{130.99}{8}=16.4.$$

Table 8

Level of factor A	Level of factor B				ΣB	Mean on lines x_q
	B_1	B_2	B_3	B_4		
A_1	14	7	13.85	13.2	48.5	12
A_2	21.85	21.85	20.16	19.08	82.94	20.8
ΣA	35.85	28.85	34.01	32.28	130.99	—
Mean of columns ($\bar{x}_n$)	17.9	14.42	17	16.14	—	—

The sum of squares of deviation Q of all observed values from their mean value is:

$$Q=\sum_{q=1}^{p}\sum_{h=1}^{q}(x_{qh}-\bar{x})^2=\sum_{i}^{2}\sum_{1}^{4}(x_{qh}-\bar{x})^2=197.$$

The sum of squares of deviations of mean values of group A from the mean value is:

$$Q_A=q\sum_{1}^{2}(\bar{x}_q-\bar{x})^2=4\sum_{1}^{2}(12-16.4)^2+(20.8-16.4)^2=155.$$

The sum of squares of deviations of mean values of group B from the mean value is:

$$Q_B=p\sum_{1}^{4}(x_h-\bar{x})^2=2\,[(17.9-16.4)^2+(14.42-16.4)^2$$
$$+(17-16.4)^2+(16-14-16.4)^2]=14.6.$$

Residual sum of the squares of deviations is given by

$$Q_R=Q-Q_A-Q_B=197-155-14.6=27.4.$$

Number of degrees of freedom of each of these sums of squares of deviations Q, Q_A, Q_B, Q_R is equal to $pq-1=(2\times4)-1=7$; $p-1=2-1=1$; $q-1=4-1=3$; $(p-1)(q-1)=(2-1)(4-1)=3$, respectively, and the unmixed estimation of dispersions of the entire set is given by:

$$S^2=\frac{Q}{pq-1}=\frac{197}{7}-28.1;\quad S_A^2=\frac{Q_A}{p-1}=\frac{155}{1}=155;$$

$$S_B^2=\frac{Q_B}{q-1}=\frac{14.6}{3}=4.8;\quad S_R^2=\frac{Q_R}{(p-1)(q-1)}=\frac{27.4}{3}=9.1.$$

Testing of the null hypothesis, i.e. the hypothesis regarding the uniformity of the set, is conducted by comparing dispersion S_A^2 between groups A and dispersion S_B^2 between groups B and by estimation of dispersion S_R^2 (Table 9), by applying F-tests.

Table 9

Variability	Sum of squares	Number of degrees of freedom	Estimation of dispersion
General	$Q=197$	$qn-1=7$	$S^2=\frac{Q}{7}=28.1$
Between groups A	$Q_A=155$	$p-1=1$	$S_A^2=\frac{Q_A}{1}=155$
Between groups B	$Q_B=14.6$	$q-1=3$	$S_B^2=\frac{Q_B}{3}=4.8$
Residual	$Q_R=27.4$	$(p-1)(q-1)=3$	$S_R^2=\frac{Q_R}{3}=9.1$

By comparing the calculated value of the F-test, namely $F_A=S_A^2/S_R^2=155/9.1=18$ with tabular values for degrees of freedom $\gamma_A=1$ and $\gamma_R=3$ and 5% confidence level ($F_{\text{tab}}=10.13$), it is necessary to assume that factor A influences the measured value in the process of the experiment.

For factor B:

$$F_B=\frac{S_B^2}{S_R^2}=\frac{4.8}{9.1}=0.526<F_{\text{tab}}=9.28.$$

for $\gamma_B=3$ and $\gamma_R=3$, the hypothesis that it does not influence the measured value undoubtedly holds good.

The technique of the processing of hierarchical classification can be considered from the example of the processing of results of the experiment on comparative estimation of productivity of three milking rigs. Each milking rig was handled by four pairs of the milkers. Observations were taken for five days. The obtained results are represented in Table 10.

For factor A: $i=1, 2, 3$, $v_1=3$; for B: $j=1, 2, 3, 4$; $v_2=4$; $k=1, 2, 3, 4, 5$, $v_3=5$,

$$\sum_{ijk} x_{ijk}^2=70607;\ \sum=2037.$$

As a result of calculations of the sum of squares, the following is obtained:

For factor A (equipment):

$$\frac{1}{v_2v_3}\sum x_i^2-\frac{\bar{x}^2}{v_1v_2v_3}=\frac{1}{20}(786^2+673^2+578^2)-69156.15=1084.3;$$

for factor B (milkers):

$$\frac{1}{v_3}\sum_{ij} x_{ij}^2-\frac{1}{v_1v_2}\sum\bar{x}_i^2=\frac{1}{5}(143^2+\ .\ .\ .\ +198^2)-70240.45=35.75;$$

for random deviations between columns:

$$\sum_{ijk}x_{ijk}^2-\frac{1}{v_3}\sum\bar{x}_{ij}^2=70607^2-70276^2=330.8;$$

Table 10

Factor A (equipment)	Factor B (milkers)	Productivity x_{ijk}					x_{ij}	x_i
1	1(1)	29	29	28	31	26	143	578
	2(2)	30	25	28	25	27	139	
	3(3)	30	27	30	27	31	145	
	4(4)	31	31	28	33	28	151	
2	1(5)	33	33	33	32	36	167	673
	2(6)	38	35	33	35	33	174	
	3(7)	37	33	33	33	31	167	
	4(8)	31	33	32	34	35	165	
3	1(9)	41	37	37	42	36	193	786
	2(10)	23	41	41	37	40	202	
	3(11)	31	43	38	45	36	193	
	4(12)	40	42	41	41	34	198	
				$\sum_{ijk} x_{ijk}^2 = 70607$				$\sum = 2037$

general

$$\sum_{ijk} x_{ijk}^2 - \frac{\bar{x}^2}{v_1 v_2 v_3} = 70607^2 - \frac{2037^2}{60} = 1450.85.$$

By comparing F_{tab} and F_{actual} (Table 11), one sees that the effect of factor B on productivity is not considerable, while the effect of factor A (on the results of experiments) is substantial.

Table 11

Source of changes	Sum of squares	Degrees of freedom	Mean squares	F(actual)	F(tabular) at p, equal to 0.05	0.01
Factor A (equipment)	1084.3	2	542.15	$\frac{542.15}{3.97} = 136.6$	4.26	8.02
Factor B (milkers)	35.75	9	3.97	$\frac{3.97}{6.85} = 0.58$	2.12	2.85
Random deviations	330.8	48	6.85	–	–	–
General	1450.85	59	–	–	–	–

Along with the estimation for the significance of effect of any factor on the dispersion of the parameter under study, it is often necessary to determine the size of this effect and its contribution in the cumulative effects of all factors. Earlier, some researchers had held the viewpoint that the estimation of the proportions of effect can be obtained from the percentage of the sum of the squares. But such a simple method turned out to be incorrect because the sum of the squares is not a measure of dispersion. Dispersion can be judged only from the mean squares S_i^2, obtained by dividing the sum of the squares by the corresponding number of degrees of freedom.

As shown above, the principle of dispersion analysis is based on the use of the F-test, which is merely a quotient of division of one average square by another. It is obvious that the absolute value of the mean squares, S_i^2, is not sufficient for the judgement about the quantitative action of factors A, B, . . ., because the value of each S_i^2, as a rule, consists of two or more components.

Thus, for single-factor analysis

$$F=\frac{\tilde{S}_i^2}{\tilde{S}_2^2}=\frac{\sigma_i^2+n\sigma_A^2}{\sigma_i^2}.$$

Numerator S_i^2 consists of two components and is bigger than the denominator by a value $n\sigma_A^2$. Hence, the effect of factor A is determined by value $n\sigma_A^2$.

In this case,

$$\sigma_A^2=\frac{S_1^2-S_2^2}{n}.$$

After establishing the value of σ_A^2, it can be substituted in the sum of σ_A^2 and σ_C^2, which constitutes the entire dispersion formed due to standard and other random factors. In that case, the proportion of effect A will be:

$$P_A^n=\frac{\sigma_A}{\sigma_A^2+\sigma_C^2}.$$

Analysis of the results of the two-factor classification is carried out on the same principle

$$\sigma_A^2=\frac{S_1^2-S_2^2}{n};\ \sigma_B^2=\frac{S_2^2-S_3^2}{n};\ \sigma_{AB}^2=\frac{S_3^2-S_4^2}{n}.$$

The denominator for determination of the contribution of effect of each factor appears as a sum

$$\sigma_A^2+\sigma_B^2+\sigma_{AB}^2+\sigma_C^2.$$

But if the role of some factor A or interaction between them (AB) is absent, then this component should not be included in calculations.

While establishing the contribution of the effect of factors in the dispersion complex, built on the hierarchical scheme, it is necessary to remember that effects are successively subtracted, one from the other.

Then

$$\sigma_A^2 = \frac{S_1^2 - S_2^2}{bc}, \quad \sigma_B^2 = \frac{S_1^2 - S_3^2}{c}. \tag{42}$$

The computation of the contribution of the effect is performed by the same method which is used for the general two-factor analysis but in case of the hierarchical scheme, it is not possible to extract the component of interaction. Therefore, the denominator consists only of three components: $\sigma_A^2 + \sigma_B^2 + \sigma_C^2$; if the action of one of the factors is not proved, then it consists of two components.

In those cases when, by the methods of dispersion analysis, the effect of the factor under study is proved, and it is necessary to find out the confidence differences between the values of the parameters under study in individual groups, analysis of the results of the experiment can be completed by testing the hypothesis about the existence of the difference between the average values of two populations by the help of Student test, the F-test or the Duncan test [48, 52, 57, 69, 79, 89, 90].

15. REGRESSION ANALYSIS

Dispersion analysis is extremely convenient for those cases when the factors can accept only discrete values, and it is not necessary to determine optimum conditions for the process to take place under continuously changing independent variables. For an estimation of the quantitative relationship between the indicators (criteria of effectiveness) of agricultural technology under estimation and the factors which determine them, plans and methods for the processing of results of regression analysis are used.

The traditional method for solving this problem consists in changing by turns each of the causative factors at fixed values of others in order to establish their effect on the result of the experiment. However, the inadequacy of such an approach has been demonstrated in at least two respects [53].

Firstly, the classical approach presumes the investigation of the so-called well-organized systems [53]. For real systems and, in particular, for system relating to agricultural production, which are badly-organized diffused systems, the alternate variation of factors does not help to quantitatively evaluate the effects of interaction. Without such an estimation, it is not possible to establish the relationship between the causative factors, and this leads to incorrect results. These effects can be studied only by varying the factors simultaneously.

Secondly, during the period of alternate factor variation, a new experimental factor unknown and unaccounted for may appear, which may influence the efficiency index under study to an equal or even a greater extent. The influence of this factor will be added to that of the combination of variables at a given moment, which is unjustified in practice.

In the modern theory of the design of experiments, therefore, it is recommended that during experimental investigations, all factors should be simultaneously varied and the sequence in which individual experiments are performed should be randomized.

The principle of randomization can be stated as follows: if while conducting an experiment, the experimental subjects are assigned to treatment groups, then it should be done randomly with the use of equal probabilities.

In this way, the effect of unaccounted experimental factors is averaged out. Since in real experiments, it is not always possible to fully randomize the experiment, a block design is used. In essence, all experiments are block experiments.

The aforementioned approach toward the designing of experiments is related equally to plans of regression as well as of dispersion analysis.

The most common testing experiment for the comparative estimation of the test and the etalon machines for its quality of fulfilling technological processes under different regimes (for example, rate of motion) will now be considered.

In the formulation of the problem, the speed of motion of the aggregates is related to the factors which qualify dependent variables.

While arranging laboratory or field experiments on seeding machines for accurate sowing or during laboratory or field group testing of a combine, individual segments are distinguished according to the state of the soil, its microrelief, crop-yield, etc. The effect of such uncontrollable factors, in the given experiment, on its results can be averaged out by using the corresponding method of randomization.

Taking into account the labor required to conduct each experiment, one is limited to three observations on three levels of the speed factor for the machine under test and for the etalon machine. Then the problem reduces to the factorial experiment of the 3×2 type, where three levels (speed of motion v_1, v_2, v_3) are quantitative, while the remaining two (the machine under test and the etalon machine) are fixed qualitative factors.

In the case of beet-harvesters with known methodological requirements and the selection method of samples (limited areas and length of considered section, necessity of preliminary unaccounted for traverse), a use of the method of full randomization without limitations is practically unrealizable.

For a seeding machine, the method of full randomization is entirely acceptable. If six experimental conditions (2 machines $\times$ 3 levels of speed) are numbered as 1, 2, 3, 4, 5, 6, then it is possible to decide, on a random basis, which of these six conditions should be realized first, second, third, etc. (Fig. 17). The scheme for conducting the experiment is shown in Table 12.

For beet-harvesters, which takes into account the abovementioned limitations on randomization, the plan of experiment can be laid out in the following way.

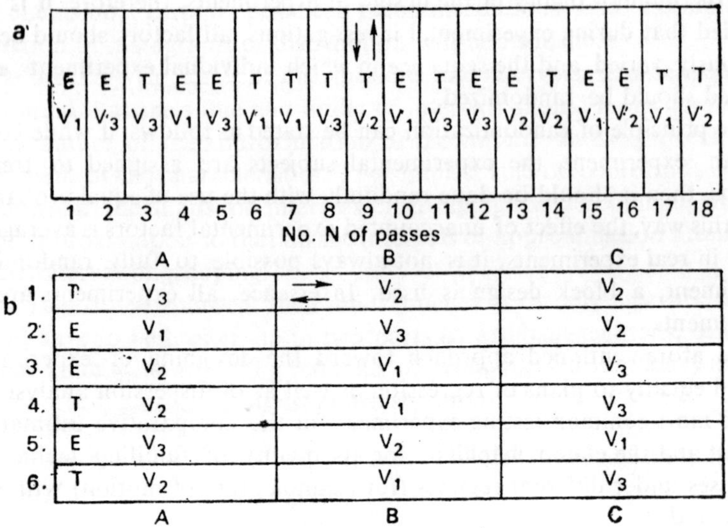

Fig. 17. Scheme of distribution of plots a) without limitation and b) with limitation on randomization.

Table 12

Machine	Speed		
	v_1	v_2	v_3
Under test T	6	9	3
	7	14	8
	17	18	11
Etalon E	4	1	2
	10	13	5
	15	16	12

Assume that in testing of combines, it is necessary that three sections (repetitions) are arranged along rows (Fig. 17, b). Then it is necessary to randomize the order of conducting the experiment for machines (blocks) and the order of repetitions *A*, *B*, *C* for each pass (within the blocks). The scheme of conducting the experiment is given in Table 13.

The problem of regression analysis consists in the estimation of coefficients on the basis of the results of the experiment of the equation of regression, describing the relationship being studied as proposed by the experimenter; testing of the statistical hypothesis of significance for the obtained coefficients, and the accuracy of the obtained equation of regression from the results of the experiment. This is just as in dispersion analysis, when the

Table 13

Machine	Machine in order	Speed		
		v_1	v_2	v_3
Under test	1	*B*	*C*	*A*
	4	*B*	*A*	*C*
	6	*B*	*A*	*C*
Etalon	2	*C*	*C*	*B*
	3	*B*	*A*	*C*
	5	*C*	*B*	*A*

mathematical apparatus of regression analysis is based on assumptions on the fulfillment of which depends the realiability of the obtained results. In regression analysis, it is assumed that:

(1) results of measurements $y_1, \ldots, y_N$ are independent and normally distributed with homogeneous dispersion. Homogeneity in dispersion snows that at different combinations of independent variates (factors) dispersions of measured values are identical;

(2) independent variates $x_1, x_2, \ldots, x_k$ are measured with a negligible error compared to the error in determination of y.

In practice, while setting up the experiment, it is assumed that these assumptions are fulfilled, and if while interpreting the results there is no contradiction between the experiment and the hypotheses put forward for its description, then the validity of the assumptions of regression analysis for the performed experiment cannot be doubted. In the alternate case, various methods are used which provide independent and normal distribution of results of measurements y_i, the uniformity of their dispersions, etc. by an increase of the technical level of the experiment, randomization and also by selection of transformation functions for transition to new variables, etc.

The apparatus of regression analysis is sufficiently well-developed for linear regressions with respect to estimated parameters, i.e. for those regressions for which the parameters are estimated by the method of least squares obtained on solving linear equations.

Assume that there exist N results of observations on the value of y, which depends on k independent variates $x_1, x_2, \ldots, x_k$; results of observations should be represented by a polynomial of the order d.

In the practice of machine testing, most of the relationships between the criteria of effectiveness and the operations and conditions of the working of machines can be described by polynomials of the first and second order. Therefore, the problem of regression analysis for cases of interest reduces to the computation of coefficients of the following equations:

$$y = b_0 + \sum_{i=1}^{k} b_i x_i; \tag{43}$$

$$y = b_0 + \sum_{i=1}^{k} b_i x_i + \sum_{i>j=1}^{k} b_{ij} x_i x_j + \sum_{i=1}^{k} b_{ii} x_i^2 \cdot \tag{44}$$

Equation (44) is easily derived from the Eq. (43) by replacing nonlinear members by fictitious linear variables. For example, equation

$$y = ax^2 + bx + c$$

on replacing $x^2 = x_1$ and $x = x_2$ will have the form $y = ax_1 + bx_2 + c$, linear equations can be derived from a series of equations using the respective transformation functions. For example, as a result of taking logarithms of both sides of equations and the use of fictitious variables, the equation $y = ae^{bx}$ also reduces to an equation which is linear in parameters, $\ln y = \ln a + bx$ or when $\ln y = y'$ and $\ln a = a'$ so that $y' = a' + bx$.

For the regressions linear in parameters, unknown estimations are found from the equation:

$$B = (XX^*)^{-1} X^* Y, \tag{45}$$

where X is the matrix of independent variables, Y—the vector of measurements (observations); the asterisk denotes transposition.

For cases with an unequal number of measurements at points of factorial space, the following relationship is used

$$B = (XR^{-1}X^*)^{-1} X^* R^{-1} Y, \tag{46}$$

where

$$R^{-1} = \begin{vmatrix} n_1 & 0 & 0 \ldots 0 \\ 0 & n_2 & 0 \ldots 0 \\ \cdot & \cdot & \cdot \quad \cdot \quad \cdot \\ 0 & 0 & 0 \ldots n_k \end{vmatrix},$$

where $n_1, n_2, \ldots, n_k$ are the number of measurements.

Residual dispersion is determined by the following formula:

$$S_R^2 = \frac{\sum_{i=1}^{N} (y_i - \hat{y}_i)^2}{N - k - 1}, \tag{47}$$

where k is the number of parameters under estimation.

If the dispersion of measurements $yS^2(y)$, which characterizes error of the experiment, is known, then the ratio F is found as follows:

$$F = S_k^2 / S^2(y)$$

and the hypothesis of adequacy of representation of results by the given equations of regression is verified.

In real situations $S^2(y)$ is generally unknown. For its estimation at the selected combination of independent variables, it is necessary to conduct a series of measurements, assuming that the conditions of experiments do not change. In a passive experiment, when independent variables are uncontrollable and change from measurement to measurement, estimation of the dispersion $S^2(y)$ is not possible. In such cases, the estimation of the completeness of description of experimental results from selected equations of regression is realized from the correlation coefficient

$$R=\sqrt{1-\frac{\sum_{i=1}^{n}(y_i-\hat{y}_i)^2}{\sum_{i=1}^{n}(y_i-\bar{y})^2}},\ 0\leqslant R\leqslant+1.$$

The significance of coefficient R is determined by the F-ratio, i.e.

$$F=S_k^2/S_R^2.$$

For the degree of freedom $f_k=k$, one gets $f_R=N-k-1$; where

$$S_k^2=\frac{\sum_{i=1}^{N}(y_i-\bar{y})^2-\sum_{i=1}^{N}(y_i-\hat{y}_i)^2}{k};$$

k is the number of parameters under estimation.

The sequence of calculations for the processing of the active experiment by methods of regression analysis is as follows:

Dispersion of the arithmetic mean for each variate of experiment is calculated

$$S_i^2=\frac{\sum_{\lambda=1}^{n}(y_{i\lambda}-\bar{y}_i)^2}{n-1};$$

where
$$\bar{y}_i=\frac{\sum_{\lambda=1}^{n}y_{i\lambda}}{n};$$

n—number of parallel observations in each modification of the experiment (number of repetitions).

The uniformity of calculated dispersions is checked by Cochran's test:

$$g=\frac{S_{i\,\max}^2}{\sum_{i=1}^{N}S_i^2},$$

where $S_{i\max}$ is the maximum value of dispersion, selected from all those calculated here.

If calculated values of g, for the degree of freedom $f=n-1$, are less than tabular, then the hypothesis that the dispersion of the dependent variable is homogeneous, holds good.

If g is greater than the tabular value, then the hypothesis that dispersion is homogeneous cannot be accepted, and it is necessary to establish the reasons for such inhomogeneity and to eliminate them. There are no formalized rules for the selection of methods for the elimination of inhomogeneity. In practice generally, methods for preliminary statistical analysis of experimental results are used or also transformations are applied to the results so as to make dispersions homogeneous.

If the hypothesis about the homogeneity of dispersion is accepted, then the dispersion of replication of the entire experiment

$$S_y^2 = \frac{\sum_{i=1}^{n} S_i^2}{N},$$

where N is the number of treatments of experiments.

Coefficients of regression, linear in parameters, are calculated by formulas (45) and (46). If the regression is nonlinear, then for a sufficient number of experiments, the coefficients of regression are determined by the method of least squares from the conditions of minimum of function:

$$F = \sum_{i=1}^{N} [y_i - f(\bar{a}_i, \bar{x}_i)]^2.$$

Residual dispersion (dispersion index) is calculated by formula (47)

$$S_{\text{ad}}^2 = \frac{\sum_{i=1}^{N} (y_i - \hat{y}_i)^2}{N - d - 1},$$

where y_i is the value of y, which is determined experimentally, $\hat{y}_i$—the value of y, calculated from the equation of regression for values x of the ith treatment of experiment, d—the number of determinable coefficients of regression.

If the calculated value of the Fischer index $F = S_{\text{ad}}^2 S_y^{-2}$ for $N - d - 1$ and $(n - 1)$ degrees of freedom is less than the tabular value then, for a selected level of confidence, the hypothesis of the correctness of regression equations can be accepted. In the alternate case, just as in the case of the nonhomogeneity of dispersion y, the possible reasons of inaccuracy are analyzed. If the dispersions y are homogeneous, but the equation is not accurate, i.e. the selected equation of regression very roughly describes the relationship under examination, then it is necessary to select a more successful model (for example, increase the order of the polynomial if the model is polynomial, decrease the range of variation, etc.). But these recommendations are acceptable only in those cases when the experiment can be repeated or extended. In the conditions of agricultural production, in most of the cases, it does not seem possible to repeat or carry out the experiment within a reasonable period. Therefore, in the selection of a model (equation of regression), it is necessary to use *a priori* information and, is the absence of it, it is necessary to perform exploratory experiments and use designs which permit the verification of two or three equations of regression.

16. PRINCIPLES AND METHODS OF OPTIMIZATION

Optimization problems (or problems of synthesis) are an integral part of the systems approach; however, in the testing of the agricultural technological process, they did not get proper coverage. In laboratory field and operational testings, generally, limitations arise because of the optimal tuning of machines by visual and, naturally, subjective techniques. When tuned in this manner, the indexes of machine operation can be determined with a known error and from time to time, regression relationships are established (the polynomial type for agrotechnical and energy estimations or the exponential type for testing of grain-harvesting combines, etc.). Generally, the response functions are given *a priori* in the analytical form, and often statistical evaluation of the measured parameters is not obtained (in this, significant systematic flaws can be discerned). Also, the estimation of the degree of optimality of basic parameters of machines and aggregates in the zonal aspect is not obtained, which makes it difficult to standardize the size-range of machines. However, it becomes necessary to solve such problems while estimating the level of technical, technological and design activities, especially during the stage of accepting recommendations from test results.

The process of searching for the best solution from among several possible alternatives is known as optimization.

Determination of optimum states and operating conditions, determination of optimum parameters of agricultural machines or aggregates and their comparison with parameters of samples under test and analogs from other countries, determination of the optimum bulk of transportation and the composition of technical personnel and also the determination of the optimum service life of agricultural machines, etc. are some of the optimization problems which require a solution at the given level of the hierarchical system (in testing of the agricultural technological process.)

In solving such problems, optimization consists in the selection of states and the operating conditions of machines, their quantitative and qualitative composition for which the minimum or maximum value of certain indexes, which characterizes the quality of the functioning of the machine and the aggregate or complex of machines, as a whole, can be achieved.

The selection of an appropriate criterion of effectiveness is one of the conditions for a correct statement of the optimization problem. A mathematically correct statement of the problem assumes the presence of only one parameter of optimization. In real problems, one is dealing not with one, but several criteria of optimization which must be solved through systems analysis.

There are no general formalized rules for the selection of some authors

[18 and others] who have given general recommendations on the selection of criteria of effectiveness that hold good even on solving the problems of testing.

Criteria of effectiveness and, consequently, of optimization are selected, in any specific cases, on the basis of the general objectives of solving the problem. For example, the optimum states of operation of agricultural machines and aggregates can be selected from conditions of the maximum observance of the agricultural requirement of achievement of maximum yield and productivity, the minimum power consumption and loss of product, the maximum comfort for service personnel, etc.

While optimizing the composition of transport personnel, the criterion of optimization can be taken as the minimum incurred expenses, number of service personnel, amount of energy sources, etc.

It can be easily seen that a number of criteria are contradictory and incompatible. If the productivity is maximum for certain states, conditions of machine operation and machine parameters, there is no guarantee that power consumption or losses in production will be a minimum, etc.

For a concrete statement of the problem of optimization, selected criterion should correspond to the ultimate aim which has been put forward for the machine.

The criteria of effectiveness for a complex hierarchical system can be represented in the form of a list of objectives of the system (see Fig. 1). For any machine aggregate under test or a complex of machines, the partial or basic (chief) indexes of effectiveness which correspondingly characterize the degree of achievement of aims, set before the subsystems and system as a whole, can be distinguished. The chief indexes of effectiveness synthesize the contribution of individual elements, operations and processes to the efficiency of the subsystems, which stand at a higher level, for example, the contribution of the plowshare group of a seeding machine to the yield and of a seeding machine to the gross yield of grains in a farm, the effect on the decrease of labor expenses or the increase of productivity, etc.

Partial factors (criteria) of effectiveness of the functioning of individual subsystems witness the effectiveness of a given subsystem in comparison with agrotechnical requirements. They do not indicate the contribution which the given subsystem brings into the achievement of the objectives before the system as a whole (seeding machine, farm).

Hence, in systems analysis, it is recommended to consider the absolute as well as the relative increment in the indexes of effectiveness of functioning of a system due to the functioning of the subsystem under estimation (for example, the effect of a plowshare group or a seeding machine on yield) as basic indexes of effectiveness of individual subsystems [70]:

$$Q_{\Omega}^{j,i} = W_{\Omega}^{P(j,i)} - W_{\Omega}^{0(j,i)};$$

$$V_{\Omega}^{j,\,i}=\frac{W_{\Omega}^{P(j,\,i)}-W_{\Omega}^{0(j,\,i)}}{W_{\Omega}^{u(j,\,i)}-W_{\Omega}^{0(j,\,i)}}\,,$$

where $W_{\Omega}^{0(j,\,i)}$ is the index of effectiveness of the system for the condition that the subsystem under estimation does not take part at all in the increase of the effectiveness of the whole system; $W_{\Omega}^{u(j,\,i)}$ is the factor of effectiveness of the system for the condition that the subsystem under estimation ideally fulfills its functions.

A methodical approach for selection of a single criterion [18] consists in the following: one out of all criteria is selected as the criterion for optimization, while the remaining are considered as constraints. For example, the following problem is to be considered: such parameters of agricultural aggregate should be selected so as to achieve the maximum productivity for a quality of work and an energy expense not below the given values.

In some cases, investigators set up a mixed or generalized criterion which represents the same function of criteria as should be preferably taken into account while solving problems on optimization [55, 70]. In works on designing of experiments, such a function is called the 'desired' function.

In setting up a 'desired' function or chief indicator of quality, the character and quantitative expression of the relationship between the chief criterion of effectiveness for the system as a whole (productivity, labor expenses or derived costs, etc.) and partial criteria which are used for the estimation of different technological, energy and industrial processes performed by machines under test, are established.

The most widely used function of the relationship between these criteria for solving practical problems in a linear function of the following type:

$$W_c=\sum_{j=1}^{s}\lambda_j W_j.$$

The selection of weightage coefficients λ_j, in the general case, is made in correspondence with qualitative assumptions about the ralative importance of each of the abovementioned criteria.

The most favourable case for the derivation of the desired function is when partial criteria W_j can, by the use of some transformation f, be reduced to one discussion, which has some physical meaning. Then, the generalized criterion is given by:

$$W_c=\sum_{j=1}^{s}f_i(W_j).$$

So as to bring partial criteria of effectiveness to one unit, often the following rating is used

$$\xi_j=\frac{W_j}{|\,W^{\text{bas}}-W^{\text{br}}\,|}$$

or

$$\xi_j = \frac{|W^{\text{bas}} - W^{\text{br}}|}{W_j}.$$

Then indexes are summed up while considering the weightage α_j of each of them.

It should be noted that such an operation does not introduce considerable changes in the abovementioned linear function, because in this case, the weightage coefficient λ_j will be equal to

$$\lambda_j = \frac{\alpha_j}{|W^{\text{bas}} - W^{\text{br}}|}$$

or

$$\lambda_j = \frac{|W^{\text{bas}} - W^{\text{br}}|}{\alpha_j}.$$

A common drawback of the component criteria is that the low effectiveness with respect to one parameter may be compensated because of another.

If the sum of weighted logarithms of relative partial criteria are used as the linking function, then to some degree, this drawback can be avoided:

$$\ln W_c = \sum \beta_i \ln \xi_i \quad \text{or} \quad W_c = \prod_{i=1}^{s} \xi_i^{\beta_i}.$$

On the whole, an unjudicious use of composite criteria of any form can result in incorrect recommendations.

An alternative way is also known for solving the optimization problem by considering criteria which involve compromise solution. This is known as the method of successive compromises. In this case, while starting from practical considerations, some compromise is made on any one of the criteria that can be permitted. For example, such a measure may be adopted to bring direct expenses to a minimum. This method of optimization, in the presence of several criteria, is useful in that here it can be directly seen as to how, at the cost of relaxation in one index, there is a gain in another.

Such a situation can often appear when it is necessary to make a decision about recommendations with consideration of the prospects of development of a given design of the machine. In these cases, it is worthwhile to reduce the number of criteria of effectiveness by breaking them into two groups: criteria that assume, as a rule, the role of constraints (labor conditions, mass, etc.) and criteria, which determine through basic parameters the effectiveness of the system as a whole (productivity, agrotechnical factors, etc.); and thereafter, the problem is solved by the method of compromises.

If the main criterion of effectiveness (optimization) of the system is selected, and a model which connects the main criterion of effectiveness with the parameters of the system and the conditions of its functioning is developed, then the next step of optimisation is the selection of a method for solution, i.e. determination of the extremal value of the so-called aim function $f(x_1, x_2, \ldots, \alpha_1, \ldots, \beta_1, \ldots)$ in some region of variation of conditions of function x and parameters $\vec{\alpha}$, $\vec{\beta}$, of the system or subject. The region is generally described by the system of inequalities of the type:

$$g(x_1, x_2, \ldots, x_n) \geqslant 0 \quad \text{or} \quad g(x_1, x_2, \ldots, x_n) \leqslant 0.$$

Depending on the complexity of the problem, optimization can be realized by different methods [5, 38, 87].

Numerical methods appear as basic methods for solving optimization problems. As is known, they consist in the use of preceding information for obtaining improved solutions for the problem with the help of iteration procedures.

One of best known and widely used numerical methods is the method of steepest (gradient) descent or ascent, the essence of which lies in approaching the optimum through a successive determination of the gradient:

$$g = \left(\frac{\partial t}{\partial x_1}, \frac{\partial f}{\partial x_2}, \ldots, \frac{\partial f}{\partial x_n}\right),$$

i.e. of the direction, in which the function f decreases or increases most rapidly. There exist many different modifications of the gradient method, which are fully described in special literature [91 and others]. However, each of them appears as most effective only for a certain class of problems and for the search of the optimal of concrete type of functions.

Methods of optimization of the linear aim function with linear constraints have been developed, and these have found wide application in practice. Problems of the search for the optimal in this case are solved by methods of linear programming.

If solution of the problem of optimization consists in the necessity of selecting the best, in the sense of the optimal value, the criterion of optimization in the sequence (strategy) of certain operations (industrial and technological operations, stages of development, testing, application, operational technique), the methods of dynamic programming are used [5].

In order to optimize the functioning of complexes of interacting operations, which in agricultural production may appear as operations of production in agricultural produce and stock-breeding, the method of network designing, known in applied mathematics as that which makes use of network graphs, is applied.

As it has been mentioned earlier, the processes of functioning of agricultural machines and complexes are statistical, and the negligence of this characteristic while developing models and their optimization can lead to an incorrect solution.

The statistical solution of the problem [51, 58, 59, 60, 61] more reliably reflects the reality but requires massive information and complicated computational procedures.

For the solution of problems in this case, two methods are used in testing: artificial reduction to a deterministic scheme and optimization relative to the probability criteria or optimization with respect to the average value.

The first method is developed on the basis of methodologies and OST (standards) developed in the system of testing; the second is used for solving various problems [55].

It is also seen that the replacement of the probability consideration by the deterministic, when the random factors are replaced by their mathematical expectations, is allowed only during a rough tentative computation, when the range of variation of random values is small.

In optimization with respect to the average value, the maximum probability of finding the criterion of effectiveness in the desired region or in the tolerance zone is determined (Fig. 18):

$$\overline{W} = MW = \iint \ldots \int f(x_1, x_2, \ldots \alpha_1, \alpha_2, \ldots \beta_1, \beta_2 \ldots)$$

$$\times P(x_1, x_2 \ldots \alpha_1, \alpha_2 \ldots \beta_1, \beta_2 \ldots)\, dx_1\, dx_2 \ldots d\alpha_1 d\alpha_2 \ldots d\beta_1\, d\beta_2 \ldots$$

The solution obtained in this way is optimized with respect to the average value.

Such a compromised solution, not being strictly optimized for specific conditions, will hold good over the entire range of conditions.

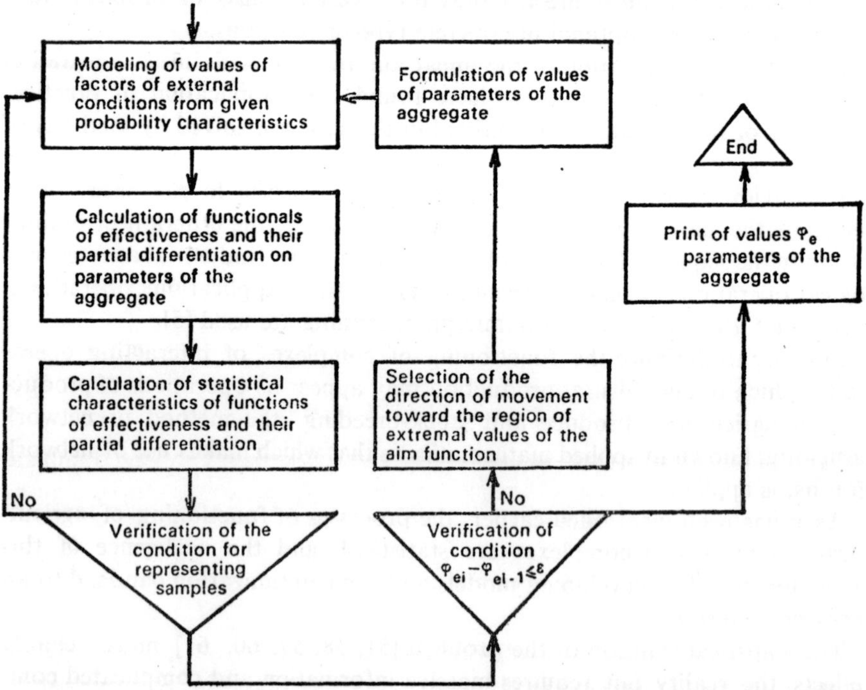

Fig. 18. Flow chart for optimization of parameters of machine aggregates.

For solving optimization problems at random values of factors, the methods of stochastic programming, modeling of operations on the pattern of Markov processes, modeling by the method of statistical testing, etc. have been developed and are used [39, 62].

Graphical methods of optimization are based on the graphical presentation of aim functions in relation to optimized variates. Limitations of these methods in respect of the number of variates (two variants on one graph) are compensated by explicity and simplicity in the search for the optimum function obtained from its graph.

In solving optimization problems with the help of factors by the methods of statistical testings on computers at machine-testing stations, the graphical methods of optimization should be considered the most acceptable as the computation of the value of the function for one iteration takes up to 5 to 7 h of machine time. The general flow chart of the algorithm for the solution of the problem for this case is presented in Fig. 19.

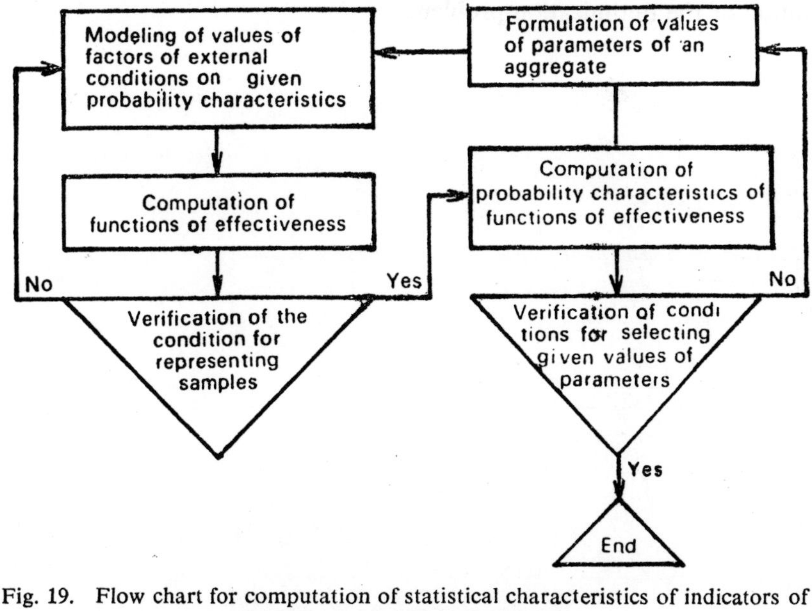

Fig. 19. Flow chart for computation of statistical characteristics of indicators of effectiveness of machine aggregate.

Specific importance is given to experimental methods for solving optimization problems with the use of highly effective methods of optimum designing of optimization experiments.

In testings of agricultural technological processes, the use of experimental methods for the solution of the multifactor (three or more factors) optimization problems is limited because of brief agrotechnical periods and often by the long lapses between the conducting of the experiment and working out its results, which makes it difficult to use sequential designing of experiments.

The results of multifactor experiments, in testings of agricultural machines, can be usefully employed in the development of models on the level of partial criteria with the objective of determining the basic criteria of effectiveness and optimization of the system as a whole.

In optimization of actual agricultural machines (operations) or systems (complex of machines, industrial processes and others), a number of difficulties appear. The criteria of optimization, in many cases, appear to be less sensitive to the change of parameters of the subject or the conditions of its functioning. Because of this, it is difficult to succeed in determining a sharp extremum. Since the subjects of testing broadly appear as crude systems, it is, therefore, advantageous to determine not a point, but a region of optimum values of factors or conditions of operation of the subject (system), in which the criterion of effectiveness changes within tolerable limits [58].

In conclusion, it must be emphasized that the process of accepting recommendations on the results of testings should also be considered as the solution of the optimization problem.

CHAPTER 5

Mathematical Modeling in Agricultural Technological Process Testing

17. BASIC METHODOLOGICAL CONCEPTS

In using the modeling method, the real process or subject is replaced by its analog, which may be realized with the help of a structure (physical model) or a mathematical description (mathematical model). A mathematical model can be developed in the form of a system of equations or an algorithm that may describe the real subject or process to a certain approximation.

As already mentioned, a complete concurrence of the model to its original can be denoted by number 1 and a complete nonconcurrence by 0. All real values of the level of 'concurrence' will lie within the limits of 0 to 1, and in these limits, the model exists. In limiting cases (at values of 'concurrence' 0 to 1), the model and modeling lose their significance. At $M=0$, the model does not contain any information, and at $M=1$, it does not differ from the modeled subject with the modeling losing its significance.

In correspondence with methodological principles of testing, the general problem involving the modeling of characteristics of an agricultural technological process under test is divided into two parts: (1) the search for a relationship of the type $y=f(x)$; (2) the validity of the obtained relationship in the entire range of parameters and conditions for machine application.

The range of problems, which can be solved under the first category, relates to the internal modeling of processes or to modeling at the microscale. Problems of second category are related to external modeling or to modeling at the macroscale [13].

The example of the problem of the external modeling of a process will be considered. Let it be established that the extent of losses in the working of haymaking machines depends on crop-yield, the type of crop and the number of harvests in the following way:

$$\Pi = a_1 y^{\alpha a} + \Pi_2 + \Pi_3,$$

where αa are the coefficients which take the effect of crop-yield on losses into account; Π_2—the traction of losses because of the type of harvested mass; Π_3—the component of losses that depends on the harvest (first or second).

Such relationships can be given in a tabular form, graphically, analytically or by some other method.

For the purpose of predicting the losses Π, when one of the machines fulfills a given volume of work in a given zone, the most acceptable method is that of numerical modeling that makes use of statistical testings (Monte Carlo). Knowing the distribution of values y, Π_2 and Π_3 from the program of modeling, specific realizations of these values are extracted from the memory of the computer and

$$\Delta\Pi_i = a_1 y_1^{\alpha a_j} + \Pi_{2k} + \Pi_{3l},$$

is calculated, where i, j, k, l are the notations of specific realizations of random values Π, y, Π_2, Π_3.

Values $\Delta\Pi_i$ are summed up to find the total volume of work which corresponds to the annual load on the machine, and $\Pi_i = \sum_i \Delta\Pi_i$ are obtained.

By repeating similar testing, a very large array of values Π_i can be accumulated, mathematical expectation and dispersion can be estimated and the law of distribution established.

The treatment of general questions involving the modeling of operational processes of agricultural machines and the solution of concrete problems are based on the following initial assumptions:

1. modeling is most effective only in combination with mechanicomathematical, logical-mathematical and physicochemical methods of operational process analysis;
2. a model can only reflect the most essential features and the logical structure of the process. In its application, it is necessary to use a set of assumptions which are directed toward the systematization of the process. The consideration of all the factors that affect the process is not possible, and the selection of basic factors is not unique;
3. any process performed by an agricultural machine should be divided into elementary acts (working functions) for building a model. These elementary acts should allow a description with the help of simple arithmetic and logical operations;
4. the algorithmic description of processes performed on agricultural machines is not unique. The design of the model depends on the conditions of the process as well as on the objectives of modeling; and
5. the essential properties of operational processes of agricultural machines are reduced by their structural (functional) characteristics and are invariant in relation to external conditions.

As a rule, the results of modeling have been of a very specific character, i.e. they are related only to those problems (or class of problems) for which the model has been developed. Therefore, a scientific generalization of the modeling results of a specific process or a group of processes assumes a very important role.

The functions of models applicable to individual problems of agricultural mechanics and also to agricultural technological process testing are different and, in fact, can be determined by the problems, for the solution of which they have been developed (Fig. 20).

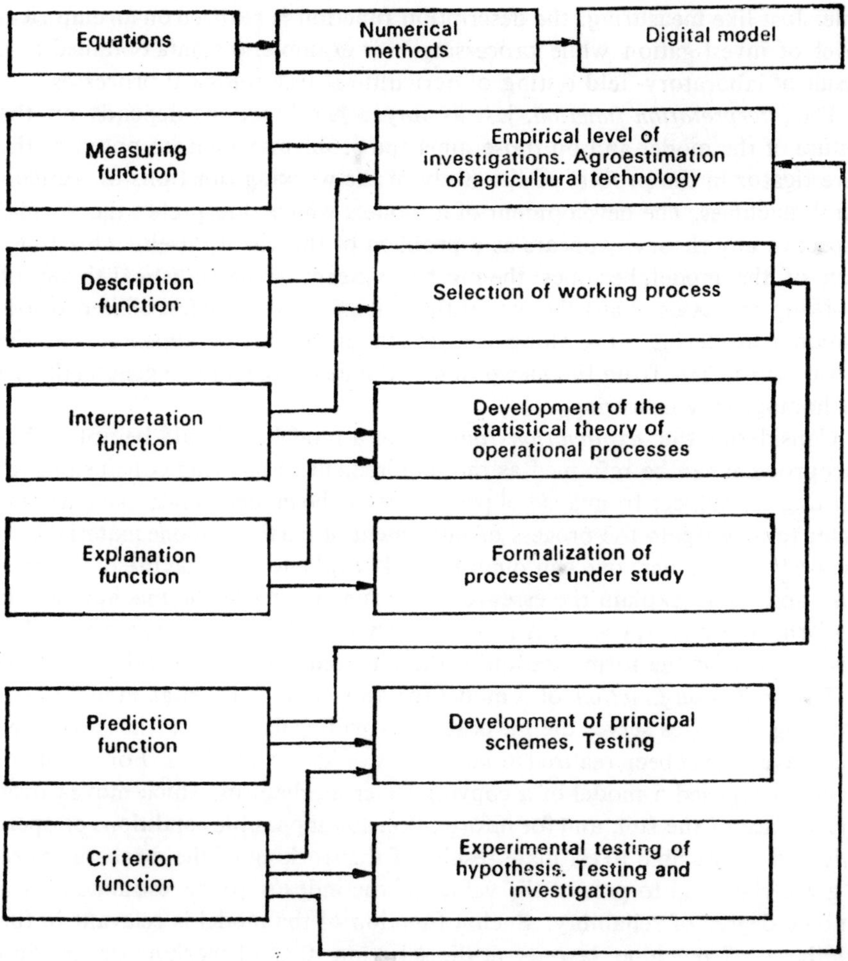

Fig. 20. Levels and functions of mathematical models.

The measuring function predominates at the empirical level of investigation. By model measurement, an operation is implied for which measurements are performed—not on the subject of interest but on its adequate model. So instead of collecting statistical data which characterizes the deviation of the working organ of the cultivator from the row of plants, statistics are collected, the elements of which are calculated by an algorithm describing the process. Models having only a measuring function are rarely encountered during testing of agricultural technological processes.

A model of the working process can fulfill the role of a *description function*. As a rule, the description function creates preconditions which are necessary for the solution of basic problems of investigation playing an auxiliary role. Just like measuring, the description function is realized on an empirical level of investigation while processing, for example, the data obtained as a result of laboratory-field testing of agricultural technological processes.

The *interpretation function*, just as any other function, depends on the nature of the model and on those aims and problems which are posed by the investigator in the process of the study of the working functions of agricultural machines. The development of a model, which interprets the formal aspect of any theory, appears as a problem of this theory itself. This function of the model becomes the most significant in the statistical theory of working processes. Thus the model that interprets the process of seed distribution considering slips along the length of the furrow appears to be the most appropriate from the viewpoint of the development of a general theory of homogeneity in sowing.

Considering the *explanation function* of a model, it should be noted that this problem can be reframed as the requirement of drawing conclusions regarding the subject from a set of properties of the model. While using a computer to investigate the process of movement of particles of one material into the body of another, one can quantitatively evaluate the structure of the process and, thus explain the essence of the functioning of the machine which performs the given process. An explanatory model usually appears as a distinct means for the formalization of the phenomenon under study.

The *prediction function* of a model for application to digital modeling of working processes serves the purpose of providing data about the processes that have not yet been realized in laboratory or field conditions. For example, having proposed a model of a copying feeler mechanism, which moves over the surface of the soil, and for information about possible conditions of operation, it is possible to estimate results of the working of the mechanisms of a given class and to predict the values of the indexes to be estimated to a known degree of reliability. Such a function of the model is relevant in the synthesis of machine layouts at the selection stage of working organs and design.

The *criterion function* of a model serves for testing of the verification of information regarding the original. This quality of model is very essential for testing machines. On the basis of some knowledge about the subject of interest, the investigator formulates hypothetical opinions about this subject, other, still unknown properties and also about the possibility of bringing more clarity into the information experimentally obtained. In such a case, the model appears in a role of verification of information about the modeled subject.

In a general case, digital models of working processes are multifunctional and depending on the aims of the investigations, one or the other function dominates. In solving problems connected with testing, the criterion function dominates; in selecting the principal scheme of the machine, the prediction function is important; in the stage of development of the theory of process, the interpretation and explanation functions of the models are significant (Fig. 20).

In the testing of agricultural technological processes within the domains of the aforementioned functions, the following problems can be solved by various modeling methods:

(1) the problem of analysis of the effectiveness of the subject under test: having known input parameters of the subject (external conditions) and known actions to determine the output parameters as the characteristics of effectiveness of the subject;

(2) the problem of the synthesis of the subject on given relations between the input and output parameters; to establish parameters of the subject for prediction of possible directions toward its improvement;

(3) the problem of optimization: the search of such a collection of factors for which the output parameters attain a given value;

(4) the problems of interpolation: determination of the indexes of effectiveness in the range of values of the input parameters or the parameters of the subject that are limited by the conditions of the experiment;

(5) the study of possible values of change in the working parameters for variation in the working conditions of the machine over wide limits; and

(6) prediction of the work quality of machines without field testing for a change in working conditions, design parameters which determine the quality of fulfillment of the working of the process, a simultaneous change in the working conditions and design parameters.

A mathematical description of the operational process of the agricultural machine (as of any other complicated system) can be given by means of mathematical models of three basic types.

Mathematical model of a system of equations investigated by analytical methods. In the given case, either explicit formulas for the parameter to be determined are derived or equations are transformed into a known form or else equations are investigated by qualitative methods. In the practice of agricultural machine testing, the application of such models is limited due to the following basic reasons:

As a rule, the analytical description of processes can be based on the assumption that random disturbances can be neglected which, as it is known, can considerably distort the real picture of the process; besides, it takes more time to compile and solve equation systems.

Description of a process with the help of equation systems which provide a numerical solution, particularly with the use of a computer. A major drawback of such a method of investigation lies in the fact that the mathematical models of complicated processes in their initial form do not always prove to be useful for the application of numerical methods; and the transformation of mathematical models to a form useful for a solution, as a rule, turn out to be just as complicated as in the case of analytical investigations.

The role of computers lies only in making the calculations automatic, i.e. to provide an automatic solution by numerical methods.

Modeling with the help of digital models preserves only the logical structure of phenomena and requires transformation of the mathematical model into a specialized model algorithm.

Digital modeling is the most suitable method for the evaluation of the quality of operation of agricultural process because of its universality, its correspondence to working process properties, the clarity and reliability of results (ensuring that machine testing stations be provided with computers which help in realizing the digital models), the complete diversity of theoretical bases of process modeling and systems and in the most general form, it is also applicable to problems of other industries.

The determination of a digital model assumes its basic realization without a computer, but with the help of operations available to digital computer technology. The problem of modeling is reduced to obtaining values of output parameters in terms of input parameters and the peculiarities of the functioning of the design which realizes the process under investigation. To date, a number of works are known in which the method of digital modeling are fairly affectively used for the investigation of industrial, technological and operational agricultural machines and the optimization of their design and technological parameters [10, 15, 58, 59, 60, 72].

Examples of the design of the mathematical models discussed in the aforementioned works and the models themselves are successfully used for estimating the effectiveness of testing techniques.

Limitations on the testing periods and increasing demands toward their speedy implementation and the greater reliability of obtained data make it possible to propose the following general scheme of using digital models for the evaluation and prediction of the effectiveness of agricultural machines:

1. To develop a zone mode in which the machine will be used characterizing work conditions in terms of statistical indicators. General principles for building a mode of the zone will be described in the subsequent sections of this book. To make a model more precise, it is repeatedly revised on the basis of accumulated information regarding the entire zone or its individual characteristic requirements.

2. To develop a model of the operational process or the mechanism which performs this process.

3. To state the hypothesis on the nature of quantitative interrelationships between indexes of the fulfillment quality of the working process and factors

which determine the working conditions and the operation plan.

4. To model characteristics of effectiveness with varying parameters and machine operation conditions.

5. To estimate the adequacy of the model and to arrive at a decision regarding the reliability of the obtained data and also for the necessary degree of supplementation of the modeling data with the data of field testing.

Realization of the model plan already considered makes it possible to substantially step up the testing of individual machines, complexes and systems for mechanization and automation of processes in agricultural production.

18. A GENERAL SCHEME FOR OPERATIONAL PROCESS MODELING

The operation sequence performed in the modeling of working processes generally consists of the following blocks (Fig. 21):

(1) statement of the problem to be investigated;
(2) selection of and basis for parameters which characterize the fulfillment quality of a given process;
(3) meaningful description of the process;
(4) working out a formalized scheme;
(5) development of a mathematical model;
(6) setting up the algorithm;
(7) realization of the algorithm; and
(8) interpretation of results.

The sequential order of the foregoing operations is preserved only as a whole; the individual blocks can sometimes be placed in a different sequence. In particular cases, some of these may even be omitted. So the selection and basis of the parameters, which characterize the fulfillment quality of the process, can often precede the meaningful description. However, in the development of a formalized scheme and mathematical model, they rarely do so. For instance, without examining the general problems of copying, the basis for selecting the ordinate of two real random processes as the output parameter of differences, according to scheme $Y_{\text{diff}} = Y_2 - Y_1$, cannot be explained. Here, Y_2 is the surface from which the depth of the trajectory of the working part is measured and Y_1 is the trajectory of the actual working part.

The general scheme of digital modeling assumes the concentric nature of transitions between blocks. Thus, while performing operations connected with the contents of Block 2, the necessity to return to Block 1 may arise, so as to draw the additional information required for making the complex of the desired parameters more precise.

While formulating the problems of modeling of operational processes of agricultural techniques, it is recommended that two stages are distinguished. The first stage is that of rendering the general objective of more precise

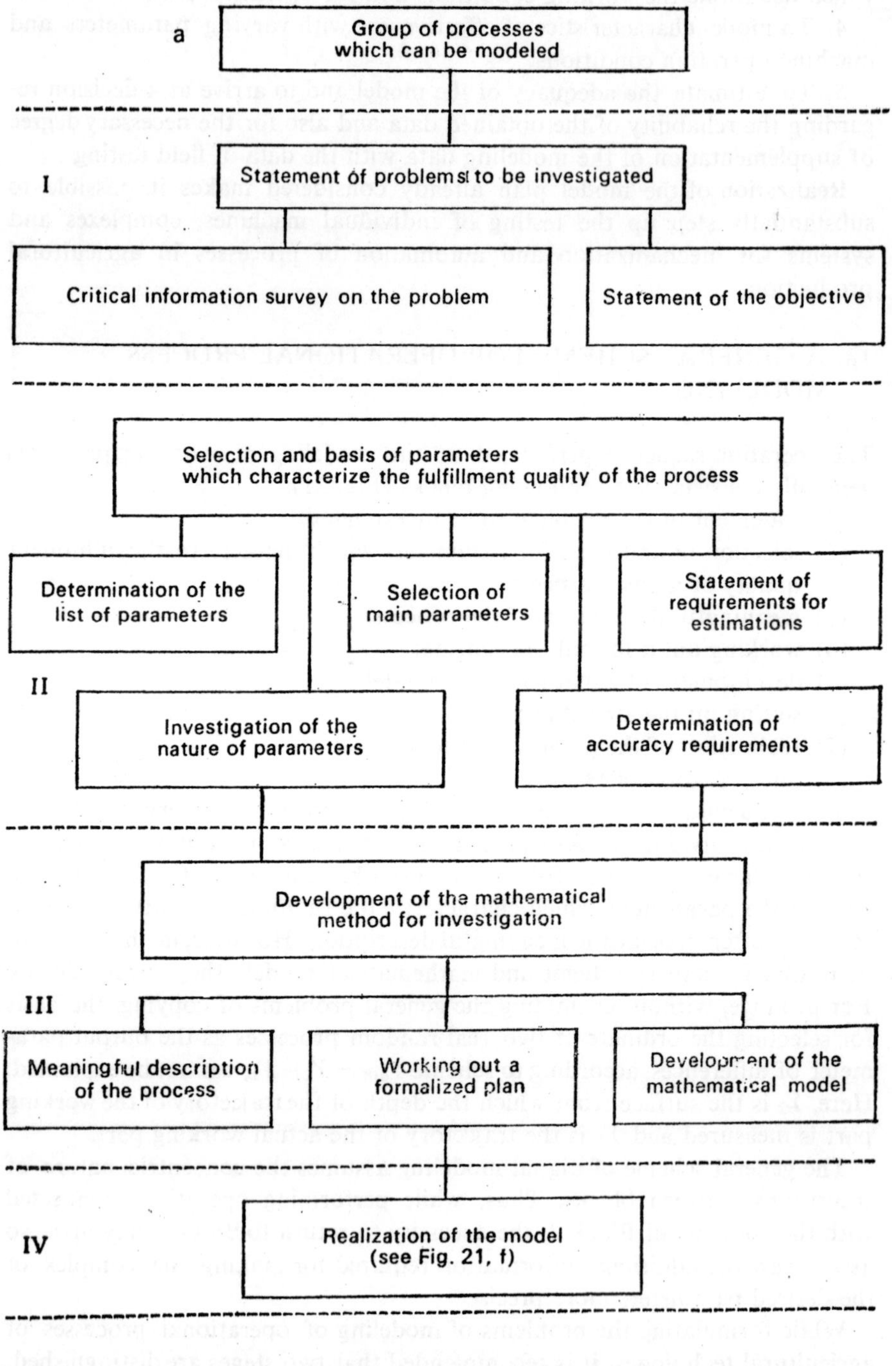
a
Group of processes which can be modeled
I
Statement of problems to be investigated
Critical information survey on the problem
Statement of the objective
Selection and basis of parameters which characterize the fulfillment quality of the process
Determination of the list of parameters
Selection of main parameters
Statement of requirements for estimations
II
Investigation of the nature of parameters
Determination of accuracy requirements
Development of the mathematical method for investigation
III
Meaningful description of the process
Working out a formalized plan
Development of the mathematical model
IV
Realization of the model (see Fig. 21, f)

Fig. 21.

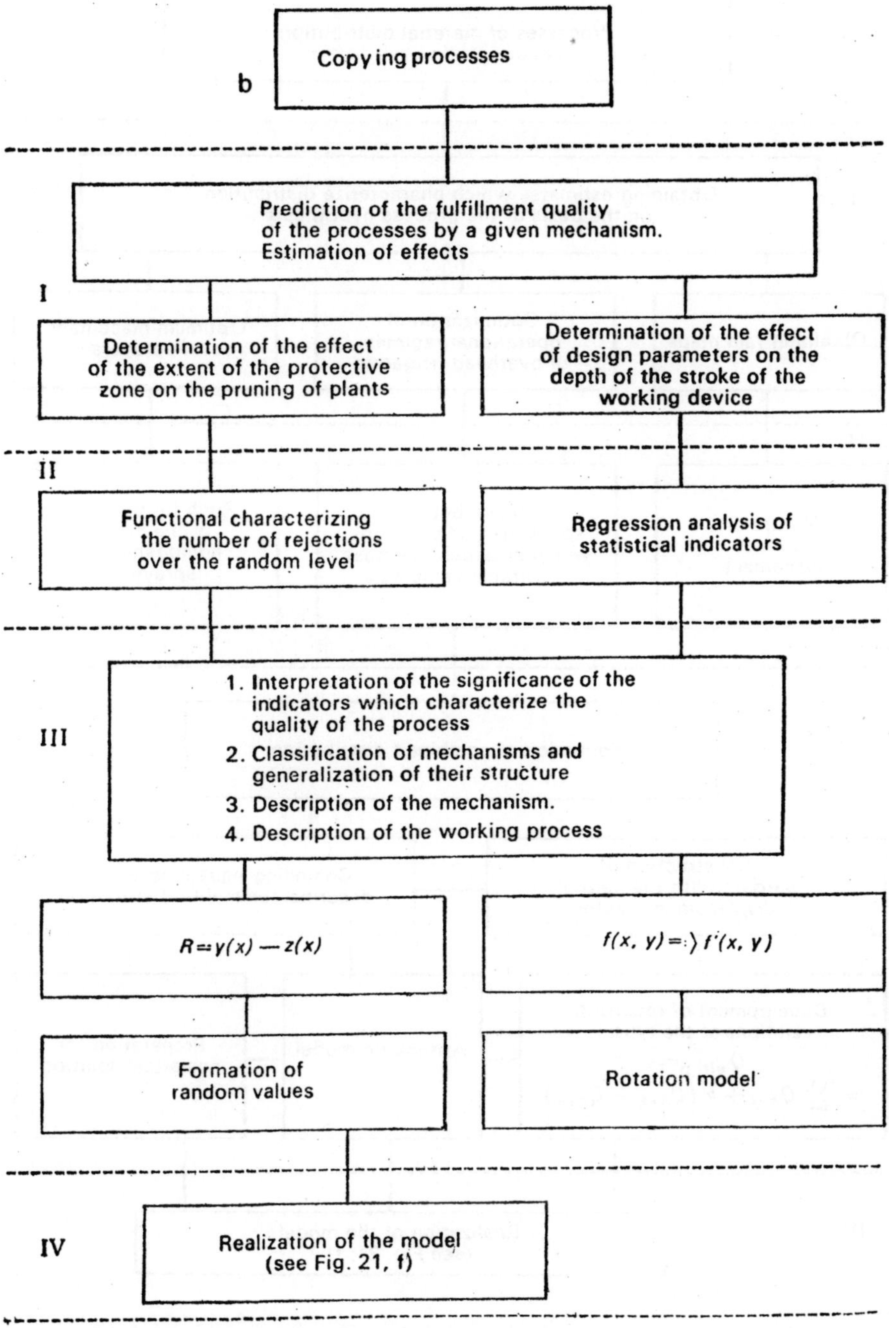
b
Copying processes
Prediction of the fulfillment quality of the processes by a given mechanism. Estimation of effects
I
Determination of the effect of the extent of the protective zone on the pruning of plants
Determination of the effect of design parameters on the depth of the stroke of the working device
II
Functional characterizing the number of rejections over the random level
Regression analysis of statistical indicators
III
1. Interpretation of the significance of the indicators which characterize the quality of the process
2. Classification of mechanisms and generalization of their structure
3. Description of the mechanism.
4. Description of the working process
R = y(x) — z(x)
f(x, y) =:⟩ f′(x, y)
Formation of random values
Rotation model
IV
Realization of the model (see Fig. 21, f)

Fig. 21.

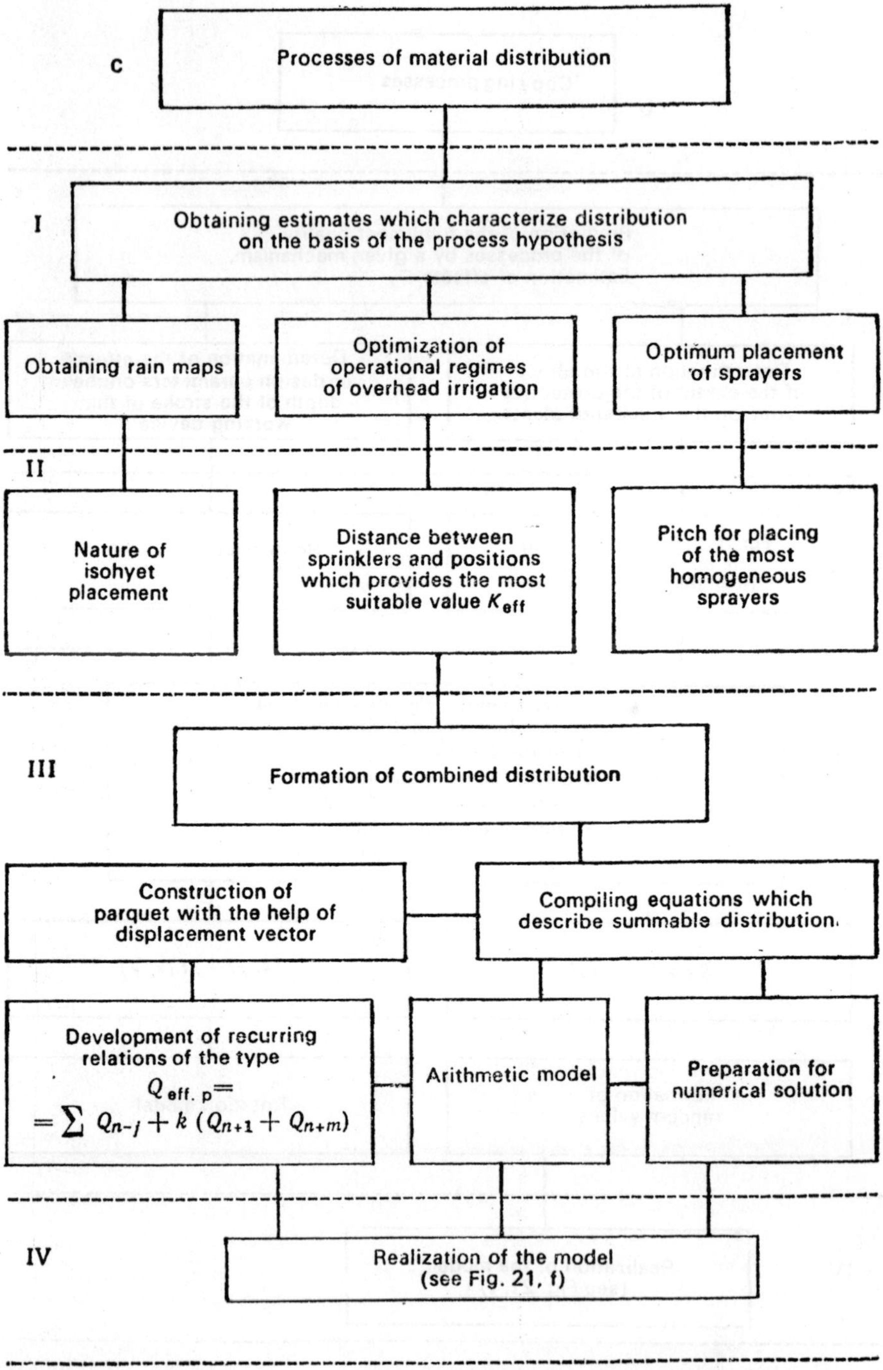
c
Processes of material distribution
I
Obtaining estimates which characterize distribution on the basis of the process hypothesis
Obtaining rain maps
Optimization of operational regimes of overhead irrigation
Optimum placement of sprayers
II
Nature of isohyet placement
Distance between sprinklers and positions which provides the most suitable value K_{eff}
Pitch for placing of the most homogeneous sprayers
III
Formation of combined distribution
Construction of parquet with the help of displacement vector
Compiling equations which describe summable distribution.
Development of recurring relations of the type $Q_{eff.\,p} = \sum Q_{n-j} + k\,(Q_{n+1} + Q_{n+m})$
Arithmetic model
Preparation for numerical solution
IV
Realization of the model (see Fig. 21, f)

Fig. 21.

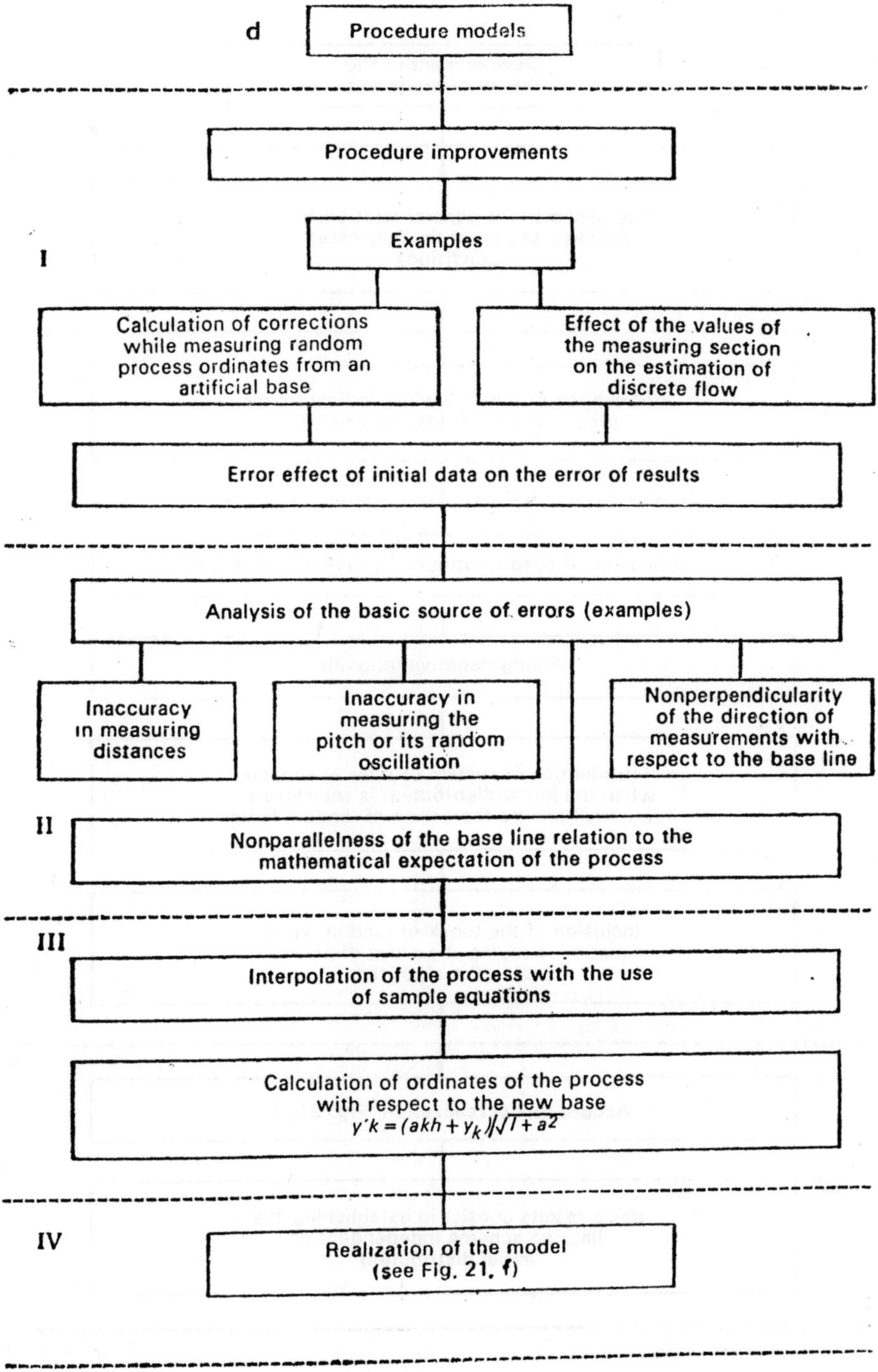
d
Procedure models
Procedure improvements
I
Examples
Calculation of corrections while measuring random process ordinates from an artificial base
Effect of the values of the measuring section on the estimation of discrete flow
Error effect of initial data on the error of results
Analysis of the basic source of errors (examples)
Inaccuracy in measuring distances
Inaccuracy in measuring the pitch or its random oscillation
Nonperpendicularity of the direction of measurements with respect to the base line
II
Nonparallelness of the base line relation to the mathematical expectation of the process
III
Interpolation of the process with the use of sample equations
Calculation of ordinates of the process with respect to the new base
y'k = (akh + y_k)/√(1 + a²)
IV
Realization of the model (see Fig. 21, f)

Fig. 21.

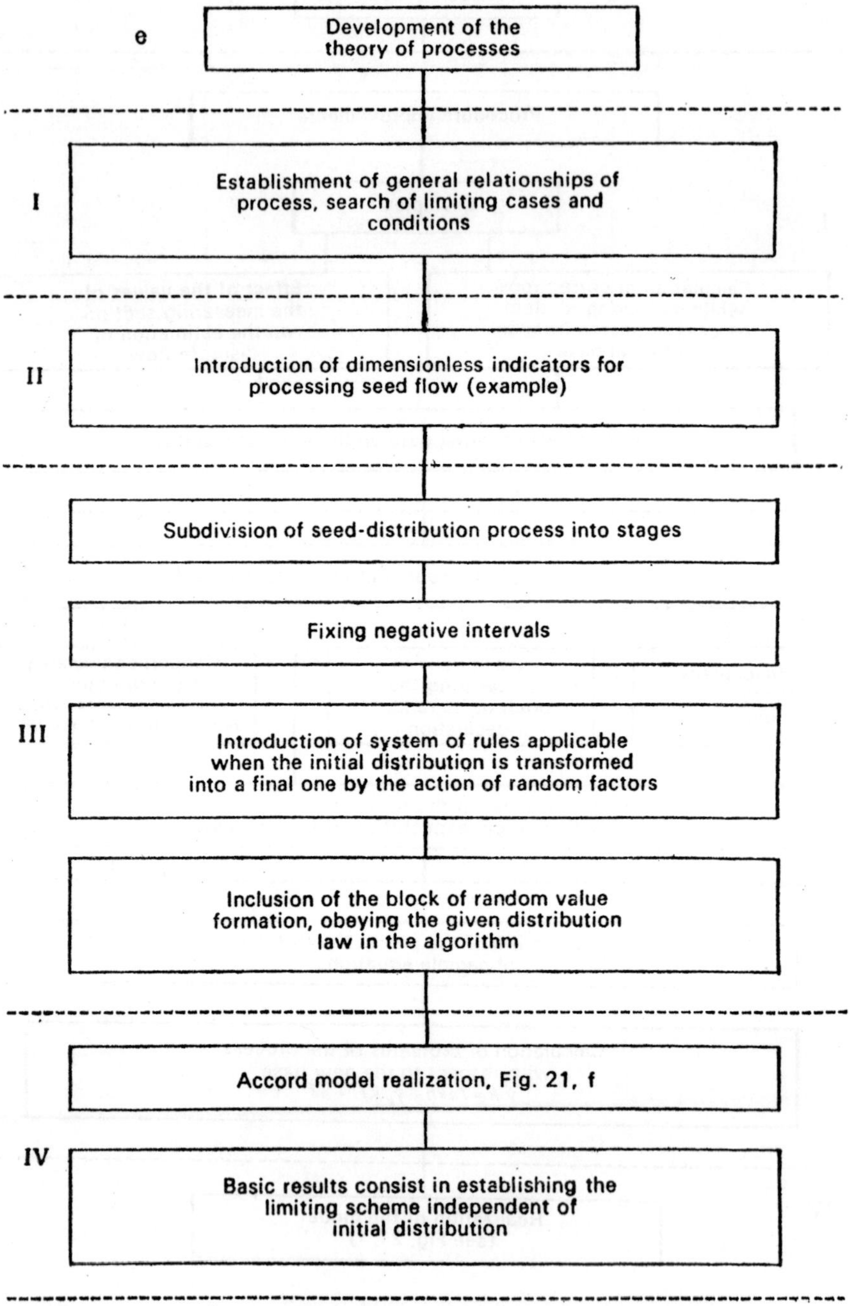
e
Development of the theory of processes
I
Establishment of general relationships of process, search of limiting cases and conditions
II
Introduction of dimensionless indicators for processing seed flow (example)
Subdivision of seed-distribution process into stages
Fixing negative intervals
III
Introduction of system of rules applicable when the initial distribution is transformed into a final one by the action of random factors
Inclusion of the block of random value formation, obeying the given distribution law in the algorithm
Accord model realization, Fig. 21, f
IV
Basic results consist in establishing the limiting scheme independent of initial distribution

Fig. 21.

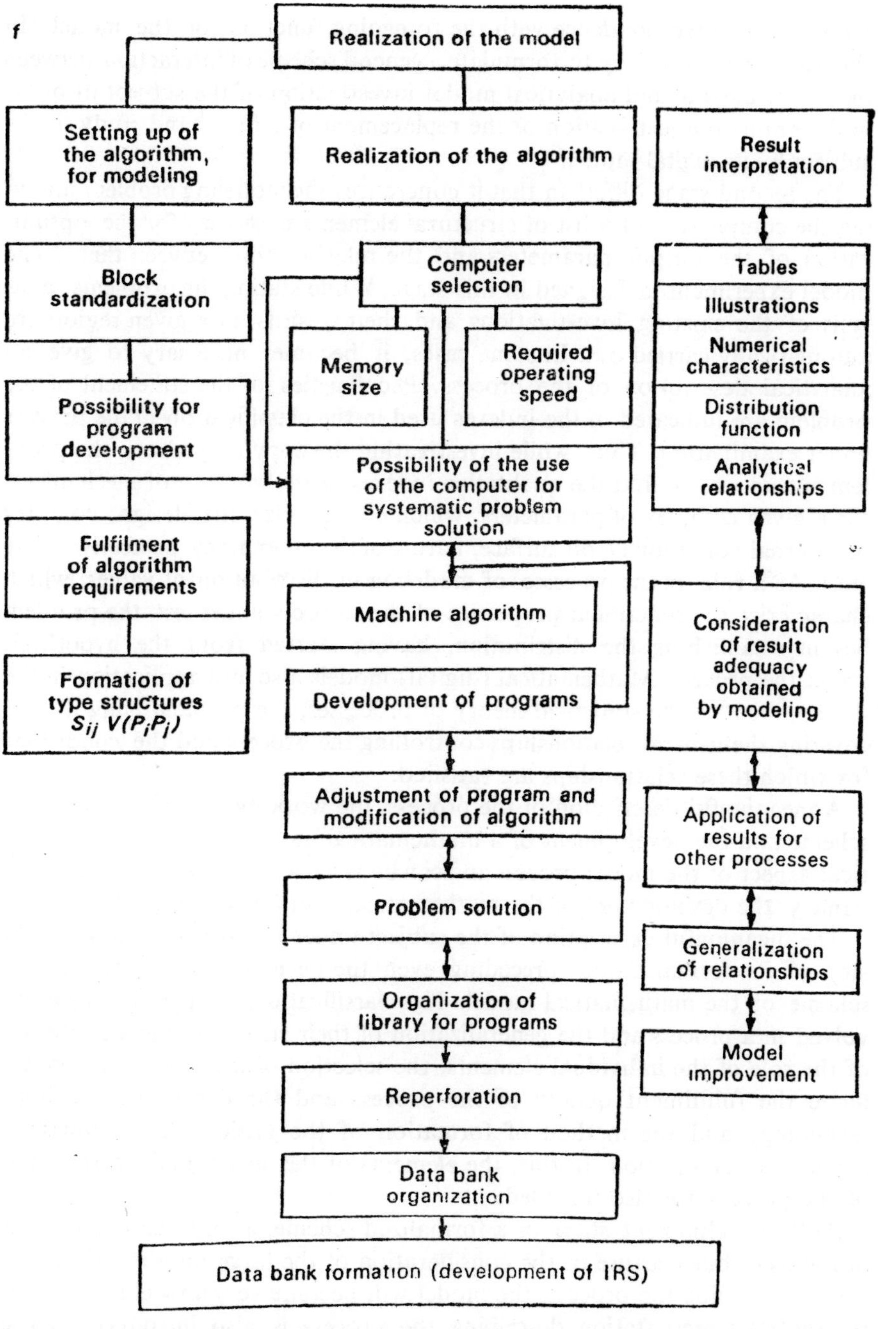

Fig .21 .Modeling scheme of operational processes of agricultural machines: a—general scheme; b—modeling of copying processes; c—distribution process of materials; d—procedure modeling; e—developing a theory of working processes using the sowing process as an example; f—approximate work content at the level of realization of the model; I–IV—levels for solving problems.

modeling in correspondence with the foregoing functions of the model. In this case, it is necessary to formulate a general scheme of interaction between the experimental and analytical model investigation of the subject in order to determine the justification of the replacement of a first-hand study of the subject by its digital model.

The second stage differs in that it concretizes the modeling problem allowing the compilation of a list of structural elements necessary for the optimization of the output parameters and the relationships between them. The model experiment is designed in this stage. While stating the problems, analysis of the existing investigations and their results in a given region are automatically carried out. In some cases, it becomes necessary to give an analytical description of the process. Peculiarities of the statement of the problem are indicated in the indexes used in the classification of the process under examination. Thus, while investigating the copying process, the problem put up was to find the qualitative indexes of the working of mechanisms for a given complex of parameters, which characterize the design, complex of external conditions (soil surface, nature of the other array, protective zone, etc.). As a rule, in the processes of modeling of distribution of values, which characterize the functional properties of the subjects under test, the problem lies in establishing the distribution, having started from the hypothesis about the process. Mathematical (digital) models also find application in the development of the statistical theory of processes, the problem lying in discovering the general relationships controlling the process and the conditions for which these relationships are satisfied.

A meaningful description of the process, the working out of a formalized scheme and the development of a mathematical model represent the theoretical aspect of the model process and can be given a combined nomenclature, namely, the development of the mathematical tool for investigation.

The meaningful description of the subject under investigation is the first step of its formalization, preceding even the development of a formalized scheme of the mathematical model. The classification of the mechanisms involved in a process and the generalization of their structure, the dercription of the role of the individual elements, the selection of indexes which characterize the fulfillment quality of the process and the description of both technology and the method of formation of the basic index comprise a meaningful description. In this, the elements of the algorithmic description of the process are also included.

At the development stage of a formalized scheme, a set of assumptions is introduced, but because of the consideration of the large number of factors which determine the process, the model will become very unwieldy. A general formal representation describing the process is also introduced, basic relationships are recorded and the type and nature of input and output variables in accordance with the general principles are clearly established.

A formalized scheme, along with the mathematical model, sum up the study and verification of the process and provide the basis for moving over to the setting up of the algorithm, i.e. a procedure that is dependent on the mathematical model of the process. At the same time, by taking the earlier procedures, considerable effort for the national organization of the calculations and the possibility of further model improvement without introducing radical change is required.

The basic requirements for digital model algorithms of working processes of agricultural machines consist in the following:

(1) standardization of blocks from which models of different processes are formed, i.e. use of universal blocks. Thus, for statistical modeling of processes, it is convenient to distinguish between blocks for the formation of multidimensional random variables, blocks for statistical processing of data, blocks for comparison, etc.;

(2) while setting up an algorithm and a program for calculations, it is necessary to take into account the possibility of subsequent additions of programs without their radical reworking; and

(3) operators used in the construction of flow chart should correspond to the division of the process into individual logical blocks and must also possess clarity and be easily realizable through elementary operations.

In conclusion, it is necessary to mention that the effectiveness of mathematical modeling in the testings of agricultural technological processes depends primarily on the quality of the model used. Formalized rules for developing models do not exist and experience in this direction is achieved gradually.

The general viewpoints presented here and also the examples of formulating and investigation models of a different complexity for various processes, described in the following sections, constitute the basis for further development and application of mathematical modeling in estimating the effectiveness of agricultural technological processes at the testing stage.

CHAPTER 6

Methods for Forecasting the Effectiveness of an Agricultural Technological Process in the Test Stage

19. BASIC DEFINITIONS AND METHODS OF FORECASTING IN AGRICULTURAL TECHNOLOGICAL PROCESS TESTING

The reliability of decision-making on test results and, hence, the conclusions arrived at from testings and the reduction in the completion period and the implementation of new agricultural technological processes in production are mainly determined by the quantity and reliability of alternative variants of the solution. These can be prepared from an analysis of the testing information obtained regarding technological, technical, operational-economic and other characteristics of the subjects under test. Taking the form of forecasting, on the basis of which opinion about the possible effectiveness of tested technological process after implementation into production are worked out, a successful preparation of such variants can be made with the help of modern scientific methods of forecasting. These methods combine analysis of retrospective information and information about the subject's behavior at the moment of forecasting with the manifestation of possible effectiveness from its application beyond the limits of values of the observed parameters and the conditions of functioning of the subject under test. Development and application of methods of forecasting in the practice of machine testing help in objectively estimating the utility of the designs of machines under test and in formulating an impression about the tendencies of their development in the near future.

In view of the scarcity of experimental information the methods and techniques of the mathematical modeling of operational processes of machines and complexes on computers and also the methods of physical modeling on actual designs appear as the basis for predicting the effectiveness of agricultural technological processes at the test stage.

Determination of possible characteristics beyond the limits of existing conditions, and deciding the regimes of operation and the machine parameters (on the basis of the accumulation of information in the testing of serial machines and experimental samples with the use of modeling) opens wider possibilities for the prognostic estimation and synthesis of aggregates and industrial processes in testing. This is possible by the determination of techno-economic and technological indexes of various potential variants of aggregates and industrial processes with certain probabilities in correspondence with perspective requirements. Forecasting of techno-economic factors is essential for individual complicated, expensive machines because it prevents testing errors. However, the methods of forecasting do not find a sufficient application for estimating the effectiveness of agricultural technological processes. In this connection, extending and improving efforts in this direction should help in the further development of the methodology and enhancement of scientific and technical levels of testings.

In literature, the concept of forecasting is defined as a prediction about the unknown (at the moment forecasting), it really is the possible states of prediction of the subject, which can be defined within the framework of scientific theory on the basis of the known laws of behavior of the subject under initial conditions and instantaneous conditions, respectively. Forecasting occupies the place of conclusions in the sequence of deduction and when it is confirmed, then the already available theoretical information is also established. On the other hand, the proposal for a forecast is directed at creating attempts for achieving the objectives at the human level of practical activity.

From a comparison of the given definition and the problems that are solved in testing (to forecast the effectiveness of machines when applied in production or to seek a method of improving the machines in order to achieve the desired effectiveness), the necessity of application of forecasting methods in the estimation of machines at the test stage [62, 84, 85, 87] follows next.

An analysis of the requirements presented for forecasting purposes shows that for the application of the principles of forecasting in testing, it is necessary to use methods that ensure the solution for the following problems:

1) determination of the effectiveness of the subject under test for the duration of its service life and in the zone of its application, while considering the statistical nature of the conditions of functioning (requirements of limiting the forecast with respect to space and time [45]);

2) decision-making (recommendations, conclusions) on the basis of formalized rules which represent a set of calculation procedures. These procedures ensure the verification of the forecast by running over the operational model of the subject while considering retrospective information (requirement for the verification of the forecast);

3) determination of the probabilities of the guaranteeing of the indexes of effectiveness by the object under test if these are not known from the test results (requirements of the estimation of forecast probabilities); and

4) experimental verification of the forecast after introduction of the tested subject into production or after determining its forecasting conditions through the comparison of predicted and actual indexes of effectiveness (requirements of verifying the realization of prediction).

In the estimation of technological processes under test, problems of forecasting include two groups of problems: static and dynamic. The first is the problem of forecasting of the indexes of effectiveness without taking the time factor into account (the estimation of agrotechnical, energy, techno-economic and other factors depending on machine parameters, states and work conditions and also industrial situations).

The second group includes problems in which one of the variable components is time. These problems are more typical for the forecast (the scientific method of prediction) because with them, the question about future change in the predicted criterion is solved. These include problems of forecasting new directions in the development of agricultural technological processes, the service life of the machine under test, its technical state and its effectiveness over the duration of this period.

An estimation of the effectiveness of tested technological processes with consideration of zone conditions should be included in the static problem. This is solved probabilistically because the forecast of the temporal characteristic of conditions of the functioning of machines is related to a long-range meteorological forecast, which still remains a sufficiently complicated problem. One of the most important conditions for solving this problem is the availability of a probable mathematical model of the zone of machine-testing station or of the proposed zone of machine application.

In correspondence with the general scheme, the process of forecasting includes three stages: retrospective analysis, diagnosis and prediction.

The first stage begins from the moment of obtaining the task on realization of a forecast. This is completed with the creation of an n-parametric model of the subject for forecasting. For the system of machine testings, the planning of the state 'test' comprises the beginning of the retrospection stage, while the problem solved at this stage is that of design for a model of the subject. Besides, singular predictions are formulated while designing the subject. For static problems of testing, these predictions take the form of opinions about agrotechnical power, operational and the technological, economic, ergonomic effectiveness of the subject under test. For dynamic problems, the form of prediction is the technical conditions of the base components or the systems, depending on the degree of the subjects detailing while drawing the structural flow chart.

In the first testing of the machine (complex), information about analogs or models of analogs and a meaningful description of the process being performed by the machine and the results of sampling are used for designing a model. Empirical formulas or exposed relationships in the behavior of the object are used as models in the retrospection stage. The regions for introduction of models at this stage include programs and methods of testing.

In the second stage of forecasting (diagnosis), the actual values of characteristics of the subject are determined. The model is corrected and made more precise. Possible methods of forecasting are analyzed and the most rational method is selected. Methods for estimation and checking of the reliability of the forecast are also determined and modified.

The third stage (proper forecast) consists of the application of selected methods of forecasting to each of the chosen structural units of the subject with a subsequent synthesis and correlation of obtained singular predictions.

Comparing the given scheme for the forecasting process and the operation sequence established in testing, it is easily seen that the stage of retrospective analysis, in fact, includes the development of standards regarding the subjects and the methods of their testings. For concrete subjects of a similar kind (technologies, complexes, machines), operational models of their testings are developed and mathematical tools and experimental methods for the determination of various characteristics of the subject are evolved. The diagnosis stage is the proper testing itself, when the operational model of testings is modified (the technological chart of testing a sample of a specific machine is drawn), the model is improved, while considering the design as also other modification and the results of trial runs and, finally, testing is carried out, i.e. characteristics (parameters) of the subject as foreseen by the model are determined, and control calculations of effectiveness of the test sample are conducted. Testings are presently completed at this stage, i.e. decisions are made on the basis of information which is only obtained in testing under conditions of machine-testing stations. However, in order to increase the problem-solving potential and to decrease the duration of testings, the estimation of the subjects under test should be completed at the stage of prediction of their effectiveness. In this, in correspondence with the selected method of forecasting, the effectiveness of technological process is determined beyond the limits of its service life conditions, regimes of testings and machine parameters; all possible alternatives analyzed, the decision made, and its reliability analyzed. For a correct forecast, it is important to have a feedback system relating to the forecast results while comparing its results with the actually achieved indexes.

While fulfilling this requirement for control samples, it is necessary to compare results of testing and inspections with that of the forecast on the basis of which the machine was put in production. If the actual effectiveness differs considerably from the predicted effectiveness, then the errors in forecasting are analyzed, the model and characteristics are corrected and the forecast is suitably revised. On the results of these modifications, a corresponding decision is made: if the effectiveness is lower than the predicted one, the design is modified and reliability increased; if it is higher than the techno-economic factors of the machine need to be modified. In experimental sample testing, the prediction can be checked on the test results in subsequent years in other conditions and after introduction, on the operation results in industrial conditions.

The prognostic approach discussed here toward the estimation of technological processes under test, foresees constant improvement of the test system in the presence of a feedback link. This is because the model of the subject for testing (methods, standards) is modified in correspondence with the design, technological and operational specifications of the processes under test, the results of comparison between predictions (for experimental samples) and the actual data of control testings and inspections under industrial operation conditions.

Methods of forecasting are determined by the nature of the subject. More than a hundred methods are listed [92] that can be subdivided into general scientific (observation and experimental analysis and synthesis, assumption and hypothesis, etc.) interdisciplinary (methods of expert estimation, interpolation and extrapolation, modeling, theory of probability, mathematical statistics, etc.) and semiscientific (morphological analysis, method of envelopes, method of trial and error, etc.).

For the prognostic evaluation of agricultural technological processes in testing, the most acceptable methods include expert estimation, interpolation, extrapolation and modeling [20]. The method of expert estimation is used for ranking indexes and for the determination of their weightage in the development of the method of comprehensive estimation of machines; the method of interpolation is used to estimate the carrying capacity of a grain-harvester combine in accordance with state standards in individual cases; the effectiveness of machine operation is estimated while considering the zonal conditions by mathematical modeling methods.

The methods of service-life prediction and the indexes of machine reliability on the basis of a physical model are also developed and are applicable to methods of accelerated testing. The method of determination of the service life of machines by the application of computers, which has been developed and introduced in practice testings at the Central Machine-testing Station, is an example of forecasting by extrapolation.

Methods of forecasting in agricultural technological process testing are often used in connection with the solution for the following problems:

1) designing of mathematical models of subjects under test which reflect the interrelationships of the indexes of the effectiveness of functioning with their design, technical, technological, ergonomic parameters and the parameters of external environment and processed material;

2) measurement and estimation of indexes of effectiveness with the necessary reliability and parameters of designed models, the development of measurement methods and the estimation of measurement errors;

3) prediction of effectiveness of the subject under test and estimation of the reliability of the forecasting; and

4) checking of the reliability of the forecast.

After subsequent development and application to concrete subjects, the methods described here can be used for solving prognostic problems for static models.

For dynamic models, it is necessary to use certain other preconditions, which are based on dialecic arguments of general methodological principles of systems analysis and the estimation of technological and techno-operational parameters from modern scientific consideration.

20. TECHNO-ECONOMIC CONSIDERATIONS FOR FORECASTING TRENDS IN THE DEVELOPMENT OF AGRICULTURAL MACHINES

In order to gather general impressions about the possible channels of past development of any type of machine and also about trends in their future development, it is necessary to ascertain the nature of most of the essential relationships and the mechanico-technological principles of accomplishment of the operational processes. The nature of adopted technical and organizational decisions with operational and economic indexes should also be established.

The economic impact of new technological processes arises mainly from factors or sources of economy: by increasing productivity, by reducing expenses on power and materials (fuel lubricant, maintenance, etc.), by decreasing manpower and by increasing the agrotechnical and general technological factors of operations (harvesting efficiency, higher quality, technical reliability, etc.).

In increasing the role of the factors of productivity and economy, as shown by analysis [88], three criteria of effectiveness have a dominating influence, namely, labor expenses, direct or reduced costs and productivity. These criteria are related to the basic parameters of the aggregates: the clow width, power saturation, metal content and the requisite capital investments.

The relationship between the synthesizing criterion of effectiveness and the parameters of aggregates is expressed by the function. This describes operational costs and reduced costs over the entire service life U_i, on the unit productivity during a certain time period T:

$$J_{U_i} = \int_0^T f(m_a, W, N_{sp}, k_{sp}, K_t)\, dt,$$

where m_a is the mass of the aggregate, K_t—the complex of agrotechnical factors, W—the productivity, N_{sp}—the power saturation, K_{sp}—the specific capital investment, etc.

The maximum impact is generally obtained after completion of the introduction and the mastering of new, highly productive technological processes and machines (for example, after the introduction of the accurate sowing of beets with a mechanized thinning of sprouts, the harvesting of grains with smaller chaft, the continuous harvesting of beets and their live-storage with simultaneous harvesting of stalls, etc.). The replacement process of old machines and related technologies with the optimization of parameters and indexes for the given level of development of science and technology should be examined from time to time and this relationship should be estimated from the economic viewpoint.

The time factor is introduced to predetermine the payback period (the earlier the effect can be obtained, the more useful it is).

In [73], it is proposed to bring in economic effects at a specified moment of time with the help of the so-called profit function

$$\overline{R}=\frac{1}{T}\left[\sum R(Q)t+\int_{0}^{T} rQ(t)\,dt\right],$$

where $\overline{R}$ is the reduced value of the mean profit for the period T, R—the one-time income (expense) at the moment t and r—the current income (expense) in unit time at the moment t.

Academician L.A. Trapeznikov considers that it is worthwhile to determine the profit function $Q(t)=(1+\alpha)^{-t}$—with the help of compound interest with a normative coefficient of capital investment payback, $\alpha=0.1$ to 0.25 as foreseen by state standards.

The cumulative savings of operational costs, derived at the moment of complete introduction of the machine in production in the given farm or zone, is determined by the following expression [73]:

$$\sum_{t_b=1}^{n} r_i=\int_{0}^{t_b} U_r(1+\alpha)^{-t}\,dt+\frac{0.5U_r}{\tan\gamma},$$

where $t_b=1/\tan\gamma$ is the time for introduction, γ—the intensity of introduction in fractions of a year (for example, proportional to time in the general case and can be a nonlinear function of time) and U_r—the savings on operational costs over a year.

For the evaluation of relationships between indexes of effectiveness of machines and basic technical and economic factors, the utilization of curves of development or the technical progress [20, 92] has proved most worthwhile.

Integrated curves, which fully describe the natural course of development of various completed phenomena, were given the utmost importance by V.P. Goryachkin, who basically used them for the analysis of biological and other subjects. While criticizing supporters of the empirical laws, he considered it appropriate to represent integrated curves in the form:

$$y=A\exp\left(-\int_{x}^{x}\frac{d\ln y}{dx}\,dx\right),$$

where $\frac{d \ln y}{dx} = \frac{dy}{y\,dx}$ is the relative increment parameter to be predicted.

Depending on function $(d \ln y)/dx$, it is possible to obtain different equations:

$$\frac{d \ln y}{dx} = \frac{1}{x}, \qquad y = Cx;$$

$$\frac{d \ln y}{dx} = \frac{m}{x} - a, \quad y = Cx^m e^{-ax};$$

$$\frac{d \ln y}{dx} = \frac{B}{x^2}, \qquad y = C e^{-x^{-1}}; \qquad \frac{d \ln y}{dx} = \frac{B_a}{x^{a+1}},$$

$$y = C e^{-Bx^{-a}}.$$

In modern works on technical forecasting, the following types of functions are used [84]:

$$y_t = a_0 + a_1 t + a_2 t^2 + \ldots + a_\lambda t^\lambda;$$

$$y_t = ab^t; \qquad y_t = ab^t c^{t^2}; \qquad y_t = k + ab^t;$$

$$y_t = ka^{bt}; \qquad y_t = \frac{k}{1 + be^{f(t)}}.$$

Having obtained the necessary information and type of integrated curve (function), it is possible to plot the curve of economic effectiveness which reflects the dependence of the obtained effect on time or the change of the basic parameters and capital investments on the farm for acquiring and introducing new types of machines. The payback period for a new technological process is determined from the expression.

$$T_{comp} = C_0/U,$$

where C_0 is the cost of the new technological process and its introduction and U—the direct costs on unit work.

The payback period should be necessarily related to techno-economic parameters of machines:

$$T_{comp} = \frac{K_{sp.in} - K_{sp.etal}}{U_{dir.ex} - U_{dir.in}} \leqslant T_r, \tag{48}$$

where $K_{sp.in}$, $K_{sp.etal}$ are the specific capital investments on the unit work of introduced and etalon machines, in rubles/ha; $U_{dir.ex}$, $U_{dir.in}$—the corresponding direct costs on unit work in rubles/ha and T_r—stipulated maximum payback period.

From the payback period, it is possible to sufficiently reliably determine a rational moment for the replacement of old machines and the introduction of more progressive machines.

In a general form, the process of replacement of the outdated and the introduction of new techniques is represented by curves which reflect the dynamics of the indexes of effectiveness E as a function of expenses K_{ex} (Fig. 22). At some moment of time, t_0, which precedes the achievement of the maximum possible effect from the corresponding techniques or technologies, the introduction of a new machine (complex) or technology begins. The replacement of machines or technologies in agricultural production is encountered with great inertia because it demands the introduction of a complete succession of new machines—especially when working on huge agricultural complexes. The introduction of new techniques begins in the face of old machines, the expenditure on which still continues. Considering that the rapid introduction of new machines is limited and the cost of new technology is generally high compared to the old, the pace of increase of general effectiveness (zone A, see Fig. 22) temporarily decreases, and then, with the saturation of farms with new technology, it increases and follows the indicated pattern (zone B). With a comprehensive adoption of new techniques and an improvement in technology, the maximum effectiveness is achieved (zone G).

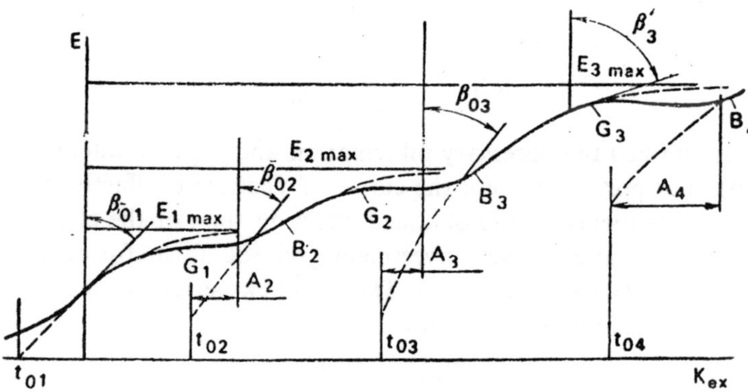

Fig. 22. Approximate course of a set of integrated curves.

Finally, the curve reaches the saturation level on account of the objective equilibrium. Further capital investments give fewer returns, which only confirms the exhaustive potential possibilities of a given design or technology and the necessity of moving over to a more perfected design. Saturation appears as a result of highly limiting factors, such as the limitation on working speeds in agricultural requirements, the limitation on the increase of productivity because of insufficient engine power, the lesser width of the claw, etc.

In this way, the curve describing the dynamics of effectiveness bears a smooth, but complicated nature with the variable values of the coefficient of intensity and the continuously repeating cycles.

Combining a certain sequence of cycles on plane EK and following the assumptions laid out in [73], a field is obtained which contains the set of progress curves (Fig. 22). Each cycle, which characterizes the lifetime and techno-economic parameters of a given type of machine or technology, determines the limiting attainable effect for the given conditions. The rhythm and intensity of change of the progress curves are variable, and the trajectory describing this phenomenon bears a zigzag nature with a certain accelerated deviation (arbitrarily assumed to be linear) from the curve representing the attainment of the 'ceiling' on possibilities to a new, steeper trajectory. This transfer will certainly characterize localized oscillations of effectiveness.

By using a three-dimensional frame of reference, the progress curve can be treated as a spiral, and the plot clearly explains the wavy nature of the curve, which is basically a projection of the spiral on the indicated coordinate axes (Fig. 23).

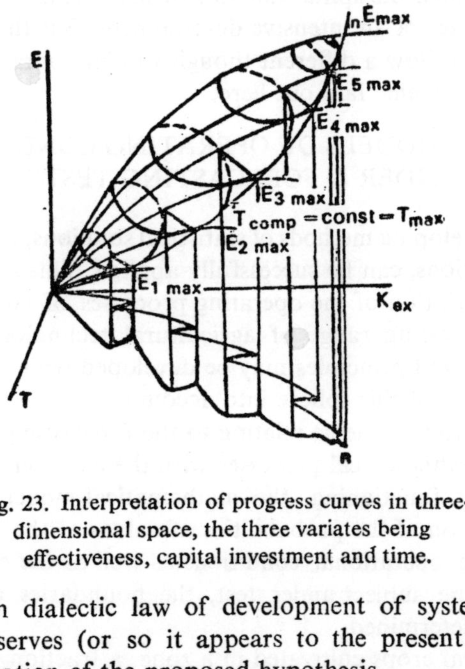

Fig. 23. Interpretation of progress curves in three-dimensional space, the three variates being effectiveness, capital investment and time.

The well-known dialectic law of development of systems in space and time on a spiral, serves (or so it appears to the present authors) as the objective conformation of the proposed hypothesis.

For forecasting, it is important to select parameters of the machine in such a way so as to provide the maximum effectiveness in the time period T_{comp}: $E_k \rightarrow E_{max}$ for the existing economic possibilities of farms and other limitations of industry.

In this, the problems of forecasting in the general case will be concerned with the study of the principles of transforming the parameters of the spiral into a function of parameters of the aggregates $\tan \beta = F(B, v, G, N_{eh}, K_{sp})$ and minimizing, at known limitations K_{ex}, the achievement time of planned and scientifically based levels of effectiveness for the maximum possible integrated effect.

The spiral should be converging, and its parameters should conform to the economic level of the given industry and even the national economy as a whole. In the initial period, at small values K_{ex}, progress curves are more gently sloping, while the periods of machine replacement are large. A steeper course of curves and the converging nature of the spiral can be ensured with an increase of economic potential. For the given conditions, on achieving a certain limiting position P when the spiral seems to be compressed into a point and further capital investments in the setting up of new techniques for a given technology seem to be ineffective, the attained level of effectiveness will oscillate with respect to a stable level (see Fig. 23). Under these conditions, a radical change in technology and the methods of production is inevitably objectively called for, i.e. an industrial revolution, in the process of which the spiral can drastically change its parameters and attain the character of an intensive development. But the process of technical progress will follow a different though similar course, which is in conformity with the scheme laid out here.

21. DESIGN OF MODELS OF OPERATING CONDITIONS OF SUBJECTS UNDER A FORECASTING TEST

New, rapidly developing methods of statistical solutions, related to the theory of random functions, can be successfully applied. This is done through the algorithmic description of the operating processes of complicated dynamic systems, in the testing range of agricultural technological processes. The general methods and principles may be developed while taking the peculiarities and problems of this sphere into account.

For solving static problems relating to the forecasting of the effectiveness of agricultural technological processes with the consideration of the zonal conditions of its functioning, the mathematical model should include the laws of distribution of the probabilities of values of the factors. These factors characterize operational conditions as well as the effect on indexes of effectiveness of the subject under test, the boundaries of the sphere, for which they are determined.

The agricultural crops cultivated in a zone, a fraction each in the general sowing area of the zone, the technological operations (plowing, sowing, harvesting, etc.) and agrotechnical periods of their performance are some of the characteristic zone signs for the location of an MTS. Each technological operation is characterized by the state of the working material. This determines the effectiveness of the performance of the technological process by the machine, by the carrying and adhesion properties of the soil. On these properties depend the effectiveness of the working of the propellers of the traction device. The meteorological conditions also affect the earlier indicated factors and determine the effectiveness of the physical protection of servicemen from the harmful effects of the external environment.

Conditions of operation of agricultural technological processes are also characterized by other environmental factors, which affect the working efficiency of machines but are not associated with specific crops or operations that may or may not depend on time. Peculiarities of the territorial allocation of farms in the zone (relief, dimensions and configuration of fields, distances for transportation, etc.) and material and labor resources (provision for transport, mode of energy, servicemen) are determined by these factors, acting mainly on operational or economic indexes of the subject under test. The scope of determination and the probability of values of these factors are established on the basis of a retrospective analysis of the statistical information comprising the data from zonal experimental stations and also from the results of systematic collection and recording of such information in the study zone.

The processing of information regarding the characteristics of external conditions for the solution of static problems consists in (a) the estimation of the probabilities of the values of different variables and (b) in the sufficiently accurate description of the obtained statistical data in terms of an analytical expression convenient for modeling on a computer.

Characteristics of the external environment are real, and as a rule, are limited by physically possible (minimum x_1 and maximum x_2) values. Therefore, the most acceptable distributions for their description are truncated distributions with the probability density

$$f_{tr}(x) = \begin{cases} 0, & x < x_1; \\ Af(x), & x_1 < x \leqslant x_2; \\ 0, & x > x_2, \end{cases} \tag{49}$$

where $f(x)$ is the probability density of the initial distribution,

$$A = \left[\int_{x_2}^{\infty} f(x)\, dx - \int_{-\infty}^{x_1} f(x)\, dx \right]^{-1}$$

the normalized quotient. Parameters $f(x)$ are determined by the method of moments from the system of equations

$$\left. \begin{aligned} \hat{x} &= \int_{x_1}^{x_2} Axf(x)\, dx; \\ \hat{\sigma}^2 &= \int_{x_1}^{x_2} A(x-\bar{x})^2 f(x)\, dx; \end{aligned} \right\} \tag{50}$$

where $\hat{x}$, $\hat{\sigma}^2$ are the estimated average value and the dispersion of the characteristic to be described. If there are more than two parameters, then an additional equation is included in the system of equation (50).

In the general case, a mathematical model of external conditions of functioning of an agricultural complex is described by a system of numbers, n, of random variates with a sufficiently complicated distribution function determining the probability of combined accomplishment of n inequalities of type $X_i < x_i$ [16]:

$$F(x_1, x_2, \ldots, x_n) = P[(X_1 < x_1)(X_2 < x_2) \ldots (X_n < x_n)].$$

The density of distribution of such a system is the nth mixed partial derivative of function $F(x_1, x_2, \ldots, x_n)$ of the first order in each argument [16]:

$$f(x_1, x_2, \ldots, x_n) = \frac{\partial nF(x_1, x_2, \ldots, x_n)}{\partial x_1 \partial x_2 \ldots \partial x_n}.$$

In practice, the distribution function of characteristics describing external conditions is unknown, and its estimation in the n-dimensional space appears to be a fairly complicated problem—both from the viewpoint of the collection of statistical information as well as from that of its subsequent processing.

A complication in solving this problem arises because of the fact that in the laws of distribution of the individual values related to in the system, it is not possible to find the law of distribution for the system in a general case [16]. In order to completely describe the external conditions (system of random variates), it is also necessary to know the relationships between them as well as the laws of distribution of each characteristic. This relationship is characterized by an arbitrary law of distribution [16], i.e. by the law of distribution of random variates, which are calculated for the given conditions that other random variates have assumed definite values. If the arbitrary law is known, then for the system of two random variates (x_1, x_2), the density of distribution is given by

$$f(x_1, x_2) = f_1(x_1) f(x_2/x_1),$$

where $f(x_1)$ is the density of distribution of variate x_1 and $f(x_2/x_1)$—the relative density of distribution of x_2.

It is known that for an independent random variate, the density of distribution of the system

$$f(x_1, x_2, \ldots, x_n) = \prod_{i=1}^{n} f_i(x_i).$$

Therefore, the problem of estimating the density of distribution for a set of characteristics describing the external conditions becomes substantially simplified if they are classified as dependent and independent, ignoring the trivial relationship between individual characteristics.

In test practice for the description of machine operation conditions for cultivation and harvesting of agricultural crops in the activity zone of the MTS, the model used includes the following factors: the agricultural crop, the crop-yield of agricultural operation, the agrotechnical period for its accomplishment (from the beginning to the end), the duration of the agro-period, the state of the working material (hardness, moisture of soil, moisture of grain mass, ratio of grain to straw, etc.); the type of soil in respect of the mechanical contents, relief, clogging up by stones and the field area.

While designing a model, it is necessary to determine the territorial boundaries of the zone in order to compile a list of the crops which are bred in the zone and to determine the areas covered by these (which are dependent variables because the common soil under cultivation in a zone is practically constant and a change of an area for the sowing of one crop leads to a change of an area for the sowing of other crops). Analysis show that the soil structure under cultivation in the zone varies insignificantly; therefore, the soil under cultivation for different crops in different farms varies only within insignificant limits. This is caused by the rotation of crops and the dimensions of fields; the soil can, therefore, be considered as an independent variate. If these changes are considered over sufficiently large durations of time, it can be assumed that the distribution of the area for different crops follows the normal law of distribution:

$$p(x_{ij}) = \frac{1}{\sigma_{ij}\sqrt{2\pi}} \exp \frac{(x - \bar{S}_{ij})^2}{2\sigma_{ij}^2},$$

where $\bar{S}_{ij}$, σ_{ij}^2 are the average value and dispersion of the area under the ith crop in the jth farm of the MTS zone.

Over the entire zone, the soil under cultivation for the ith crop in farms can be distributed according to various laws, depending on the characteristics of that zone. So, for example, the distribution law for land under beet cultivation in the forest-steppe zone follows the logarithmic normal law [55]; for the area under a grain crop—the truncated normal distribution.

In order to design a model for each crop, statistical data about crop-yield for either all or randomly selected farms is collected. If in the activity zone of the MTS, it is possible to pick out groups of farms which differ by crop-yield because of factors (plains, foothills, hills; chernozem soil, sandy soil, etc.) essentially independent of the farm, then the selection of farms for information gathering is made randomly in each group, and the distribution of crop-yield in the zone is estimated while considering the weightage of each group.

On the completion of each year, this information is revived, keeping in view the obtained crop-yield in the said year. For a more reliable description of the zone conditions, it is necessary to draw attention to the presence of the limiting values of the crop-yield (or other factors), characterized by a sharp deviation from the norm (for example, very moist conditions for drought-prone regions, sporadic changes in crop-yield etc.). This type of abnormality should be studied with the application of distributions for extremal values of random variates:

for maximum values

$$f(x) = \frac{1}{\sigma} \exp\left[-\frac{1}{\sigma}(x-\mu) - \exp -\left(\frac{1}{\sigma}\right)(x-\mu)\right];$$

and for minimum values

$$f(x)=\frac{1}{\sigma}\exp\left[\frac{1}{\sigma}(x-\mu)-\exp-\left(\frac{1}{\sigma}\right)(x-\mu)\right].$$

The characteristics of working materials used for designing a model of the zone should completely reflect that state of material on which the quality of machine operation and its operational-economic indexes depend. For soil-processing and seeding machines, such a characteristic may be the moisture content and hardness of the soil; for beet harvesters, it is the hardness and moisture of the soil and the yield of both beet tops as well as roots; for grain harvesters, it is the moisture of harvested fields, the yield of grain and straw and the moisture and soil hardness. Long-term experimental investigations have confirmed that the hardness and moisture of the soil encountered during agricultural operations followed the normal law of distribution fairly accurately.

Analysis shows that characteristics of working material such as hardness and moisture, the yield of beet tops and roots, grains and straws, the soil moisture and moisture of the straw mass are correlated in pairs. Therefore, for the estimation of the probabilities of distribution of these characteristics, two-dimensional truncated distributions (preferably the normal distribution) are used, and the probability density of these are determined by formulas (49) and (50) for the two-dimensional case:

$$f_{tr}(x, y)=\begin{cases} 0 & \begin{matrix} x\leqslant x_1 \\ y\leqslant y_1 \end{matrix} \\ Af(x, y) & \begin{matrix} x_1<x\leqslant x_2 \\ y_1<y\leqslant y_2 \end{matrix} \\ 0 & \begin{matrix} x>x_2 \\ y>y_2 \end{matrix} \end{cases}$$

where

$$f(x, y)=\frac{1}{2\pi\sigma_x\sigma_y\sqrt{1-r^2}}\exp\left\{-\frac{1}{2(1-r^2)}\left[\frac{(x-m_x)^2}{\sigma_x^2}-\frac{2r(x-m_x)(y-m_y)}{\sigma_x\sigma_y}+\frac{(y-m_y)^2}{\sigma_y^2}\right]\right\},$$

$$A=\left[\int_{x_2}^{\infty}\int_{y_2}^{\infty}f(x, y)\,dx\,dy-\int_{-\infty}^{x_1}\int_{-\infty}^{y_1}f(x, y)\,dx\,dy\right]^{-1},$$

(r being the correlation coefficient for x and y).

While evaluating the seeding area in the zone under examination in respect of the soil type and relief, the stretch of land having each of these characteristics is determined. Soil charts and other sources can be used in collecting such information.

Information on the indicated characteristics is represented in the form of a series of distribution, which can be approximately described by a suitable law.

The abovesaid approach for designing a mathematical model for the zone is acceptable for the solution of forecasting problems with the help of models present in an analytical description of the criteria of effectiveness in the function of the speed of movement, the duration of technological operations, the length of the strip, the crop-yield, dimensions of the fields, etc. The mathematical apparatus of the theory of random functions is used for predicting statistical characteristics of the speed of movement of the aggregates in terms of their parameters. This is done with the help of dynamic models for describing the physical state of the soil and the processed biological masses during harvesting.

Characteristics of the external environment and working material can be a certain quasi-steady process (such as a sufficiently large set of stationary pieces), governed by some law of probability (see Section 6).

Collection of information for the estimation of the characteristics of the aforementioned pieces is, from the technical viewpoint, a rather complicated problem. It requires continuous or a regulated discrete step recording of the characteristics over sufficiently large intervals of time or path.

The general methods for collection and processing of the realizations of random functions and also of the methods applicable for evaluation of machine loading characteristics are given in a series of works indicated [6, 8, 55].

For estimation of the characteristics of hardness and soil moisture, grain mass and straw in bales, etc. as random functions over the path, one is limited to realizations with 150–200 measurements.

In the case of the estimation of low-frequency components, sliding average values of measured characteristics (for example, the mass of the bale over a limited interval) are used

$$\bar{x}(t)=\frac{1}{\Delta t}\int_{t_1}^{t_1+\Delta t} x(\tau)\, d\tau.$$

In order to avoid errors in processing the obtained realization of a limited path length, because of unsteadiness it is necessary to use control tests on steadiness and ergodicity [6, 8].

In conclusion, it is necessary to point out that the use of methods of forecasting for the estimation of technological processes under test does not replace and certainly does not exclude the experiment and opinions of specialists. But, it permits only to obtain within the limits of probability, the quantitative estimation of the possible alternative solutions within the framework of those hypotheses and laws laid down as a base for the designing of models of external conditions, objects under test and their use for forecasting. Prediction depends on the undertakings of an agricultural technological process and increases the reliability of accepted recommendations on the test results.

PART TWO

Methodological Principles for Types of Agricultural Machine Estimations

CHAPTER 7

Agricultural Technological Evaluations

22. GENERAL CONSIDERATIONS

Specific problems for the determination of indicators of effectiveness of machines under test are related to the type of estimation [12] (agrotechnological operational and technological, power, engineering and technical, sanitary and hygienic, economic, etc.) as well as to the general methodological principles, which assume the complexity of estimations and the possibility of their combination with integrated indicators, as for example, the economic indicators. Diversity in the types of estimation arises because of the complexity of processes that can be accomplished by agricultural machines. A sample classification of technological and industrial processes to which different types of estimations can be applied is shown in Fig. 24.

Along with the development of a system of agricultural technological process testing, a method of machine rating has been developed, their nomenclature expanded and provisions of instrumentation also enlarged. Beginning with recommendations of V.P. Goryachkin and also V.I. Nagibin, M.Kh. Pigulevskii, K.G. Shindler and many others, the methods of functional, reliability, power and other ratings are formulated as directions, recommendations and other assessments.

Even now at the machine-testing stations headed by V.P. Goryachkin, tests are conducted according to outmoded methods, which take into consideration soil preparation and fertility, the repetition of experiments, etc.

At the state machine-testing stations, recommendations on conducting various types of evaluation were initially included in test methods according to the types of machine. The industry standards were then evolved, keeping in view the type of rating and its development. More recently, standards have been evolved for practically all types of machines and ratings. Among various types of evaluations discussed in the book, agrotechnical evaluation has a significant place. The agrotechnological estimation of agricultural machines is necessary for a comparative determination of the totality of indicators that characterize the accomplishment quality of the technological process.

In view of the variety of machine types, the indicators for machine complexes number more than a thousand in field-crop cultivation. Specific

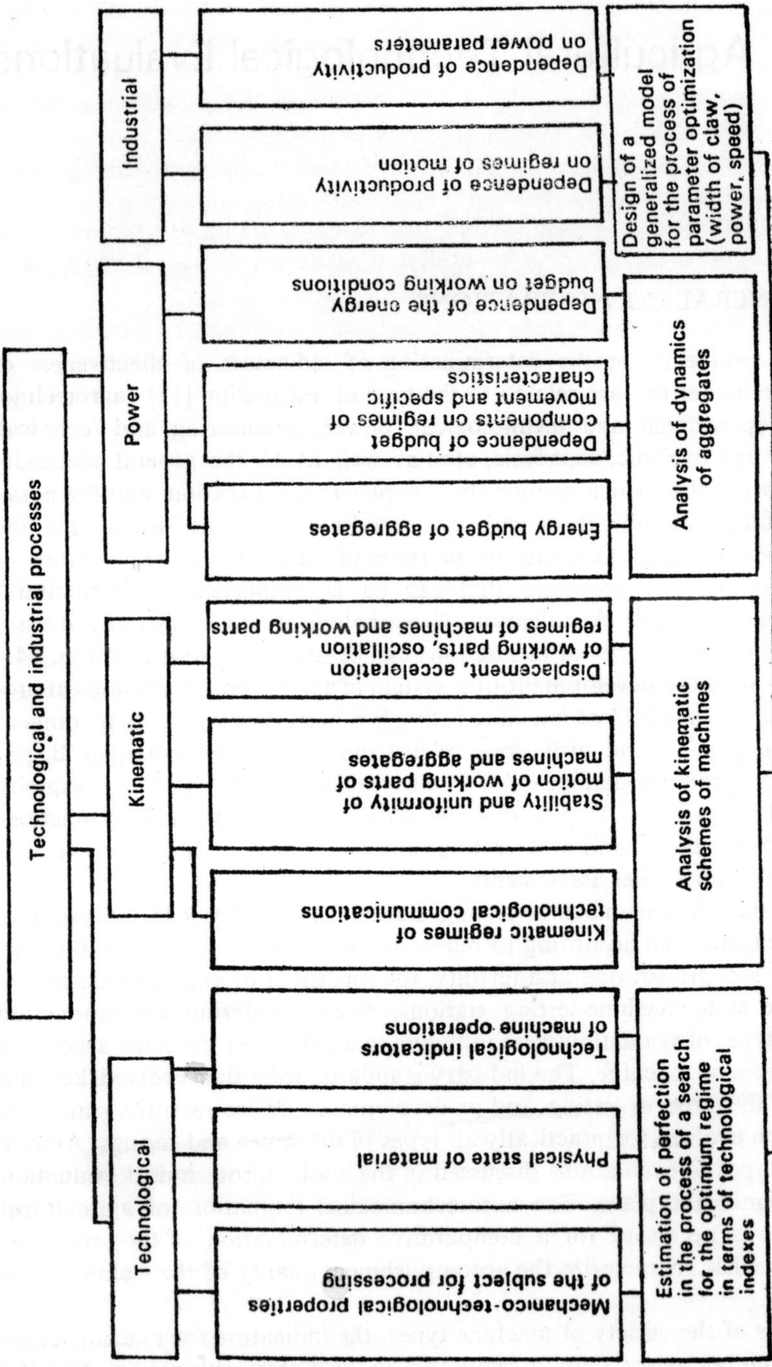

Fig. 24. Chart of relationships among rating types.

methods are developed in test standards of individual types of machines. Agrotechnological indicators can be provisionally classified into several groups.

The *first group of indicators characterize* the technological possibilities of agricultural machines. The indicators of the preceding crops, operational parameters and technological characteristics, for which a technological process can be satisfactorily realized, belong to this group (soil moisture and other inputs, obstructions and soil hardness, crop yield, lodgeability of the grain crop, speed of motion, load-carrying capacity, norm for applying inputs, etc.).

The *second group of indicators characterize* the quality of machine operation, mostly in favorable and other conditions, typical for the zone under optimum and limiting parameters. These indicators are varied and stipulated by agrotechnological requirements and standards on the test methods of each type of machine. For standard groups of machines, the following characteristic indicators can be named:

1. *For soil-processing machines*, the indicators named include the depth of processing, the cutting of weeds and covering up of plant residues and the breaking of soil and leveling of its surface.

2. *For sowing and planting machines*, the indicators included are the quantity of seeds (plants) on a unit area, the depth and quality of embedding, their distribution over the area (in lines, clusters) the field-germinating capacity for seeds, the acclimatization of plants, the dynamics of germination, damage to seeds of plants (seedlings and saplings), the thickness of the soil layer between the seeds and fertilizers, etc.

3. *In machines utilized for the care of saplings*, the indicators are: weeds destruction, damage of plants, the degree of herbicide and pesticide coating of leaves (for sprayers and sprinklers) and the norm of applying inputs (water for water sprinklers, pesticides and herbicides for sprayers, sprinklers and fumigators, fertilizers for additional forage, etc.).

4. *For harvesting machines*, the indicators include: the losses of the basic and the accompanying produce (grain, strawberries, root crops, fiber, fruits, straw, etc.) due to the machines and through individual working parts; damage to produce (crushing, smashing), infestation of soil heaps by weeds and other admixtures.

5. *For machines used for the processing of produce after harvesting*, the indicators include: losses, damage and weed infestation, quality grading (of produce, of seeds), spoiling or preservation of nourishing components (contents of sugar, protein, starch, carotene and vitamins).

Indicators of the third group characterize the stability of technological process in space and time under varying external conditions.

The working of some types of machines is *rated with the help of a point scale*. This method finds application in the testing of sprayers and sprinklers in order to determine the thickness of the coating and the dispersability of the spray.

Point estimation can be used for the determination of the *quality of plowing* (for embedding vegetable residues, the leveling of arable land and the crumbling of the layer). Points are also used for the evaluation of such machines whose operational quality cannot be easily expressed by sufficiently accurate quantitative indicators.

Indicators concerning the quality of machine operation are obtained experimentally and analyzed by various methods. Presently, several estimation modifications of agrotechnological indicators are used such as the comparison with standards and the estimation of economic and combined criteria.

Comparison with standards foresees the comparison of the actual indicators obtained in testings with those stipulated by agrotechnological requirements, technical targets or else indicators which foresee the requirement of an international system of machines (ISM).

The amount of yield is one of the most widespread criteria of agrotechnological machine ratings and complexes for grain and crop cultivation. Crop losses and the corresponding damage of harvesting machines are easier to determine than that of plowing and seeding machines, tillers, stubble plows, fertilizer distributors, spraying devices, broadcasters, sprinklers and other machines. The effect of these machines and complexes on the yield can be established by arranging field experiments from the start to the completion of the harvest. This procedure is very lengthy and labor consuming. According to generally accepted practice, the experiment stretches over a period of three years. Only then is it possible to give appropriate recommendations. But it has not always been possible to reveal the advantages and drawbacks of machines from the date of three years experiments. The real advantage of such experimentation appears to be the prediction of yield at the various stages of plant vegetation.

While *testing watering devices*, a mathematical relationship between the yield and the distribution of showers, which is characterized by the coefficient of effectiveness of watering, was established [55].

On the basis of the analysis of *agrotechnological ratings of mineral fertilizer distributors*, it is possible to derive the mathematical relationship between the amount of wasted crop and the nonuniform application of fertilizers. These relationships are generally characterized by variation coefficients. In most cases, these relationships are represented by polynomials containing second order or, at best, third order terms [55].

Mathematical models, similar to that described here, are obtained for beets and corn seeding machines [55, 59].

Thus the quantity of produce or the deterioration of its quality is an important economic criterion for machine ratings. It is used basically for harvesting machines and also for machines used for post-harvest processing and the preservation of agricultural produce.

Expenditure for upgrading the quality of the product to match the agrotechnological indexes of a given level is also an important economic criterion. For example, if on using beet-harvesting machines, there is an increase in the content of the soil and beet tops in the harvest collected, then for upgrading these indexes to the standard value, it is necessary to invest additional labor in order to separate the soil from the beet tops (these operations being manually realized). The investment on these operations also appears as an economic factor in the agrotechnological indexes for these machines. In the comparative rating of thinness not only the direct expenditure for this work, but also the additional expenses for subsequent operations have to be accounted for.

Hence, in the agrotechnological estimation of various agricultural machines, it is often necessary to take the investment expenses of the means of the operational sequence of the technological process into account. But this should be accomplished only in that case when one anticipates a difference in the investments on carrying out the successive operations.

23. MODELING AND PREDICTION OF AGRICULTRAL TECHNOLOGICAL INDICATORS OF EFFECTIVENESS

In the prediction of agrotechnological indicators by the use of analytical and digital models, practice shows that the experimental determination of agrotechnological indicators of machine operation in the presence of the totality of externally conditioned indexes, different values of technological and technical parameters and working regimes, is an extremely complicated problem. Its solution by experimental methods in the presence of limited periods of testings, material and labor resources is hardly possible.

In connection with the aforementioned, the prediction of effective agrotechnological criteria by methods of mathematical modeling, which is presently based on fairly well-developed models and methods of design, plays an ever-increasing role. Apparently, in the near future, experimental methods for obtaining initial information will dominate in the designing of such models, because the biological systems and agricultural processes closely connected with them appear to be so complicated that it is difficult to have a thorough study and analytical description within reasonable periods of time. A mathematical description of the probability mechanisms of the formulation of agrotechnological criteria in combination with experimental methods is a sufficiently effective direction for designing models for prediction purposes.

While designing models of agrotechnological criteria of effectiveness, agricultural yield or any of the other factors which influence it indirectly may be used as an output variable in the most general and the detailed cases. Examples of such factors can be the deviation of the working parts from the axis of the row, the extent of damage to the plants, the quality of layer crumbling, the dispersion or variation coefficient of plant placement, etc.

Depending on the number of plants on one hectare N, the standard moisture content of produce W_x, the characteristics of photosynthesis μ_{min} and the number of parameters pertaining to the growth of plants which reflect the piling up and decaying of dry and organic substances, A.I. Budko has proposed the following formula for the estimation of agricultural crop yield:

$$Q_x = \frac{100 N g_x}{100 - W_x};$$

$$g_x = \sum_{i=1}^{i} \gamma \mu_{\min i}\, \beta_{x_i} \frac{(1 + \gamma \mu_{\min i} \beta \Phi_i - \alpha_i - \alpha_i')^{t_{i-1}}}{\gamma \mu_{\min i} - \alpha - \alpha_i'}$$

$$\times (1 - \alpha_{xi})^{t_i}\, g \Phi_0 \prod_{i=1}^{i-1} (1 + \gamma \mu_{\min (i-1)} \beta \Phi_{(i-1)} - \alpha_{i-1} \alpha_{i-1}')^{t_{i-1}}.$$

For an estimation of the effectiveness of the compared technological processes of machines, it is recommended that parameters of the derived equation are determined, the yield computed and the most effective technology or machine selected from the highest yield value. However, the use of this model in testing is difficult in view of the objective difficulties involving the experimental estimation of parameters such as photosynthesis and the growth of plants.

For the prediction of agricultural yield, the distribution of intervals between plants and the dependence of the mass of the productive part of a plant (beet roots, corn cobs, etc.) on the intervals between the plants are used.

The yield of beets on the segment of area Q can be expressed by the formula:

$$Q = \frac{S}{bx} \int_0^{\infty}\!\!\int p(x_1, x_2) f(b, x_1, x_2)\, dx_1\, dx_2,$$

where $f(b, x_1, x_2)$ is the dependence of the mass of the root on the width between lines b and distances x_1 and x_2 to neighboring roots in the line and $p(x_1, x_2)$, the density of the probability distribution of a two-dimensional random variate (x_1, x_2).

It has been established through experimental investigations that $f(b, x_1, x_2)$ is an asymptotic function and approaches the limiting value g_{max}, which can be interpreted as the integrated index for biological peculiarities of variety and for the potential possibilities of growth conditions for beets. A relationship whose parameters g_{max}, α_1, n_1, α_2, n_2 can be determined from experimental data by the method of least squares can be used for its description:

$$f = f(b)\{1 + \exp(-\alpha_1 x_1^{n_1} - \alpha_2 x_2^{n_2}) + \exp(-\alpha_2 x_1^{n_2} - \alpha_1 x_2^{n_1})$$

$$- \exp[-\alpha_2 (x_1^{n_2} + x_2^{n_1}) - \exp(-\alpha_1 x_1^{n_1}) - \exp(-\alpha_1 x_2^{n_1})]\},$$

where

$$f(b)=g_{\max}[1+2\exp(-\alpha_1 b^{n_1}-\alpha_2 b^{n_2})-\exp(-2\alpha_2 b^{n_2})$$
$$-2\exp(-\alpha_1, b^{n_1}).$$

The density of distribution $P(x_1, x_2)$, while assuming an independence of random variates x_1 and x_2 can be determined if the density of distribution of the plants in a row $p(x)$ is known:

$$p(x_1, x_2)=p(x_1)\,p(x_2).$$

A model for the loss of yield (that is most characteristic for harvesting machines) can be designed either on the basis of a study of the interaction mechanism of the working part and the harvested crop or by experimental methods on the pattern of the 'black box'.

Prediction of the accomplishment quality of the process by the lifting plows of beet-harvesting machines and the probability of extraction by the machine is given by [10]:

$$\text{Prob}\,(P_{\text{ex}}\cos\delta_2 \geqslant m\ddot{z}+P_z+Q)$$

and the probability of root damage

$$\text{Prob}\,(P_{\text{ex}}\cos\delta_x \leqslant m\ddot{x}+[P_x]),$$

where P_{ex} is the force of extraction, P_z—the force of resistance of the root against vertical displacement $[P_x]$—the assumed value of the force of resistance of the root against horizontal displacement, δ_x, δ_z—the angles between the extracting force and the corresponding axes of coordinates and Q—the root mass.

Assuming that when soil is compressed between surfaces of the lifting plow and the root, it possesses the properties of a linearly deformed body; the process itself proceeds. But in free lateral expansion, the specific pressure on the surface of the root is approximately equal to

$$p=E_n\Delta SS^{-1},$$

where E_n is the deformation modulus of the soil that is experimentally determined, ΔS—the deformation of the soil layer between the surfaces of the lifting plow and the root and S—the initial thickness of the compressed layer of soil.

Resolving pressure p into horizontal and vertical components and multiplying each of these by that corresponding area of the root's surface that experiences the pressure, horizontal and vertical components of the force needed for the extraction of the root are obtained:

$$P_x=p(d_k-h\tan\gamma)h\cos\delta_x;$$
$$P_z=\pi ph\tan\gamma(d_k-h\tan\gamma)\cos\delta_z,$$

where d_k is the diameter of the root and h—the length of the root.

The horizontal cross section of the root, which experiences the maximum bending moment on account of the horizontally applied load, has coordinates satisfying the following relations:

$$\left(-\frac{2}{3}\tan\gamma\right)z^3+(2\tan\gamma-0.5)z^2+z$$

$$+\left\{0.5(l_n-h)^2+\frac{2}{3}(l_n-h)^3\tan\gamma-[\sigma_s]^{-1}P_x\left[d_n(l_nh-0.5h^2)\right.\right.$$

$$\left.\left.+h^2\tan\gamma\left(l_n-\frac{1}{3}h\right)\right]\right\}=0,$$

where $[\sigma_s]$ is the instantaneous resistance of the soil to compression.

Stress at the critical section, which makes its appearance when the root bends, is given by:

$$\sigma=32\mid(M_{[\sigma]}-M_P)\mid\pi^{-1}(d_n-2l_z\tan\gamma)^{-3}$$

where

$$M_{[\sigma]}=[\sigma](d_n-2l_z\tan\gamma)\frac{l_z}{2}+\frac{2}{3}l_z^3\tan\gamma;$$

$$M_P=P_x\left[d_nh(0.5h+l_z)+h^2\tan\gamma\left(\frac{2}{3}h+l_z\right)\right];$$

$$l_z=l_n-h-z,\qquad d_n=d_k-2h\tan\gamma.$$

When condition $[\sigma]>\sigma$ is satisfied, the root will not be damaged under the action of the working part intended for lifting.

The relationships and algorithms described in work [10] and also the characterististics of the physicomechanical properties of roots, soil and their interaction make it possible to model the processes of the lifting of roots by a disk and share lifting plow. In modeling, the values of extracting P_z and shearing P_x components are calculated, while in modeling by the Monte Carlo method, the probabilities of lifting of and damage to roots for different values of parameters of the working parts and the characteristics of roots and soil are estimated so as to obtain the effectiveness of various types of lifting plows, the optimum selection of their parameters and to recommend possible directions for improvement of their construction.

In order to estimate the loading capacity or grain-harvesting combines, the model of dependence of the loss of grain on the flow of grain mass $y=a\,e^{bx}$ is used, the parameters being determined by the method of least squares from the results of the experiment.

Of all the models reflecting the link between the conditions and the state of the subject under test with indirect characteristics in relation to crop, the model, copying process, precise seeding and distribution of watering and rains, which can be used without considerable changes for modeling of a wide range of processes, can be considered.

Designing of a model for copying begins with detailed analysis of the mechanism and the working process performed by it.

While examining the general scheme of the feeler mechanism along with the basic design (Fig. 25), it is seen that in the place of initial relief $f_1 \mid x \mid$, it is necessary to use relief $f_2 \mid x \mid$, deformed by the cam surface. The unknown value of the technological parameter (depth of the movement of the working part) is expressed in terms of vectors, which represent the trajectory $f_2 \mid x \mid$ and $f_3 \mid x \mid$, given by $\bar{h} = \bar{y}_{p'} - \bar{y}_k$, which corresponds to a scalar relation $h = f_3(x) - f_2(x - l)$, where l is the mechanism parameter is known as the pitch of copying. As a measure of the statistical relationship between the coordinates of the copying and implementing mechanisms, the role of parameter l is determined by the correlation function of the relief.

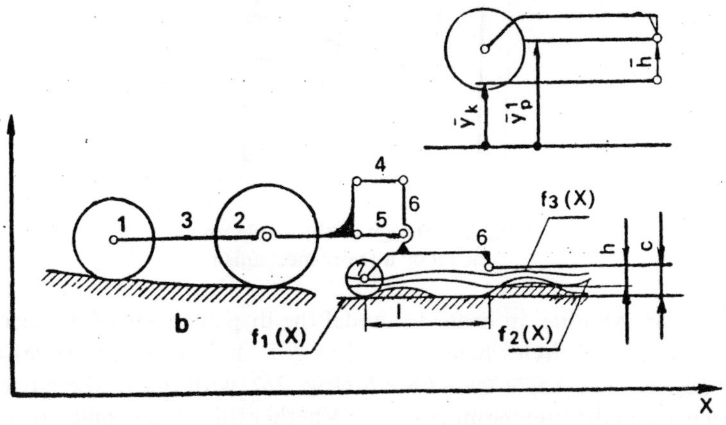

Fig. 25. Diagram for a feeler mechanism and geometry for copying. 1, 2—supporting elements of a base design; 3—base construction; 4, 5—links; 6—link of working parts; 7—feeler mechanism.

It follows from the foregoing equations that the dispersion is given by $D_h = D_2 + D_1 + 2K_{21}$, where K_{21} is the correlation moment of ordinates y_2 and y_1. D_2 and D_1 are the dispersions of ordinates y_2 and y_1.

In the particular case, when $y_2 = y_1 + \text{const}$,

$$D_h = 2D_1 + 2K_{11}, \tag{51}$$

where K_{11} is the correlation moment of the ordinates of function $f_1(x)$, which are separated by pitch l.

The dispersion is calculated in such a way that it characterizes the accomplishment quality of the process of copying, but for the calculation it is necessary to know values D_1, D_2 and K_{21} and K_{11}. These values are determined with the help of a figure model, which is based on the initial realization $f_1(x)$ or $f_2(x)$ and makes use of rules by which initial functions are transformed into $f_3(x)$. Such rules may include the construction of a geometric space of circle centers, enveloped by function $f_2(x)$ or by plotting the trajectory of some point describing the mechanism which rolls over the given surface (Fig. 26).

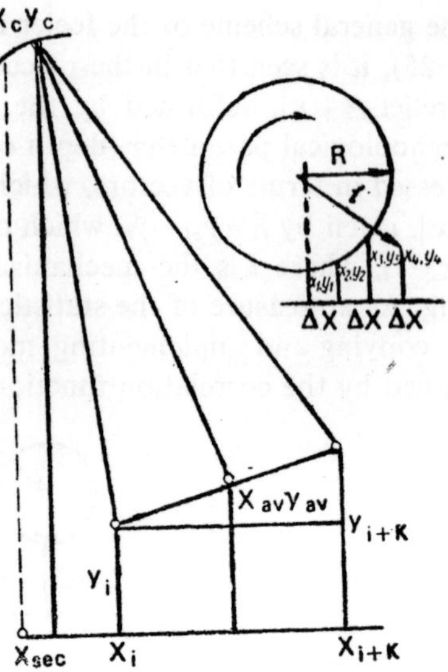

Fig. 26. Design of the displacement model of a feeler mechanism.

Thus, for example, in order to model the displacement of a feeler mechanism, a family of circles having a radius R is drawn, which passes through point (x_1, y_1): $(x_1-a)^2+(y_1-b)^2=R$ (Fig. 26), with one of the circles of the family passing through point $|x_2, y_2|$. Whether this corresponds to the actual position of the circles, i.e. whether the following relationships hold good, is then checked:

$$(x_3-a)^2+(y_3-b)^2 \geqslant R^2;$$
$$(x_4-a)^2+(y_4-b)^2 \geqslant R^2;$$
$$\cdots\cdots\cdots\cdots\cdots\cdots$$
$$(x_n-a)^2+(y_n-b)^2 \geqslant R^2.$$

Here, the number n can be selected, for example, from condition $\Delta x(n-2) \leqslant R/2$ (a, b are fixed coordinates of the center of the circle).

After checking the imposed conditions and determining the coordinates of center (x_c, y_c), coordinates (x_{sec}, y_{sec}) are found which correspond to the line which passes through the base point (x_1, y_1) or (x_1, y_c), depending on the position of the center of the feeler mechanism.

By calculating the sequence of required coordinates using scheme (Fig. 27), the necessary estimation parameters are found.

Results of the modeling of real subjects with the use of an approach toward the designing of the copying model described here shows that the dependence of the dispersion of the ordinates of the trajectory of the center of the feeler mechanism D on its diameter d is expressed by the curve which asymptotically approaches axis d (Fig. 28, a).

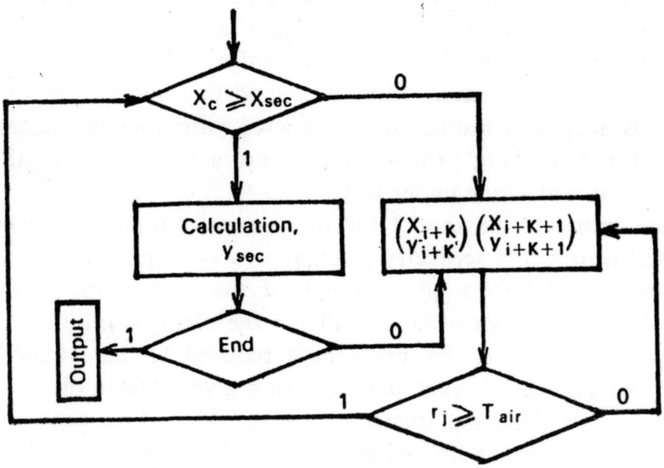

Fig. 27. Block diagram modeling of the displacement of a feeler mechanism.

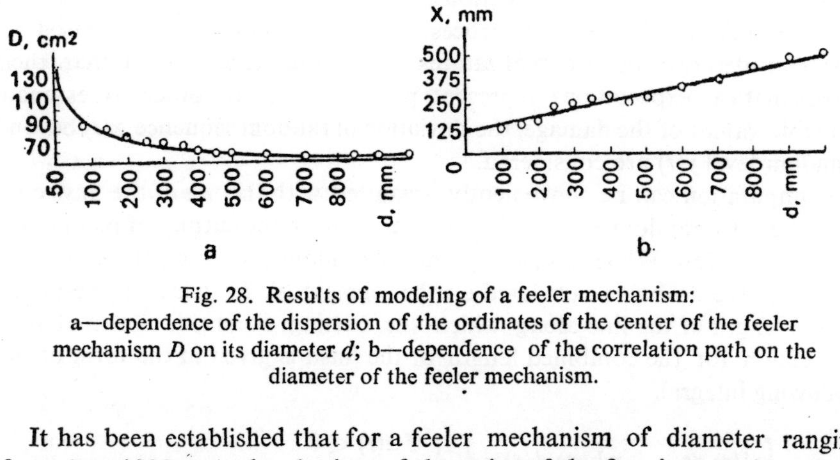

Fig. 28. Results of modeling of a feeler mechanism: a—dependence of the dispersion of the ordinates of the center of the feeler mechanism D on its diameter d; b—dependence of the correlation path on the diameter of the feeler mechanism.

It has been established that for a feeler mechanism of diameter ranging from 0 to 1000 mm, the abscissa of the point of the first intersection of the plot of the correlation function of the trajectory with the abscissa (path of correlation) depends on the diameter of the wheel approximately linearly (Fig. 28, b). It is thus possible to select the diameter of the feeler mechanism, keeping in view the design and technological limitations.

From the data of the modeling of a twin-support mechanism developed on a similar scheme, it can be established that with the increase of the base and size, the effect of the vertical coordinate on the dispersion of the unknown trajectory is reduced. In this case not only the effect of the depth to which the working part sinks is reduced, but so is the fluctuation of this depth. Also the spectral content of the process changes becoming considerably responsible for the effect of the working parts distance from the center of the supports on the factors of the quality of copying.

From the results of modeling, one can judge whether the selected parameters are in fact optimum. Moreover, modeling makes it possible to supplement the results of testings with predicted ratings of the accomplishment quality of the working process and to more fully evaluate the copying mechanism of a machine under test.

On the basis of the model for the process of interrow working by the furrow cultivator, it is necessary to plant a row of sprouted saplings while remaining confined to a given stretch of the protection zone. Because a sufficiently large number of factors affects the changeability of the stretch of the protection zone, the problem is reduced to that of predicting the possible percentage of plant damage for a given stretch of the protection zone from the statistical characteristics of the row and the path of the working part obtained over a limited segment. And then based on these, statistical characteristics which should possess the necessary reliability, an attempt to predict the quality of copying of the row with a change in the protection zone within given limits is made. It is, therefore, possible to compare indexes of several machines by identification of conditions of their use.

A formalized scheme of the process of interrow cultivation is based on the consideration of two realizations of random function so that their mathematical expectations represent parallel lines. In order to estimate possible values of the damage, the deviation of random sequence $x(t)$ beyond random level $y(t)$ are considered.

Computation can be conveniently organized with the use of the flash concept, i.e. of a random event, which is equivalent to the cutting of plants by a cultivator's claw. While comparing realized random processes, the combined length of the flashes is calculated on the segment $(0, T)$, which corresponds to the length of the line along which the cutting occurred. An analytical expression for the combined length of the flash is given in the form of the following integral:

$$\int_0^T \psi(t)\,dt, \qquad \text{where } \psi(t)=\begin{cases}1, & \text{if } z(t)\leqslant 0\\ 0, & \text{if } z(t)>0,\end{cases} \quad \text{and} \quad z(t)=y(t)-x(t).$$

In view of the fact that the processes $z(t)$ can be recognized as ergodic after considering a sufficiently large number of realizations, the following is obtained:

$$R=\frac{1}{T}\int_0^T \psi(t)\,dt \approx \frac{\sigma}{\Sigma}, \tag{52}$$

where σ is the number of values for which $z(t)\leqslant 0$, Σ—the number of considered values and R—the fraction of the combined length of flashes in the total length of realization T.

Having thus established basic concepts, one can move over to designing a mathematical model and algorithm.

A mathematical model preceding the development of the algorithm includes the following elements:

1. Initial data

$$x^{(h)}(t_i); \quad y^{(h)}(t_i); \quad i=1, \overline{n}; \quad h=1, \overline{m},$$

where h is the number of the realization for the respective processes, i—the number of discrete points and m—the total number of realization.

2. Computation of correlation matrix $K_z \mid t_i, t_j \mid$ for the process $z(t)= y(t)-x(t)$ [55].

3. From the values of the correlation matrix, the values of the process are evolved:

$$z_1=z(t_1)=C_1\eta_1+b-a;$$
$$z_2=z(t_2)=C_{12}\eta_1+C_{22}\eta_2+b-a;$$
$$\ldots\ldots\ldots\ldots\ldots\ldots\ldots\ldots\ldots$$
$$z_k=z(t_k)=C_{1k}\eta_1+C_{2k}\eta_2+\ldots+C_{kk}\eta_k+b-a;$$
$$z(t_{n*})=z_{n*}=C_{1n*}\eta_1+C_{2n*}\eta_2+\ldots+C_{nn*}\eta_n+b-a,$$

where

$$C_{ii}=\sqrt{k_{ii}-\sum_{\mu=1}^{i-1} C_{\mu i}^2}; \qquad [i=j=1, \overline{n}^*];$$

$$C_{ij}=\left(k_{ij}-\sum_{\mu=1}^{j=1} C_{\mu}C_{\mu i}\right)\Big/C_{ij}; \qquad [i>j; \quad j=1, \overline{n}^*],$$

$\eta_1, \eta_2, \ldots, \eta_{n*}$ being the independent normally distributed random numbers with a mathematical expectation equal to zero and dispersion equal to one.

After modeling, the realizations of the values of the process are checked for condition $z_k \geqslant 0$; the number of positive and negative values z_k are calculated, and then the fraction of damage in percentage is calculated as the ratio of the deviation number to the total number of values z_k.

As an example of the modeling results of the interrow cultivation process, the dependence of the extent of plant cutting by various cultivators on the width of the protection zone is given in Table 14.

Table 14

Brand of cultivator	Speed, km/h	Extent of plant cutting (%) as a fraction of the width of the protection zone			
		4	3	2	1
6-KRKh	5.6-5.8	0.35	2.50	9.78	26.00
FKSh-2.7	5.6-5.8	0.00	0.00	1.00	13.90
KRNO-2.8	5.6-5.8	1.99	6.90	16.00	31.00
6-KRKh	1.4	0.59	3.40	11.70	28.00
FKSh-2.7	1.4	0.00	5.89	15.00	31.00
KRNO-2.8	1.4	0.00	0.00	1.40	14.00

The region of optimum width of the protection zone and the rate of movement for each of the tested machines is selected from the modeling data. It is then possible to conduct fruitful testings and to compare machines at optimum values of regimes and technological parameters of their operation.

A model for the distribution of intervals between plants, which contains the quality characteristics of sowing, planting and seed embedment in furrows by a device and also indexes for quality soil preparation by ensuring the most favorable conditions for their germination, can be developed on the basis of the formalization of the foregoing assumptions regarding the mechanism of the processes taking place during the sowing and germination of seeds and by considering the nature of the indexes influencing these as probabilistic.

On analyzing the probability distribution of the intervals between seeds and assuming that the process of cell filling, in the most general case, can be described by a sequence of Bernoulli tests with two outcomes, it can be shown that the density of distribution of intervals x considering inversion [35] and double filling will be equal to

$$P(x)=\sum_{k=1}^{\infty} P_k(x) \prod_{\substack{j\neq k \\ j=1}}^{\infty} \left[1-\int_0^x P(t_t^j)\,dt\right], \tag{53}$$

where

$$P_k(x)=pp_2P_\xi(\xi)+p^2\sum_{i=0}^{\infty} P_i(i)\int_{-\infty}^{\infty} P_\xi(\xi)P_\xi[x+\xi-h(k+i)]\,d\xi.$$

If seed distribution in relation to the calculated point of unloading P_ξ follows the normal distribution with null average and dispersion σ^2, then the distribution density of the intervals between the seeds, when sowing is accurate, will be given by

$$P(x)=\begin{cases} 0, \; x<0 \\ \left\{\dfrac{1}{2\sqrt{\pi\sigma}}\sum_{k=1}^{\infty} p^{k+2}\sum_{i=1}^{\infty} C_{k+i-1}^{k-1}(1-p)^i\left[\exp\left(-\dfrac{\eta^2(x)}{4\sigma^2}\right)\right.\right. \\ \left.\left.+\exp\left(-\dfrac{\eta^2(-x)}{4\sigma^2}\right)\right]+\dfrac{2pp_2}{\sigma_0\sqrt{2\pi}}\exp\left(-\dfrac{x^2}{2\sigma_0^2}\right)\right\}\prod_{\substack{j=1 \\ j\neq k}}^{\infty} \\ \left\{1-\dfrac{p^{j+2}}{2\sigma\sqrt{\pi}}\int_0^x C_{\substack{j-1 \\ j+i-1}}(1-p)^i\times\left[\exp\left(-\dfrac{\eta^2(t)}{4\sigma^2}\right)\right.\right. \\ \left.\left.+\exp\left(\dfrac{-\eta(-t)}{4\sigma^2}\right)\right]dt-\dfrac{2pp_2}{\sigma_0\sqrt{2\pi}}\int_0^x \exp\left(-\dfrac{t^2}{2\sigma^2}\right)dt\right\}, \quad x\geqslant 0, \end{cases}$$

(54)

where

$$\eta(x) = x - h(k+i);\ \eta(-x) = -x - h(k+i);$$

p_2—the probability of cell filling by two masses and σ_0^2—the dispersion of seed distribution after liberation of cells with a double filling.

An analysis of the obtained expression shows that for inequalities $h/\sigma \geqslant 6$ $[P(\epsilon > 1/2h)] \geqslant 0.999$ and for a sufficiently high (0.95) probability of the filling of cells (a fact that is characteristic for sowing by small norms (10–12 pieces per meter) of coated seeds by accurate seeding machines), the distribution of intervals between seeds under initial conditions ($p=\text{const}$, ξ—the normal distribution law) approaches the normal law (Fig. 29). For other values of h/σ and p, this distribution differs from the normal and is characterized by considerable asymmetry.

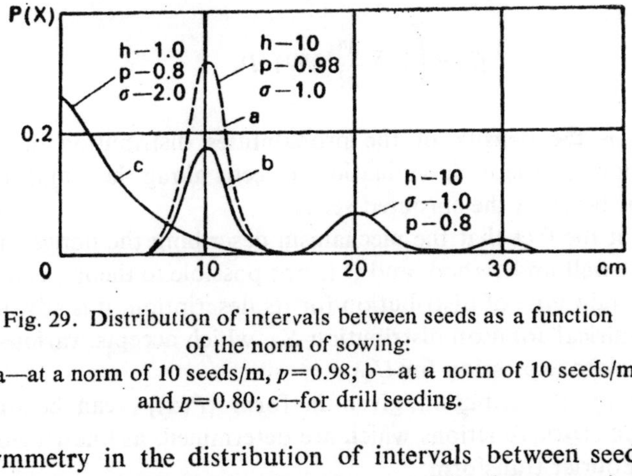

Fig. 29. Distribution of intervals between seeds as a function of the norm of sowing:
a—at a norm of 10 seeds/m, $p=0.98$; b—at a norm of 10 seeds/m and $p=0.80$; c—for drill seeding.

The asymmetry in the distribution of intervals between seeds, as the results of experimental investigations show, can be due to an imperfection of the process of unloading and seed embedment into the soil. It could also consequently be caused by the relatively large dispersion of placement deviations from the calculated sowing point, the imperfection of the process of cell filling by the seeds as well as rather low probability of their filling.

In connection with the foregoing, one can make use of the coefficient of asymmetry as an accurate index in the working of the seeding device together with the coefficient of the variation already in use.

For designing the model, which adequately describes the distribution of interval between plants, it is necessary to take into account the peculiarity of the mechanism of their formation, which is expressed in terms of the probable characteristic of field seed germination in a given row.

Analysis has shown that the interval between the neighboring plants in a row can be represented as the sum of a random number of components (intervals between seeds), the number of which (these items) is determined by the number of unsprouted seeds between them. The density of the probability distribution of such a sum [75] is given by:

$$P(x) \sum_{n=1}^{\infty} \mathcal{G}_n\{P(x)\}^n, \tag{55}$$

where $\mathcal{G}_n$ is the density of distribution of the number of unsprouted seeds between two plants, $\{P(x)\}^n$—n, the ratio of the intervals between the seeds, which are distributed according to the formula described in Equation (53).

An analysis of the probable manifestation of unsprouted seeds shows that the Poisson distribution with random variable αt appears as a sufficiently complete model of distribution, $\mathcal{G}_n$, which takes into account the changeability in the germination or nongermination probability of seeds in the soil along the length of a row [75]. Then the probability of having just n unsprouted seeds in succession will be equal to:

$$\mathcal{G}_n = \int_0^{\infty} e^{-\lambda} \frac{\lambda^n}{n!} U(\lambda)\, d\lambda, \tag{56}$$

where $U(\lambda)$ is the density of the probabilities distribution describing the manifestation of the nongermination in seeds along the length of a row in the interval between the sprouted seeds.

In view of the fact that the mechanism describing the nongermination of seeds is not well understood, and it is not possible to theoretically verify the use of a certain type of distribution for its description, it is advantageous to use the empirical Johnson distribution S_B, which accepts various forms as an approximate expression for $U(\lambda)$.

The density of distribution, given in form $\{P(x)\}^n$, can be obtained by using characteristic functions which are determined, as known from [16], by means of Fourier transform:

$$\{P(x)\}^n = \frac{1}{2\pi} \int_{-\infty}^{\infty} e^{-itx} \varphi_n(t)\, dt, \tag{57}$$

where

$$\varphi_n(t) = \prod_{k=1}^{n} \varphi_k(t); \quad \varphi_k(t) = \int_{\infty}^{\infty} e^{itx} p(x)\, dx.$$

For the most modern seeding machines, the seed distribution in a row is sufficiently close to the normal distribution. The density of distribution of the intervals between plants is described by the following expression:

$$P_p(x) = A^{-1} \sum_{n=1}^{\infty} \frac{\exp\left[-\dfrac{1}{2} \dfrac{(x-nh)^2}{\sigma^n \sqrt{(2\pi)^n}}\right]}{\sigma^n \sqrt{(2\pi)^n}}$$

$$\times \int_0^{\lambda_m} \frac{e^{-\lambda}\lambda^n}{n!(\lambda-\epsilon)(\lambda_m-\lambda+\epsilon)} \exp\left\{-\frac{1}{2}\left[\gamma+\eta \ln\left(\frac{\lambda-\epsilon}{\lambda_m-\lambda+\epsilon}\right)\right]^2\right\} d\lambda, \tag{58}$$

where

$$A^{-1} = \int_0^{\lambda_m} \frac{1}{(\lambda-\epsilon)(\lambda_m-\lambda+\epsilon)} \exp\left\{-1/2\left[\gamma+\eta \ln\left(\frac{\lambda-\epsilon}{\lambda_m-\lambda+\epsilon}\right)\right]^2\right\} d\lambda$$

—normalization factor.

Expressions (53) to (58) determine the density of distribution of the intervals between plants and include the technological characteristics of the sowing apparatus p_1, p_2, σ_0, σ, h. Function $U(\lambda)$ reflects the effect of conditions created in the soil; this function is used to solve problems of systems analysis and to evaluate the sowing apparatus, technological operations and working parts for the preparation of soil under sowing by seeding machines in testing technological processes, regulating the actual distribution of plants. From the condition of the least value of χ^2 test, namely

$$\chi^2 = \min \sum_{i=1}^{l} \frac{(m_i - np_i)}{np_i} \tag{59}$$

the characteristics of the sowing apparatus and the parameters of function $U(\lambda)$ are determined, and the degree of agreement between the actual and theoretical distributions is estimated.

The lowered value of p and the high values of p_2, σ_0, and σ and the considerable difference between the actual distribution and the theoretical only go to prove how insufficiently developed are the processes of cell filling and emptying, the embedding seeds in the soil and the presence of factors brought about by the design defects of the apparatus and the drawbacks in its manufacture, etc.

Replacing argument λ by value $q=1/(1+\lambda)$ in the function $U(\lambda)$, where q characterized field germination, distribution q along the length of the line can be analyzed and, by comparison with the etalon, judgements regarding the degree of improvement of the process of seed embedment in the soil and the technological and working parts for preparation of the soil before sowing can be expressed.

In order to analyze and estimate the quality of soil preparation for sowing and to understand the possible reasons for the high nonuniform germination of seeds along the length of the row, it is worthwhile to investigate the spectrum of the interval sequence in a row by using nonintegral values as an argument and the length of a row as $i\,\Delta l$. For this, it is worthwhile to replace the interval sequence between plants by the sequence $x(i\,\Delta l)$ with the nonrandom argument $x \in \{0, 1\}$ and $i \in \{0, 1, \ldots, N\}$, considering that

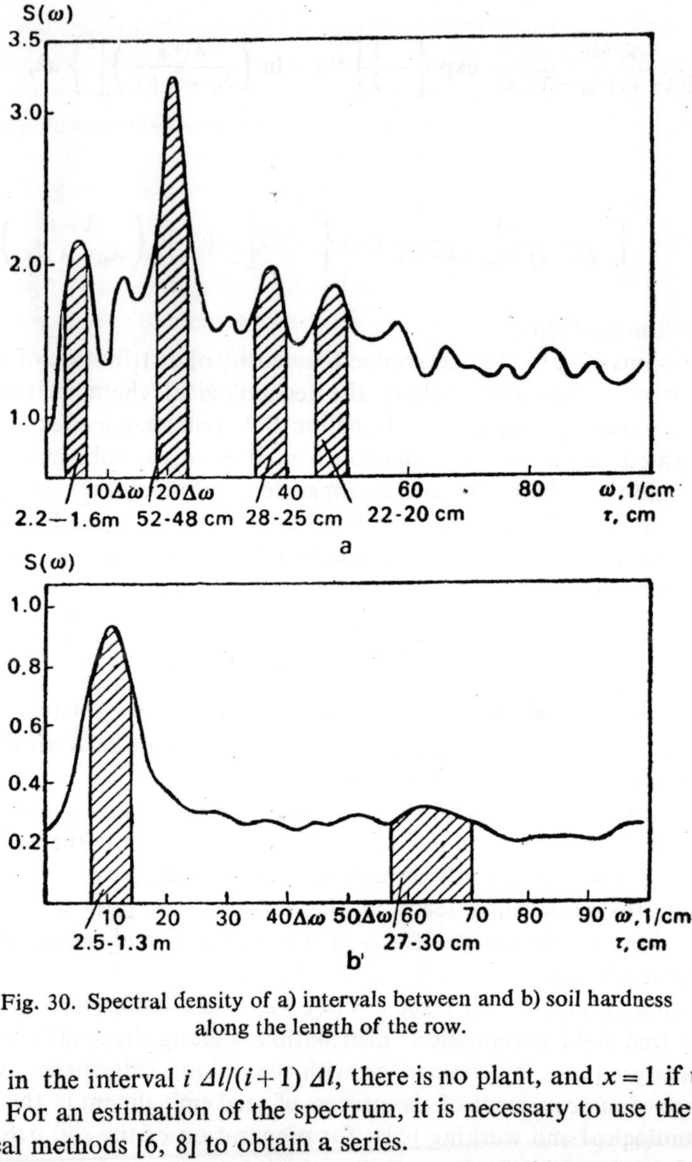

Fig. 30. Spectral density of a) intervals between and b) soil hardness along the length of the row.

$x=0$ if in the interval $i\ \Delta l/(i+1)\ \Delta l$, there is no plant, and $x=1$ if there is a plant. For an estimation of the spectrum, it is necessary to use the known numerical methods [6, 8] to obtain a series.

Analysis of the intervals between plants and of the root mass during harvest (which is accomplished with the use of the aforementioned transformation) and also analysis of the hardness of the soil along the length of a row have shown that in the spectrum of these characteristics, there are frequencies commensurable with the width of the claws of various frames and the plow as a whole (Fig. 30). In those cases, when the field was cultivated from autumn, such frequencies in the distribution of plants were not so clearly expressed.

From this analysis, it follows that on the basis of the evolved model, the development of more precise seeding machines, technological operations and working parts for the preparation of the soil, which provides the most favorable conditions for the growth of seeds and excludes the formation of sections with unfavorable soil conditions, appears as a very important direction for increasing the effectiveness of beet sowing.

While solving problems of the synthesis of technologies and technological devices for the sowing and training of plants, the criterion of effectiveness (which predicts yields) is set up, using the methods of optimization, the characteristics of the sowing apparatus, the distribution function of field germination along the length of the row and also other technological parameters (for which the maximum yield is obtained and which satisfy given limitations) can be written as follows:

$$Q = \max_{h,\, p,\, \sigma,\, n,\, U(\lambda)} \left\{ \frac{S}{b\bar{x}} \iint_0^{\infty} P(x_1)\, P(x_2)\, f(b, x_1, x_2)\, dx_1\, dx_2 \right\}.$$

Analysis and synthesis of patterns of plant training densities, manually or by mechanical implements under test, can be accomplished by using the following model:

$$Q = \max_{p(x)} \left\{ \frac{S}{b\bar{x}} \sum_{i,j=1}^{N} \hat{P}_i(x_{1i}) \hat{P}_j(x_{2j})\, f_{ij}(b, x_{1i}, x_{2j}) \right\},$$

where $\hat{p}_i(x)$ is the distribution of the intervals between plants after the training of plant densities.

The mathematical model of such generalized agrotechnical criterion of effectiveness such as yield, which reflects its dependence on the most essential factors, makes it possible to predict the direction of further improvement of the technology of crop cultivation, and machines and working parts for the accomplishment of individual technological operations [47].

Thus, for example, computer investigations of a mathematical model of a beet crop including as variates the distribution parameters and intervals between plants and the parameters of seeding apparatus, have shown that an extremum occurs for uniform placing of plants ($D_x \to 0$), although at various densities (depending on the probable values of q_{max}) (Fig. 31).

However, a similar approach toward the training of optimum plants was once only acceptable while sowing in rows with comparatively high norms (25–30 kg/ha). The application of accurate sowing with norms of 12–20 lumps per unit length (m) is coupled with a considerable increase in the nonuniformity of plant distribution. This demands a search for compromise relationships between indexes of the inhomogeneity of plant placement and the seeding density of plants on the basis of maximum yield. Under industrial conditions, with the existing methods for the preparation of soil before sowing, the accurate sowing and training of plants and the density of 75–90

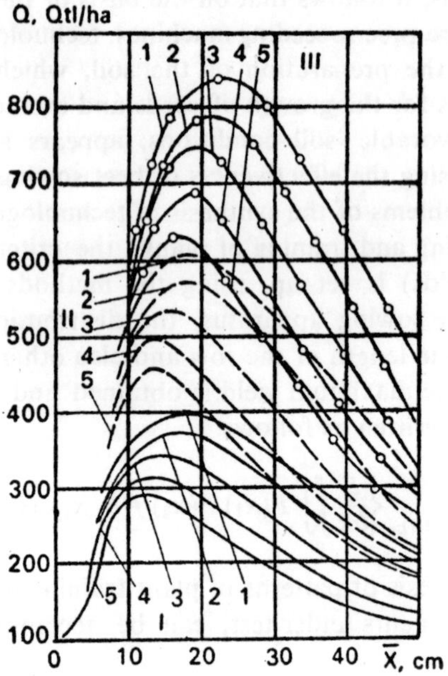

Fig. 31. Prediction for beet yield under various conditions of plant growth, density and plant homogeneity: 1—coefficient of variation of intervals between plants $v=25\%$; 2—$v=50\%$; 3—$v=75\%$ 4—$v=100\%$; 5—$v=125\%$; I—$q_{max}=500$ g; II—$q_{max}=750$g; III—$q_{max}=1000$ g.

thousand plants on one hectare (coefficient of variation 60–75%) is not optimum as the criterion for maximum yield.

Under favorable conditions of plant growth, the optimum density of plants should be 127–148 thousand plants, depending on the homogeneity of their placement. Under more favorable conditions, plantation density should be within limits of 97–127 thousand plants per hectare.

Modeling results have also shown that a tentative direction for an increase of beet yield lies in the increase of the sowing accuracy of seeds as a result of the realization of technical and technological measures (improvement in the design of sowing apparatus, etc.).

Considering the limitations of labor resources and the breakdown periods, it will be necessary to limit the level of manpower requirements up to 17–25 man-hour-hectare (5-7 hectares for one man, engaged in manual training) in the near future. This calls for the necessity of reducing plant density (not more than 8–9 plants in 1 m with blind mechanized thinning and not more than 6–7 plants for manual thinning, Fig. 32). While solving the problems for the increase of field seed germination and its stability along the length of the row in the future, it will be possible to sow beets at a specific density, i.e. without manual or mechanized training of plants. Calculations show that at constant field germination, the sowing norm should consist of 17–20 seeds per meter. With an increase in field germination to 50%, it is possible to predict a decrease in the norm of sowing to 8–12 seeds per meter. On the results of prediction using the model described here, the use of automatic thinner turns out to be a very advanced method of technological improvement in beet cultivation.

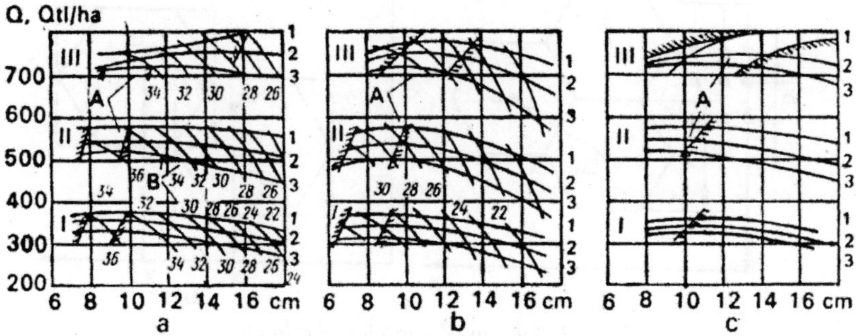

Fig. 32. Prediction of beet yield depending on parameters and the method of plant training:
a—manual training; b—blind thinning and manual check; c—thinning by automatic thinner; 1—$v=60\%$; 2—$v=80\%$; 3—$v=100\%$; I—$q_{max}=500$ g; II—$q_{max}=750$ g; III—$q_{max}=1000$ g; *A*—zone of optimum density for maximum yield; *B*—level of manpower requirement for training plant thickness, man-hour/ha.

The model for overhead irrigation is a typical example of the process of material distribution.

Modeling of the distribution of rain in testing overhead irrigation machines, permits on the basis of the information obtained in field conditions, to determine the optimum pattern of placement of overhead irrigation machines at positions which provide the extremal value of the overflow parameter, i.e. the coefficient of effective irrigation:

$$\gamma_{eff.r} = F_{eff}/F_{tot},$$

where F_{eff} is the effectively irrigated area and F_{tot}—the total irrigated area.

In accordance with the normative materials used on the testing of overhead irrigation machines, F_{eff} and F_{tot} are determined on the basis of rain charts plotted from the calculation of experimental results for a given pattern of rain gage and water-sprinkler arrangement.

While calculating the rain chart, possible overlapping as a result of the parallel trajectories of the mobile or stationary vans standing at adjacent points has to be considered.

A mathematical model of the distribution of rain formalizes the plotting of distribution maps of rain for a particular position. On calculating the charts for various positions, it is possible to select the best pattern of movement or sprinkler arrangement from the maximum value of K_{eff}.

In designing a model of overhead irrigation, the following considerations have been taken into account.

With the consideration of overlapping from adjacent positions, the rain map can be represented in the form of a hexagon, the relationship between the parameters being determined from geometric consideration (Fig. 33).

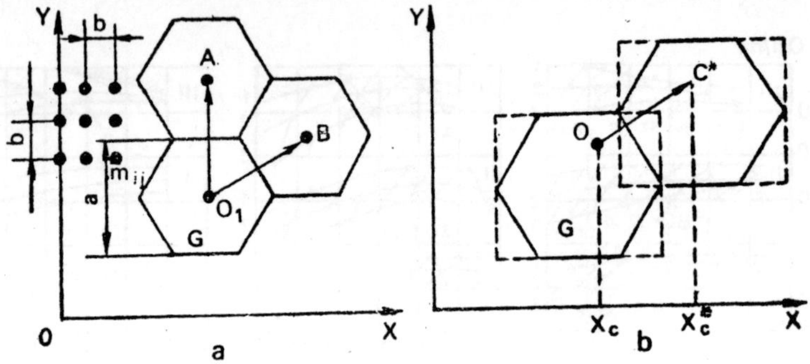

Fig. 33. Design of a model of rain chart.

As a result of displacement of the hexagon in the direction of vectors $\overline{O_1A}$ and $\overline{O_1B}$ (both being perpendicular to its sides), it passes into new polygons with centers at points A and B (Fig. 33, a).

On the displacement of hexagon G in the direction of vectors of the type described above, the hexagons will cover the entire area in a pattern seen in Fig. 33, b. Considering the hexagon G as a certain set of points, it is possible to establish that the displacement of the hexagon in the direction of the vector consists in the displacement of all points belonging to the set under consideration.

In the model considered here, the problem on the identification of corresponding points which belong to different hexagons on the plain has to be solved. Two points C and C^* of two hexagons (Fig. 33) are considered as corresponding to one another if one is obtained from the other by displacement along the vector of the linear combination of the type described here.

On the basis of the foregoing considerations, point C^* can be found to correspond to point C.

While setting up the parquet so as not to lose any points or count those which belong to common sides twice, it was considered in modeling that the upper side and two adjacent sides belong to set G, whereas the lower and the two adjacent sides do not belong to set G.

In machine realization the first part of the algorithm consists of partitioning the cycle of integers, representing the squares with respect to X and Y (blocks 6, 8, 19, 20 in Fig. 34).

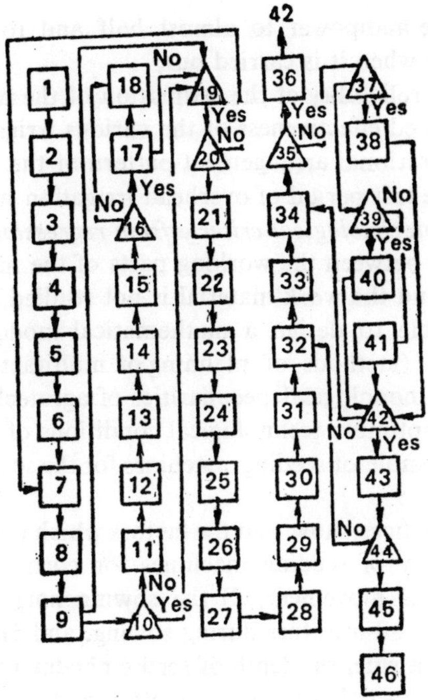

Fig. 34. Flow chart for a rain map model.

At first, it is checked whether point (X, Y) belongs to C (block 10). In case of a positive answer, the corresponding element m_{ij} of the matrix of intensities does not change.

Conversely, the value of the element of the matrix is memorized in a certain cell (block 11), following which a certain arbitrary number (a constant) is sent by block 12 to that cell where the value of the element a_{ij} is stored. Then from point C, the corresponding point C^* (blocks 13–14) and the nearest integral point C and C^* (blocks 13–18) are found. An element of the matrix of intensities, the value of which exceeds that value memorized in block 11 corresponds to this established point.

In the second part of the algorithm (blocks 21–30), ranges of normal, insufficient and excess intensities are determined.

The third part of the algorithm (blocks 31–41) formulates and prints out the rain maps. The cycle of the partition of the rows of the matrix, which characterizes the map, and the internal cycle of the partition of the element of the rows' material is constructed (block 31, 43). One of the symbols +, −, *, is printed out, depending on the range to which the element of the matrix belongs. These symbols represent normal, insufficient and excess intensity, respectively.

The use of the described model for analysis of the effectiveness of operation of the overhead irrigation machines permits:

1) to reduce the manpower to almost half and to obtain experimental results on the day when it is carried out;

2) to raise the reliability of the estimation of the machines under test in view of the estimated effectiveness of the various irrigation patterns and for the selection of a rational arrangement pattern of the working parts; and

3) to reduce the test period of overhead irrigation machines.

Forecast of agrotechnological criteria from regression models. If the interaction mechanism between the working parts of the machine and the external environment and the work material is not studied very thoroughly so as to make it possible to design a mathematical model, then experimental methods are used (methods of planning of multifactor regression experiments). Some methodological peculiarities of agrotechnological models for the prediction of optimum technological conditions of test machine operation, using the example of seeding machines for the sowing of corn and beets will be considered.

Agricultural practices show that the factors which most considerably affect the working quality of seeding machines for corn, beet and other crops are: 1) the speed of movement, 2) the sowing norm, 3) the depth of seed embedment, 4) the soil moisture during sowing, and 5) the depth of tilling before sowing. As a rule, the depth of seed embedment is selected intuitively and the establishment of a standard depth for all seeding machines, under comparison tests is in practice not possible. The depth for embedding seeds, tilling and other factors selected in this way, may be the optimum for one machine but not so for another. In this connection, machines that seem comparable actually have nonidentical conditions, and thus the reliability of their evaluation result is reduced. Often, the established and factual sowing norms for seeding machines under test differ from each other. Therefore, the results of these experiments must be analyzed for different factor values, which determine the density and homogeneity of the placement of plants and finally the criterion of effectiveness, namely yield.

In order to avoid such situations and also to obtain exhaustive information about seeding machines as the results of a multifactor experiment over the course of one year (to establish optimum working regimes, estimate the effectiveness of mechanisms for regulating the sowing norm and the depth of seed embedment, etc.), in agrotechnical estimation, a mathematical model in the form of a regression equation is obtained, and the criterion of effectiveness is predicted within the limitation of the factors influencing it. This can be dealt with while working with machines under field conditions.

In designing such experiments, the established sowing norms and the depth of seed fixing, the speed of movement, germination of seeds (for beet seeding machines), the soil moisture during sowing, the depth of tilling before sowing, etc., can be accepted as the inputs (independent factors) which characterize the conditions of a machine operation. Depending on the aim of the experiment, variable output factors are accepted.

Factors involving the working quality of seeding machines, which are foreseen in the state standards or the complex criteria of effectiveness, are taken as output variables.

Analysis of *a priori* information about the dependence of output variables on the input shows that within permissible limits and the physically realizable conditions of the regulating mechanisms of seeding machines and power devices, these dependences can be described with sufficient accuracy by a polynomial of the second order.

Considering the manpower requirement for field experiments on seeding machine testing and the very compressed periods necessary for their realization, the essential limitation in the selection of a plan for a multifactor experiment appears to be the number of experiments conducted. Therefore, in practice for three or four factors, plans B_3 and B_4 [55], respectively, with 14 and 24 experiment modifications were realized on three levels, namely, -1, 0 and $+1$.

In the realization of such a plan, each experimental variant is repeated thrice, but because of the limitations on the area to be sown, the same row can be used for three repetitions. In this case, the requirements for homogeneous field allotted for laboratory field testing increases.

The sequence of arrangement of the experimental variates in each field section is randomized, and same other models of machines take part in the experiment, then all the variates are randomized.

In selecting factor levels following the instructions for seeding machines, the medium level of the established norm of sowing g_0, is selected in such a way that it would be possible to control the upper g_{max} and lower g_{min} levels of the sowing norm with identical or nearly identical intervals ($g_0 - g_{min}$ and $g_{max} - g_0$) by adjusting the seeding machines.

The depth of seed embedment is determined directly in the field with the minimum allowable depth at which all seeds are embedded in the soil (for corn—4–5 cm, for beets 1.5–2 cm) established by the plowshares mechanism for adjusting the depth of movement. Then, using the mechanism for depth adjustment, the maximum depth allowable by agrodemands (for corn—10 cm, for beets—5–6 cm) is established.

Such a situation is not allowed when the adjustment mechanism is placed at the maximum depth, and the plowshare goes in to only a portion of the established depth. This can violate the initial conditions which were used as a basis for designing the recommended plans, and this would hinder a proper interpretation of the experiment's results.

The medium or zero level is established by the adjustment mechanism in the center between the maximum and minimum level marks.

In selecting the speed level, it is necessary to try to satisfy condition $v_0 - v_{min} = v_{max} - v_0$ and to take into account traction as well as the power characteristics of the drive for ensuring operations of the aggregate in power-consuming states (for example, at maximum depth and maximum speed).

For determining field sowing norms, sowing is conducted in an open furrow at established sowing norms with the actual movement speed selected in the experiment (three norms of sowing and three speeds in all nine variants). On the basis of this information, the coefficients of the regression equation (polynomial of the second order), which describe the dependence of the field norm of sowing on the two above-indicated factors are determined.

The speed of the aggregate's movement is measured during the run of the seeding machines, and after the run, the depth of seed embedment is measured. The depth of embedment should be measured in all three repetitions.

The experimental information is processed, and regression equations, characterizing the agrotechnical indexes of the working quality of seeding machines as a function of the established norms of sowing the speed of movement, the depth of seed embedment or other indexes included in the experiment are obtained.

For the analysis of these results, the regression equations are tabulated, and the graphical relationships are plotted from the obtained values.

The mechanism of depth adjustment, the movement stability of the plowshare group along the line of movement and the width of the claw are evaluated according to the actual values of the depth of the seed embedment as well as dispersion depending on the established depth, the movement speed, the cultivation depth, etc.

The indexes of the distribution quality of plants can also be analyzed in terms of the established values of factors included in the experiment.

As an example of the interpretation of the results of a multifactor experiment, which was conducted by specialists O.A. Makovetskii, G.I. Yakimenko and V.G. Bondarenko at the Ukrainian Machine Testing Studies, against working conditions obtained in the testing of seeding machines 2STSN-6 and 'Palm-Agromatik' for the sowing of beets SKNK-6. SUPN-8 and 'Maksikorn', for corn with the use of above-described method, are shown in Figs. 35–38.

An analysis of these results for the formation of the optimum density for plantations permitted the establishment of a unit length (at an average interval of 8 cm) provided for by 2STSN-6 seeding machine at the seeding norm of 16 seeds per meter at a speed of 7.2 km/h.

The mean-square deviation in this case is 7.4 cm (see Fig. 35).

Regression equations, obtained experimentally, also permit the determination of the maximum movement speed of a seeding machine under stipulated conditions of sowing, seed germination as well as the stipulated average interval between plants.

A prediction of agrotechnological indexes of corn-seeding machines, accomplished on regression models, permits the evaluation of comparable machine for optimum regimes, and the determination of optimum values of certain technological parameters.

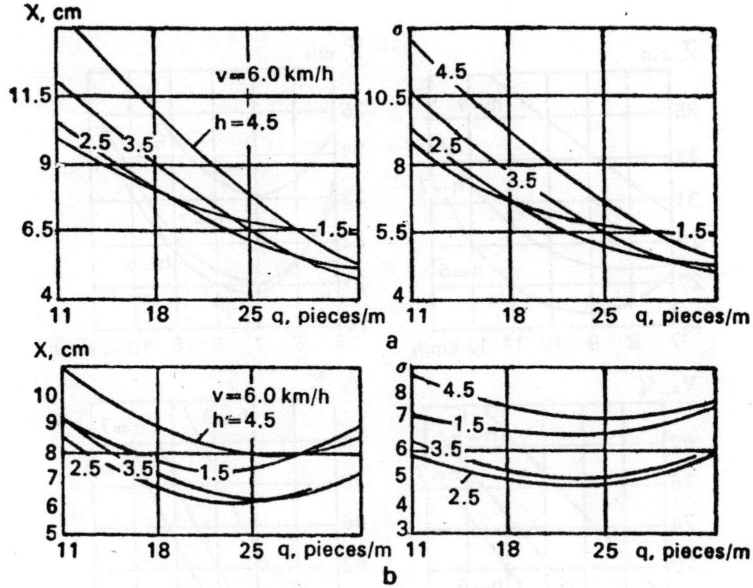

Fig. 35. Relationship between the predicted mean value $\bar{X}$ and the mean square deviation of intervals between plants for (a) 2STSN-6 beet seeding machine and (b) 'Palm-Agromatik' and the sowing norm and depth seed embedment.

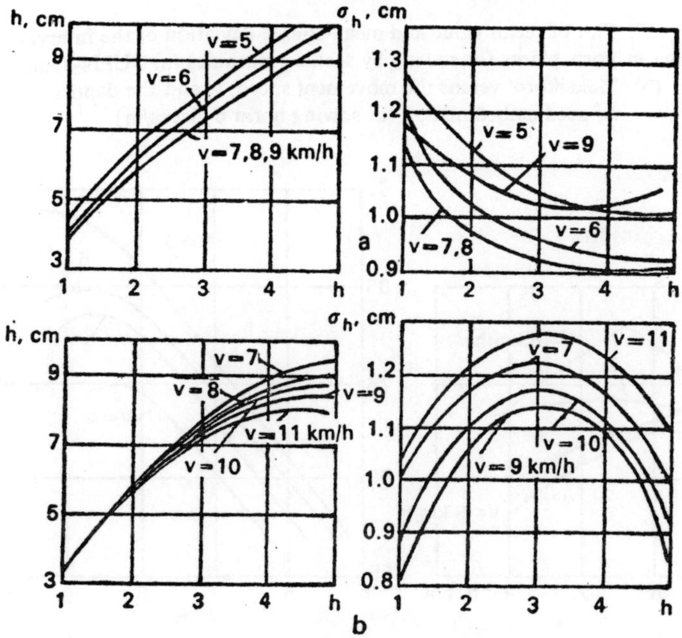

Fig. 36. Plot of mean value and mean-square deviation of the embedment depth of corn seeds versus the depth adjustment of a plowshare movement and the movement speeds of seeding machines (a) 'Maksikorn' and (b) SUPN-8.

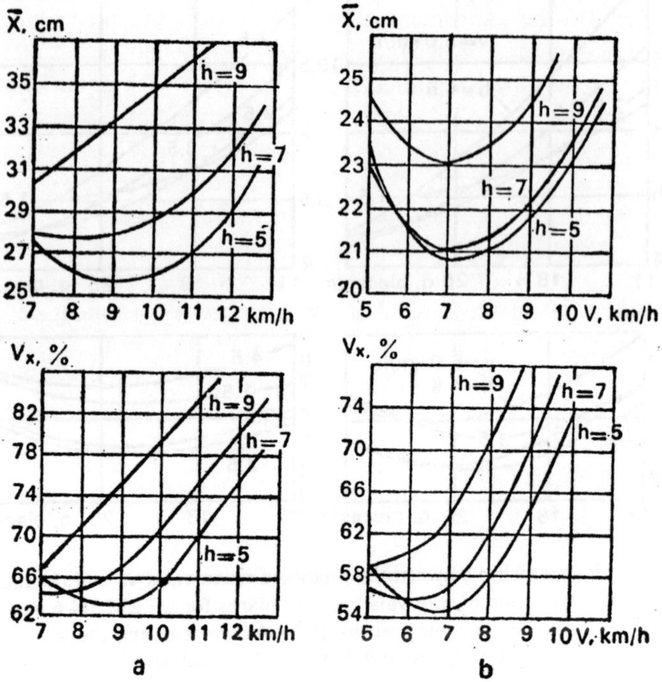

Fig. 37. Plot of mean value and mean-square deviation of the interval between corn plants for sowing by seeding machines (a) SUPN-8 and (b) 'Maksikorn' versus the movement speed, *v*, and the depth of seed embedment, *h* (at sowing norm 6 pieces/m).

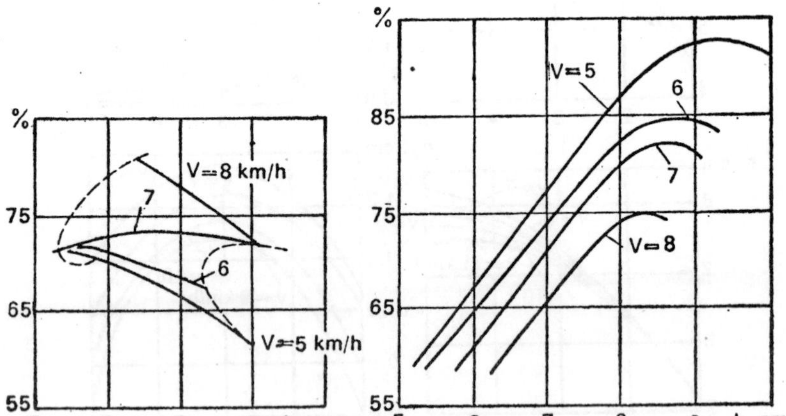

Fig. 38. Plot of predicted field germination of corn seeds versus the movement speed and depth of tillage before sowing, on sowing, by seeding machines a) SUPN-8 and b) SKNK-6.

The effect of the movement speed of seeding machine SUPN-8 on the depth of seed embedment depends on the established depth of the plowshares movements. For small depths (up to 6 cm) with an increase of speed up to 11 km/h, the depth of seed embedment remains almost unchanged. But for large depths (above 7 cm) a tendency toward a decrease in the depth of seed embedment is observed (see Fig. 36). It should be mentioned that for an average depth of plowshare movement at 7.5 cm, the depth of seed embedment is not homogeneous.

From a comparison of the predicted quality of plant distribution at various conditions of sowing, movement speeds and seed embedment depths by the seeding machine, 'Maksikorn' and SUPN-8 (see Fig. 37), a conclusion can be drawn to the effect that a better performance can be expected at the seed embedment depth of 4-5 cm under the experiment's conditions. The optimum rate of movement of the seeding machine 'Maksikorn' is 7 km/h and that of the seeding machine SUPN-8 is 9 km/h.

Depending on the depth of tillage, prediction of the field germination of corn seeds while sowing by seeding machines SUPN-8 and SKNK-6 has shown that the quality of seed embedment by the plowshare type of seeding machine SUPN-8 depends on the tillage depth to a lesser degree as compared to seed embedment by seeding machine SKNK-6 (see Fig. 38).

The optimum kinematic and technological working modes for seeding machine SUPN-8 are the cultivation depth before sowing at 6-7 cm, depth of seed fixing at 6-7 cm with a speed of 8-8.5 km/h. Probably at low speeds, the SUPN-2 plowshare type of seeding machine does not provide seed embedment on a compact bed as a result of which a very low percentage of seed germination (70-72% as against 77-80%) is obtained (see Fig. 38).

For seeding machine SKNK-6, the optimum working modes are: depth of tillage at 6-7 cm; the depth of seed embedment at 8.5-9.5 cm when the movement speed is 5-5.5 km/h.

Hence the above-discussed examples of the prediction of agrotechnological indexes on models (which are designed on the basis of multifactor experiments) only point to the utilization possibilities of these methods in testing for quality evaluation of the work of agricultural techniques in the sphere of probable values of parameters and conditions of operation. On the basis of such evaluation, that quality of work can be predicted which the machine can provide, owing to its design specifications, and also the comparison of test machines is conducted under optimum parameters and conditions of operation. The uncertainty in estimation, existing because of differences in experimental conditions, has been eliminated here.

For a complete representation of the suitability of test machines for operation under various conditions, it is necessary to include other factors as well (soil moisture during working crop-yield, etc.) in the experiment. Different levels of soil moisture can be established by changing the period of operation, whereas different levels of yield can be ascertained from harvesting in different fields. For sowing and tilling machines under significant design limitations of prediction models, multifactor experiments can be conducted on specially prepared sections (or polygons) so as to provide the possibility of controlling some factors and to estimate machines with respect to predicted yield instead of experiments on actual yield. In this, machine operation under production conditions using the same technology as was used in the MTS zone is not eliminated since it permits the checking of technology and a subsequent improvement in the design of the model.

CHAPTER 8

Power Rating of Agricultural Technological Processes

24. GENERAL RULES FOR POWER RATING OF MACHINES

The determination of the power consumption of the work performed by a machine and its working parts under test is the main objective of the power rating of agricultural aggregates along with correspondence of the latter's power consumption to the traction and power indexes of engines, while operating in the range of high speed and loading conditions, which have been established by both agrotechnical requirements as well as technical assignment.

Modern methods of power rating should provide source material for evaluating the optimum design decision and aggregate assemblage. At present, a certain discordance exists between the development of rating methods for engines and a theoretical basis, especially in the application of statistical techniques.

In testing, power rating is accomplished in accordance with industry standards, extending over the testing of prototype of modernized and industrial machine and implement samples, and also over stationary machines provided with internal combustion engines.

The power rating of agricultural aggregate is made along with the agro-zootechnical evaluation of the machine.

Conditions for carrying out power estimation, background, the number of repetitions and working parameters of the aggregate are established by industry standards specifying test methods for different types of machines.

Before the beginning of testing the control characteristics of the engine should be determined by a method of braking. The actual values of power, torque, rotation frequency of the shaft, fuel consumption and other indexes should correspond to the data in the instruction manual. Results of brakings are endorsed in the report documents.

Adjustments of the agricultural machine should be accomplished in accordance with the manufacturer's instructions on operation and should be optimum from the agrotechnical viewpoint. Pressure in the machine tires and agricultural assemblies should correspond to that indicated in the manufacturer's instructions.

In the power rating of mobile aggregates, the following basic parameters are recorded: torque on the engine shaft, the driving parts of the engine or the shaft of their drive, on the power takeoff shaft, etc., M_{ki} kgf m; traction resistance (for tractor-drawn agricultural machines)—R_t kgf; frequency of crank shaft rotation of the engine, driving wheels, power takeoff shaft—n_r rev/min; path covered—S m; duration of experiment—t seconds.

Power parameters of mobile aggregates are recorded in a steady-state operation during a working and the idle run of the aggregate. The parameters are recorded at least twice in each of the operational regimes during the movement of the aggregate in both forward and backward directions.

In order to explain the effectiveness of the new working parts of the machines under test in individual cases, an additional recording of power parameters (forces, torques, displacements, rotation frequency, etc.), which characterize their power consumption, is accomplished.

The list of working parts and the additional power parameters is provided in the industry standards describing the test methods of various types of machines.

The registration duration of one repetition of an experiment for power rating of both mobile and stationary aggregates should be as follows: for recording of power parameters by a tensiometric laboratory—not less than 20 seconds, and by a measuring laboratory of the ChEK-1 type—not less than 30 seconds.

The following indexes are determined on the results of power ratings:

1. Effective engine power while measuring torque M_{en} and the rotation frequency of shaft n_{en} of the engine:

$$N_e = \frac{M_{en} n_{en}}{716.2}.$$

While measuring the torques and the rotation frequency of the right and left drive half-axle shafts of wheels M_{wr}, n_{wr}, M_{wl}, n_{wl}, the power takeoff shaft (PTO) M_{PTO}, n_{PTO} the front wheel drive shaft of tractor M_{wf}, n_{wf} and the hydraulic motor M_h, n_h, one obtains

$$N_e = \frac{M_{wl} n_{wl}}{716.2\eta_{mw}} + \frac{M_{wr} n_{wr}}{716.2\eta_{mw}} + \frac{M_{wf} n_{wf}}{716.2\eta'_{mfw}} + \frac{M_{PTO} n_{PTO}}{716.2\eta_{m\,PTO}} + \frac{M_h n_h}{716.2\eta_{mh}},$$

where η_{mw}, $\eta_{m\,PTO}$, η_{mh}, η'_{mfw} are the mechanism efficiencies which take into account the mechanical losses in rear wheel drive transmission, the power takeoff shaft and in the hydraulic motor and front wheel drive transmission of tractors respectively.

The mechanical drive efficiency is determined by the formula

$$\eta_m = \eta_{cyl}^m \eta_{bev}^p,$$

where η_{cyl} is the efficiency of one cylindrical pair of gears, η_{bev}—the efficiency of one bevel pair of gears, m, p—the corresponding number of cylindrical and Bevel gear pairs, simultaneously engaged in meshing.

2. The power on the driving parts of the engine at the torque measurement and rotation frequency of the engine shaft, PTO shaft and the hydraulic motor

$$N_w = \left(\frac{M_{en} n_{en}}{716.2} - \frac{N_{PTO}}{\eta_{PTO}} - \frac{N_h}{\eta_{mh}}\right) \eta_{mw};$$

at the torque measurement and rotation frequency of the rear wheel drive and the shaft for the front wheel drive

$$N_w = N_{wr} + N_{fw} = \frac{M_{wl} n_{wl} - M_{wr} n_{wr}}{716.2} + \frac{M_{fw} n_{fw}}{716.2} \eta'_{mfw},$$

where η'_{mfw} is the transmission efficiency for front wheel drive (from the place of measurement to the wheels). $N_{PTO} = M_{PTO} n_{PTO}/716.2$—the power on the drive of the machines from the PTO shaft.

3. Power on the drive of the working parts of machines from the power takeoff shaft and the hydraulic motor

$$N_{PTO} = \frac{M_{PTO} n_{PTO}}{716.2}, \qquad N_h = \frac{M_h n_h}{716.2}.$$

4. The power required to overcome mechanical losses during the measurement of torques and the rotation frequency on the axles of driving wheels, on the shafts for the drive of machines from the PTO shaft and the hydraulic motor is given

$$N_m = \frac{N_w(1 - \eta_{mw})}{\eta_{mw}} + \frac{N_{PTO}(1 - \eta_{PTO})}{\eta_{PTO}} + \frac{N_h(1 - \eta_{mh})}{\eta_{mh}}.$$

5. The power required for self-driving is determined during the idle run of the engine without the agricultural attachment, using the same background as in the active run.

While measuring the torque on the driving parts of the engine, the power consumed is:

$$N_{mov} = \left(\frac{M_{wli} n_{wlw}}{716.2} + \frac{M_{wli} n_{wrw}}{716.2}\right)\left(1 - \frac{\delta_i}{100}\right).$$

6. The power consumed in overcoming skidding

$$N_{skid} = 0.01 N_w \delta,$$

where δ is the skidding of the driving parts of the engine, %.

The skidding of the driving parts of the engine:

$$\delta=\left(1-i\frac{n'_{w}}{n'_{ww}}\right)100\%,$$

where $i=n'_{wi}/n'_{Si}$, n'_{wi}, n'_{Si}, n'_{ww}, n'_{Sw} represents the number of marks of the driving wheel of the tractor and the path-measuring wheel for idle and active runs of the aggregate.

The coefficient i is determined on a road with a hard coating, at operational pressure in the tires at the time of idle running of the engine.

7. The traction of the engine is assembly with a mounted or semimounted agricultural machine:

$$N_t=N_e-N_{PTO}-N_m-N_{mov}-N_h-N_b$$

or

$$N=N_w-N_{mov}-N_b;$$

in assembly with tractor-drawn agricultural machines

$$N_t=Rv/75,$$

where v is the motion speed of the aggregate, m/s.

8. The total traction resistance of mounted and semimounted agricultural machines and implements

$$R'=75N_t/v.$$

9. Specific power requirement (on unit productivity)

$$Q_e=\frac{N_e}{W},$$

where N_e is the effective power developed by the engine for a given operation, in horse power; W—the productivity for one hour of real time of operation (ha/h, t/h, kg/s).

10. Specific resistance

$$K_{sp}=R/F$$

or

$$K_{sp}=R/B,$$

where F is the area of seam, cm^2 and B—the width of claw, m.

The power load factor of the engine is determined by the formula

$$K_{Ne}=\frac{N_e}{N'_{enom}}100\%,$$

where N_{enom} is the nominal effective power of the engine according to the instruction manual, in horse power.

11. Effective power consumed in overcoming ascents

$$N_{e\alpha}=\frac{mv_m \sin \alpha}{75},$$

where m is the total mass of the aggregate, kg; α—the maximum allowable angle of ascent at which the technological accomplishment and agrotechnological indexes for hilly regions are ensured, in degrees.

The comparative rating of mobile aggregate is mainly carried out on the specific indexes and the degree of coordination with the power potential of the engine.

In connection with the wide application of tensiometry, the selection of representative characteristics from the realization of random functions in power rating and investigations of working parts is important. The generally accepted method for oscillograph processing establishes rules for the selection of the realizations under investigation.

The length T of the dynamogram which is to be processed and the pitch for quantization of the ordinates Δ are determined by preliminary examination and calculations from the dynamogram recorded during an experiment.

By simple calculation of the oscillations of the curve, the lower change in frequency of the studied parameter f_l is determined. The corresponding angular frequency ω_l is then determined by formula $\omega_l = 2\pi f_l$.

The upper frequency of the change in parameter f_u, which the experimenter considers necessary for analysis, may be determined in a similar way.

Taking the allowable error for determination of the correlation function as 2%, it is assumed that $1/\omega_1 T \leqslant 0.02$.

The maximum value of the shift of function τ_{max}, for the determination of the correlation function, is calculated from the lower frequency of the process, i.e. $\tau_{max} \geqslant 2\pi/\omega_1$.

The length of the dynamogram under process for the corresponding frequencies $T \geqslant 8\tau_{max}$.

The pitch of ordinates Δ, in the determination of the correlation function of the process, is selected as a function of quantity f_u using the condition $\Delta \leqslant 1/2f_u$.

In this connection, a limitation in accordance with the well-known Kotel'nikov theorem is imposed on the minimum value of $\Delta \leqslant \pi/\omega_{max}$ over the realization length.

According to the aforementioned recommendations on the dynamogram (oscillogram) of the length, corresponding to the obtained value T, N ordinates are drawn (marked) over the interval of Δ (from the null line of the parameter record). The obtained numerical sequence for the parameter change under study represents the values of the initial data for calculation of all the characteristics on the computer. For mechanization of the processing of the oscillogram, it is recommended to use the automatic device,

'Siluet', which puts out a perforated tape which is useful for direct feeding into the computer.

The power estimation of machines driven by an electric motor, mainly for the processing of agricultural materials and the mechanization of processes in stock breeding, follows instructions in GOST 183-74 and OST 70.2.3-73.

In the power rating of machines, mobile tensiometric laboratories are used [55]. A mobile laboratory ChEK-1 (Fig. 39) is introduced, which permits the measurement of registration and distribution of the mean values of force and speed parameters over six independent measuring channels directly after the completion of the experiment.

Fig. 39. Mobile laboratory ChEK-1:
a—at work for the tensiometry of the harvester; and b—internal view of the laboratory.

Each measurement channel of the force parameters includes a strain gage amplifier for direct current and an integrater. The measurement of the mean values of the force parameters is achieved by integrating amplified signals from pickups during the experiment. For each measuring channel, the strain gage is fed from a separate stabilized source of dc voltage of 12 V.

Contact interrupters or induced impulse pickups are used for speed parameter recording.

System ChEK-1 provides a measurement at discrete intervals of time with the duration of experiment at 7.5, 15, 30, 60, 120 and 240 seconds.

The output of values of the measured force and the speed parameters is numerical.

System ChEK-1 is mounted in the body of automobile UAZ-452. The system is limited with the aggregate, under testing, by a multicore, unshielded wire of 100 m in length.

In recent times, equipment has been developed for the comprehensive estimation of mobile and stationary machines.

Telemetric information and the measuring complex ChEK-3-5 are meant for power estimation of tractors, combines, agricultural aggregates and any other machines in laboratory field testings under particularly complicated working conditions, when a connection of wiring between the subject under test and the secondary apparatus is not possible (Fig. 40). The complex permits the conducting of measurements, the transmitting of information on a radio channel, the processing measurement of data and the furnishing of results in the form of printouts.

The multipurpose information-measuring complex ChEK-6 is used for the estimation of external factors in the working zone of the machine operator (in the testing of tractors, combines and any other mobile and stationary subject) and their conformity to safety technique requirements and industrial sanitation. System ChEK-6 (Fig. 41) includes a complex of apparatus, which permits the measurement of vibration and acoustic noise, forces acting on the controls and temperature in order to estimate the efficiency of the braking system.

Information-measuring systems 'Urozhai-11, -12 and -13' are meant for prolonged recording and decoding of operational, power, physiological and sanitary characteristics of agricultural aggregates. The system includes the stationary equipment 'Urozhai-11' for on board recording of the parameters on a magnetic tape; the portable equipment 'Urozhai-12' for operative processing and recording of information; the device 'Urozhai-13' for feeding information into the series computer and a system of pickups.

'Urozhai-11' (Fig. 42) represents a small magnetic recording device, which is mounted in the cabins of MTZ-50, DT-75, T-150, K-700 tractors and can be used as an independent, complete set. The equipment on board consists of the following standard subsystems: magnetic storage for multiple

Fig. 40. Mobile radiotelemetric laboratory ChEK-3-5:
a—general view; b—apparatus on board.

recording of information, registered pickups mounted on the subject under test; blocks of electronic devices and supply commutators, patch board, normalizers plus a block for amplification of recording and self-control and a block of service parameters. The number of simultaneously recorded parameters is 248 and the duration of continuous recording is up to 24 hours. The supply of a direct current source on board varies between 18-31 V. In assembly, the equipment weighs 20 kg.

System 'Urozhai-12' (Fig. 43) is placed in the body of automobile GAZ-66 and consists of the following functional devices: for recording, reproduction

Fig. 41. Mobile laboratory ChEK-6:
a—general view; b—apparatus on board.

and transformation of information for visual and graphical representation. While moving from one object to another, the equipment reads out information from certain recording systems placed on board or removed from the object. The speed of reading out is 15 times higher than the speed of recording by a recording system.

The decoding results are printed on a paper tape in the form of graphs (up to 32 pieces), which show the change of parameters in the time scale or in the form of ordinates on a screen (for 16 parameters) and in decimal numeration form (for one parameter).

The 'Urozhai-13' system represents a device for conjugation with computers in batches. Each of the connecting blocks takes a code from the block of commutation and on command, transfers it into a computer and using controls signals which come from the computer.

In comparison with the existing technical means, the 'Urozhai-11, -12, -13' systems have a multipurpose application: smaller recording dimensions and higher universality; possibility of information storage for a prolonged period of observation (shift, ten-day period or month) in the presence

Fig. 42. Mobile laboratory 'Urozhai-11':
a—general view; b—apparatus on board.

of a large number of channels for obtaining it; maximum mechanization and automation of the experiment; considerable reduction of the period for obtaining finite results; and the comparability of experimental results. All this renders it broadly useful in operational and technological estimation.

25. THE STATISTICAL INTERPRETATION OF OPERATIONAL RESISTANCE IN MACHINES AND IMPLEMENTS

Statistical methods have acquired a high significance in the power estimation of machines during recent years [6, 31, 46, 55, 64].

Besides a large number of investigations on the statistical analysis of power processes, contradictory interpretations are often encountered and questions (which deal with the laws of formation of processes, their structure analysis and methods of estimation of working resistances) remain insufficiently elucidated.

Fig. 43. General view of the portable recording block of the 'Urozhai' system, mounted on the machine to be tested.

A study of the statistical mechanics of operational resistances demands the search for an adequate description of strictly random influences of the surroundings on the working parts as well as of the random reaction of a mechanical system on these influences.

The significance of the evaluation of operational resistances consists in the fact that their measurement is generally made indirectly, by fixing the reaction of working parts (system) on external loading. Since it is not possible to directly measure the input value, and the formation of processes bears a closed nature, therefore, the displacement of working parts in the surroundings under processing and resistances, originating in this way, appears to be linked on the one hand by internal connections and on the other by the feedback connection, which reflects the interaction between the energy carrier (working part) and a certain source of resistance. This gives rise to a closed circle of arguments, and while measuring reactions, brings about uncertainty in the determination of the true nature of the process. The spectral density of this process is the result of a multiplicative process, i.e., a product of the spectral function of the source and the transmission constant of the working part in the presence of disturbances.

Analysis shows that it is necessary to differentiate two basic components of the working resistances: quasi-static and dynamic [58, 66].

It is well known that a quasi-static component determines the mean level of operative resistances, slowly changing in time with a frequency which is one order below the natural frequencies of the working parts or elements of the machine. The dynamic components of resistances, which are measured on the working part, are of higher frequencies and include forces of inertia of the working parts and also of the adhered mass of the material under processing. They are observed on oscillograms in the form of fluctuations with respect to the averaged level of processes.

In his time, Goryachkin repeatedly pointed out that processes of operational resistance are related to the disorderly appearance of pulses and proposed a basic simple equation, which describes the order of action of working parts and machines on the material or soil: $P\Delta t = m\Delta v$ or $P\Delta t = \Delta\mu v$, in which force P characterizes the engine or working part, time Δt—the period of action of the force (step of a horse, time for splitting of a layer, etc.), m—the mass of the machine, $\Delta\mu$—the final mass of the processed material and v, Δv—the working speed imparted to the mass or the machine. Thereafter, Goryachkin proposed that the spontaneous action of the force and resistance are related as follows:

$$P\Delta t = \int_0^{\Delta t} R\, dt = \int_0^{\Delta S} R d\Delta S$$

and power $75N\Delta t = m\Delta v^2/2 = \int_0^{\Delta S} R d\Delta S$, which is consumed over path ΔS. In this, the value $\Delta v = dv = \sqrt{\dfrac{2\int_0^{\Delta S} R d\Delta S}{m}}$ represents the acceleration which is absorbed by the resistance of the material during the pulse.

On an elementary area of the working part, the resistance of the surroundings is created in the form of a component of the normal pressure from deformed masses and frictional forces acting on the working part as it moves along. Taking the statistical nature of the properties of soil and other agricultural materials into account, it would be correct to assume that from the microscopic standpoint, these power phenomena arise due to discrete primary phenomena during the uniform movement of the rigid working part. The discrete primary phenomena are caused by microheterogeneity and deformation peculiarities which get damped, averaged out and then absorbed by the inertial working parts in some integrated way. They may be observed (measured), even if on a macroscopic scale, since the continuous fluctuations of the forces can be simply represented as the integration of the pulse flow, which often succeeds in oscillating during period Δt (for example, 10^{2-6} 'dense noise', (Fig. 44) by a rigid body. The rapid damping of a set of pulses, which arises in the processing of the surroundings, takes place due to intense internal friction.

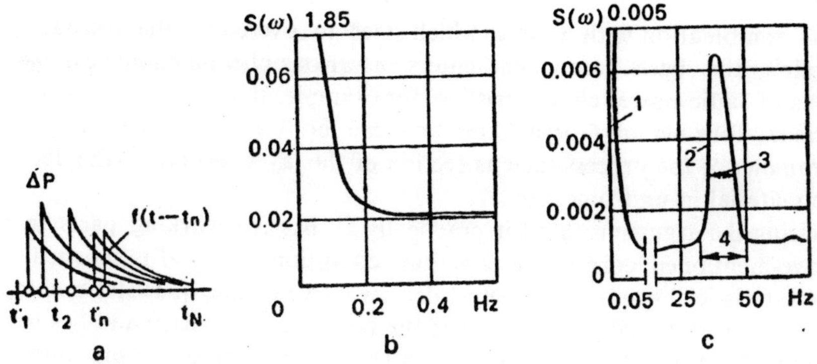

Fig. 44. Schematic representation of a—pulse process; b—spectral torque density on the grain conveyer; and c—spectral density of the working resistance of the pointed share in sand:
1—quasi-static component; 2—dynamic component; 3, 4—phase fluctuations and amplitude of oscillations, respectively.

This type of pulse process as the random superposition of certain determinate functions is given as

$$\Delta \bar{P}_{st\Delta S}(t) = \sum_{k=1}^{k} g_k f_k(t - t_k)$$

which get damped; for example, according to the law $e^{-\alpha t_k}$ is described by the correlation function of the type

$$R_x(\tau) = a^2 e^{-2\lambda|\tau|}$$

in accordance with the postulated mechanisms of the probable phenomena; here a^2 is the mean square of the pulse amplitude, λ—the intensity of the pulses which, in the general case, follows Poisson's distribution law with a variable random parameter.

From the foregoing and, in particular, on the basis of the central limiting theorem, the predominant manifestation of the normal law of distribution for fluctuations of power process is explained.

It can be assumed that a similar physical pattern is observed in the operation of all types of working parts, and it ensures the building up of an averaged level (quasi-static component) of power process, which under the influence of highly heterogeneous soil, grain rolling devices, mass feeding into the machine or a change in the modes of movement will represent a gradually proceeding (with frequency $\omega \leqslant 0.1$ Hz) random function of the path or time with the correlation function of the type:

$$K_{xi}(\tau) = e^{-\alpha_i|\tau|}(\cos \lambda_i |\tau| + \frac{\alpha_i}{\lambda_i} \sin \lambda_i |\tau|)$$

where α_i is the coefficient of the damping of the correlation function and λ_i—the predominant frequency of the load oscillation.

In combination with a force, which goes to overcome the resistance of friction, this basis component enters in various phenomenological expressions of static power characteristics (for example, the first two terms in the rational formula of Goryachkin) and can be related to any characteristic parameter of the process such as section of the layer, extent of the feed of the material in unit time, etc.

Using the argument that for practically all passive working parts in the above-mentioned correlation function, condition $3\lambda_i^2 < \alpha_i^2$ is satisfied, the characteristic shape of the spectral density curve of the quasistatic components of operational resistance is of the bell type (Gaussian) and lies in the region of 'infralow' frequencies, i.e., in the region of zero on the graph of spectral density (Fig. 45).

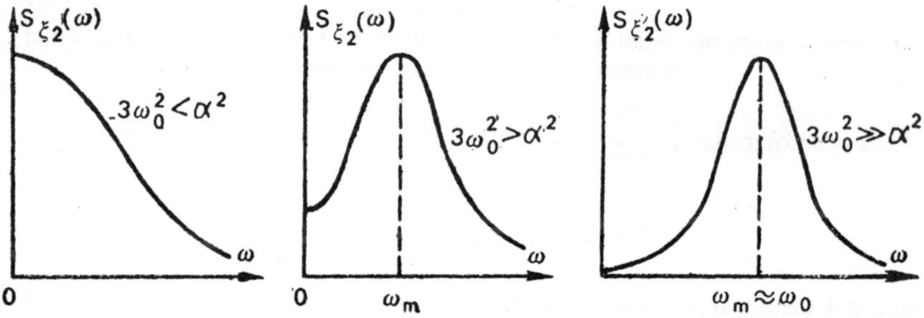

Fig. 45. Model examples of the course of curves describing spectral densities of processes.

The formation of dynamic components of resistances is also linked with the vibrations of systems under deformation (working parts) and with the stability of the movement of machine aggregates over the field as a whole. The dynamic component of the mechanical resistance of working parts describes the ratio of the increment of the resistive force of the material to the increment of the displacement seed of the working part.

Often, the problems of vibrations of the mechanical systems under random loads, which are formed in the process of their interaction with the surroundings under processing, are solved by expansion of the deformation in terms of the modes of natural vibrations of the design under consideration in the form:

$$\Delta(x, y, z, t) = \sum f_m(t)\lambda_m(x, y, z)f,$$

where Δ is the deformation, λ_m—the modes of natural vibrations of the system, and $f_m(t)$ ($m = 1, 2, \ldots$)—the generalized coordinate.

Assuming the dynamic characteristics of the working part (system) and the experimental information about the properties of the material under processing, it is possible to consider linear differential equations with the constant coefficients within the framework of the correlation theory:

$$\frac{d^n y}{dt^n}+a_1\frac{d^{n-1}y}{dt^{n-1}}+\ldots+a_{n-1}\frac{dy}{dt}+a_n y=AR_x(t)$$

and to find the corresponding spectral density of the dynamic forces of the surroundings under processing which cannot be measured:

$$S_R(\omega)=\frac{S_y(\omega)[(i\omega)^n+a_1(i\omega)^{n-1}+\ \ldots+a_{n-1}(i\omega)+a_n]^2}{A^2}.$$

In the general case, the solution of such an inverse physical problem leads to a similar correlation function, but already under a different condition

$$3\lambda_i^2 \gg \alpha_i.$$

The shape of the oscillographs and correlation functions show that the fluctuations of resistance on the working parts of a certain type of cultivator share, etc., because of their very origin, reflect the periodic nature of their movement. And its approximation can be the process, which is described by exponentially damped sinusoidal functions in each period of the fundamental mode of vibrations

$$R(t)=\sum_{i=1}^{n}[\exp(-\pi\Delta\omega_i\tau_k)](a_i\cos 2\pi f_i t+b_i\sin 2\pi f_i t)+a_0,$$

where f_i is the frequency of the ith mode; $\Delta\omega_i$—the width of the band for a random change in the frequency.

Quantity $2\tau_k=T\omega_1$, $(\omega_1=2\pi/\tau_k)$ determines the frequency of the lower mode of the working part along the interval τ_k (between the first two zeros of the correlation function).

As it has been mentioned earlier, it is worthwhile considering the processes under investigation to be steady fluctuations in terms of mathematical expectations, which are calculated as the mean integral over the 'sliding' interval of time:

$$\Delta x(t)=x(t+T)-x(t)=x(t)-\frac{1}{T}\int_{t-T/2}^{t+T/2}x(t)\,dt.$$

The nature of approximate expression and the spectral densities of the processes under discussion, as it appears from the set of experimental data, generally show several peaks (Fig. 46). This corresponds to the solution of the differential equation of at least the second order, which characterizes the vibrational requirement of the working parts with a frequency of the lower mode in relation to the slowly changing stable positions.

Such a reaction shows a very useful aspect of the problem, namely, that the designs of the working parts of agricultural machines possess filtering properties and actually appear as narrow-band dynamic systems, which makes it possible to neglect the correlation between the various degrees of freedom and generally leads to normalization of operational resistances.

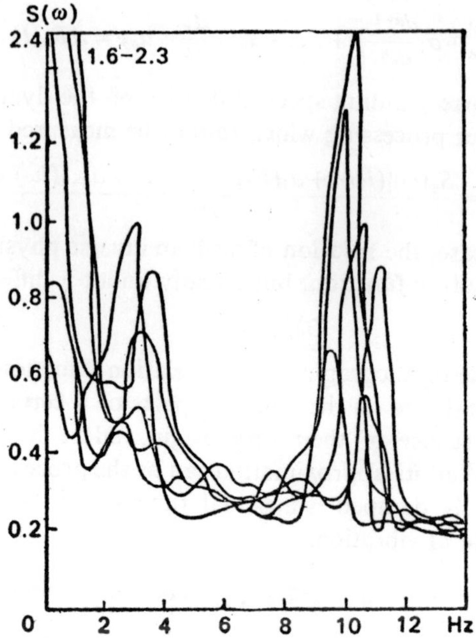

Fig. 46. Spectral density of torque on the driving sprocket of tractor T-150 in an aggregate with a beet combine SKD-3.

The aforesaid can be illustrated by the shape of the curve of spectral densities of resistances for some working parts (see Fig. 44).

The abovesaid hypothesis regarding the mechanism of the formation processes of operational resistances become fully confirmed under ideal conditions (for example, the working of a cultivator share in sand, a conveyer in grain).

Spectral analysis of operational resistances shows that the operational mode of cultivator shares, which is close to that of natural vibrations (self-excited vibrations) in ideal conditions (Fig. 47), loses its periodical nature in real conditions under the influence of random disturbances. This is related to the fact that with an increase in the intensity of random disturbances of soil as compared to sand, the self-excited vibrational mode of the motion is disturbed, and a considerable shift of the peaks of spectral densities toward the side of lower frequencies takes place, with the shape of the peaks becoming symmetrical. At the same time, the main condition for the existence of proposed self-excited vibrations no longer holds good.

In real conditions, when the soil is highly heterogeneous, and there is considerable effect of the dynamics of the machine aggregate, the structure of functions of the spectral density of operational resistance gets somewhat

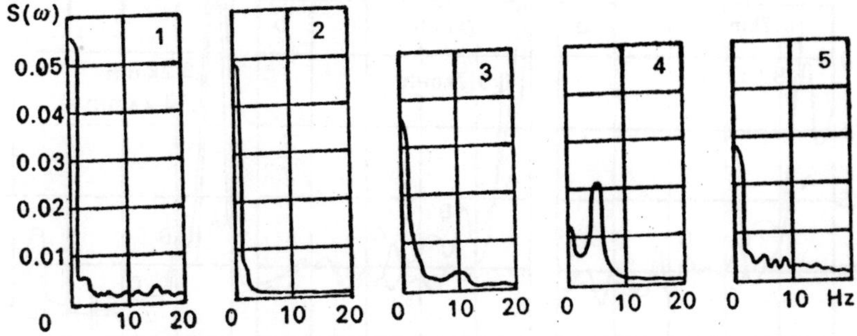

Fig. 47. Spectral densities of working resistances of cultivator working parts mounted on one frame and on an identical depth for the processing of soil: 1—pointed share of cultivator KPS-4; 2—bit of cultivator 2KRN-2, 8M; 3—spring-controlled share of cultivator 'Kongskild' (Denmark); 4—spring-controlled share of cultivator M-200 (Canada); 5—spring-controlled share of cultivator KPS-4.

altered. Thus, the vibrational movements of the aggregates leads to a localized change of speed of the interaction of the working parts and the actuating devices of machines with the soil and masses under processing, thereby creating additional dissipative forces of a nonconservative nature. The contribution of these to the general dispersion of processes attains considerable value and assists in the formation of flashes of spectral density in the frequency range of the longitudinal and vertical vibrations of aggregate (see Figs. 46, 47).

This is explained by the fact that in the presence of vibrations of the aggregate, the gradients of the deformation speed of the soil by the working parts considerably increase ($1.5\text{-}2v_m$), so that there is a considerable dependence of resistance on the derivatives (at least the first derivative) and the dissipation of energy is sufficiently high. Such parametrical phenomena are called modulation [66]. Modulation determines the multiplicative character of formation of fluctuations of operational resistances which are transmitted to the hook, axle shaft and engine shaft. During measurements, it does not separate the effects created because of the surroundings, machines and engines. The source of slow modulation is the field relief and other slowly changing physical factors (physical state of the soil, distribution of the crop along the field, etc.), and the source of fast modulation is basically the three-dimensional vibrations of the aggregate.

The results described here are confirmed by the data obtained by Solov'ev in the nonchernozem region [55]: here, curves *A* (Fig. 48) characterize the correlation between the ordinates of the surface profile of a stubble field of winter wheat on plots with ordinary, uneven and even backgrounds.

From the curves of the correlation function, it is seen that the length of the path of correlation connection for the field with an uneven background

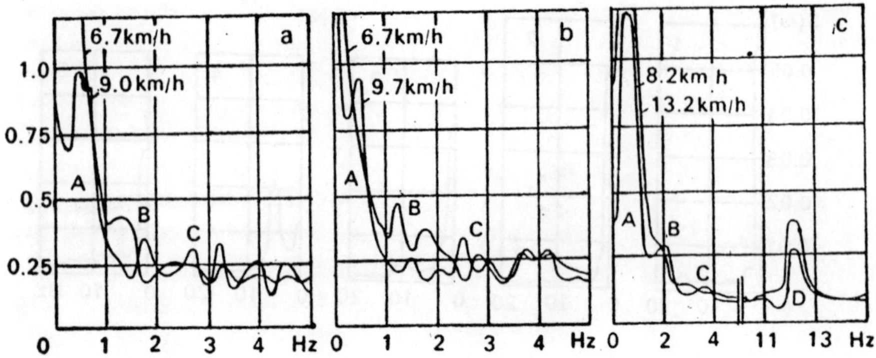

Fig. 48. Spectral densities of torque on axle shafts of tractors YuMZ-6L in an aggregate with (a) roller plow GEF-3-35, (b) plowshare PN-3-35 and of tractor T-150 in an aggregate with (c) coupling SG-21.

is equal to 1.66 m and with an even background, it is more than 5 m. If the interval of the path of the correlation connection is small, then higher frequencies dominate in the curve of the surface profile of the field, and the values of the coefficients α and β are higher in the approximate expression.

For the curve of an uneven background, the values of these coefficients are $\alpha = 1.22$ l/m, $\beta = 0.94$ l/m; while an uneven background, $\alpha = 0.45$ l/m, $\beta = 0.31$ l/m.

It is also established that the surface profile of the field shows the effect on the formation of the curve of the torque on the axle shafts of the tractor and other self-propelled machines [24, 55, 58]. A machine-tractor aggregate, consisting of tractor MTZ-5M and plow PN-2-35, was investigated on two backgrounds of different evenness of a stubble field of winter wheat.

The correlation function of the change of the ordinates of the torque curve while working on the machines aggregate on an even background shows that the correlation time connection for this curve is equal to 0.41 s and for working in an uneven background, it is 0.17 s.

From this data, it appears that the maximum energy during the working of the aggregates on an even background is transmitted in the range of frequencies 0-61 s^{-1} and during the working of an aggregate on uneven background—0-11 s^{-1}.

Since the disturbances, which influence the actuating device and the working parts, possess a continuous spectrum, while the dynamic (partial) systems themselves only possess certain discrete spectrums of natural frequencies, therefore, their reactions differ by narrow local splashes of the spectrum and lie on a lower frequency (vibrations of elastic elements of the actuating unit) as well as in a higher frequency region (axle shafts, shanks, spokes, frames, etc.).

Modulated vibrations have a random nature and preferably carry one and the same physical content: their amplitudes and phases considerably change only beyond the time, which is more than the periods of natural vibrations of elements of design or characteristic changes of influencing factors.

Modulated vibrations are expressed in the quasi-harmonic form:

$$\xi(t) = A(t)\cos(t) = A(t)\cos[(\omega_0 t - \theta(t)],$$

where $A(t)$, $\theta(t)$ are the functions of time, which are slow in comparison with frequency ω_0, which is characteristic of the given type of movement of the processes.

In the manifestation of the sources of modulation, which, in a general case, have random intervals of time, it is possible to schematize each elementary event which leads to excitations of vibration by using the known stochastic scheme of creation and annihilation. This can be conveniently adopted to the characteristics of external conditions.

In fact, in Fig. 48, it is possible to discover a distinct manifestation of the stochastic modulation of the process of operative resistances. In region *A*, flashes of spectral density of torque on axle shafts of tractor YuMZ-6L in the aggregate with plows PN-3-35 and GEF-2-35 and of tractor T-150 with coupling SG-21 are linked with the sign-variable displacements of bodies of machines in the presence of longitudinal and angular vibrations. The first natural frequency of tires and other elements of the actuating unit (region B) also influences the power processes, while the second (C) is weakly manifested. Sign-variable vibrations of the corresponding design elements of the tractor are transferred on its connecting link or mounting mechanism, and these induce fluctuations of the traction resistance of the machines at corresponding frequencies.

For track-mounted tractors, when coupled with connected or semi-mounted implements, a peak is observed in the spectral density of the torque on the axle shaft in the region which corresponds to the frequency of impact of the ridges of the caterpillar chain on the teeth of the driving sprocket and the nonuniformity of the rewinding of the chain (D).

The transmission of harvesting machines and power engines, the nonuniformity of the moment of driving forces as well as of the resistive forces generates cyclic loads, which are aggravated by the errors of universal-joint, chain, shear and other drives.

Interaction between the machine aggregate and the surroundings can be estimated more thoroughly with the help of mutual correlation functions (Fig. 49). The delay in the change of moments of resistance particularly on the axle shaft and connecting link, in relation to the engine torque and, on the contrary, the lag in the change of the increment of the speed of motion in relation to the engine torque, characterize the ability of the aggregate to absorb the engine power and to consume it in overcoming the external disturbances. A delay of the order of 1.5-6 s, which is found from mutually-

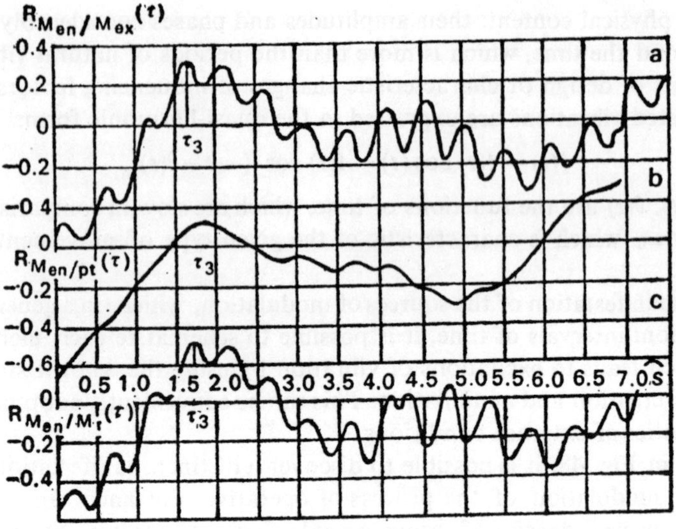

Fig. 49. Mutual-correlation functions of the torque on the engine shaft of tractor T-150 in an aggregate with a beet combine SKD-3 and the torque on (a) axle shafts; (b) traction resistance; and (c) a combined resistance moment.

correlated functions, depends on the parameters of the aggregates and is caused by the inertial development of the processes for the consumption of a fraction of the kinetic energy and its replenishment by increasing the rotation frequency of the engine.

If the aggregate could ideally reproduce an input signal from external disturbances, i.e., $M_{en}(t) = kM_r(t)$ then the mutual correlation function would be as follows:

$$R_{M_{en}/M_r}(\tau) = k[M_r(t)M_{en}(t-\tau)] = kRM_r(\tau)$$

and the phase shift and temporary delay would be absent. But for this, an engine with an extremely high power would be required. The mean time of delay, which is determined from the first maximum of the mutual correlation function, physically corresponds to the intervals of the dissipation of fluctuation in kinetic energy of the aggregate as it executes a nonuniformly, 'small', translational motion. The reduction of this interval is the result of the rise in the operating speed of the aggregate and, consequently, its power saturation.

It should be borne in mind that filtration of the relatively higher frequency external disturbances, which is accomplished by the inertial parameters of the aggregate, possesses a simple physical meaning; the system narrows the spectrum of input action and index correlation at the output, in comparison with the input. From this point of view (increase in stability of motion), it is worthwhile to increase the total mass of the machine aggregate and the speed of its movement, and this increase can be limited mainly by the trafficability on the field and in the space between the rows as well as by design considerations.

The formation patterns of power processes (examined here) are sufficiently general for a large class of machines and implements. The spectral structure of power processes can be schematically represented in the form of a combination of a number of characteristic components (Fig. 50). The area under the narrow-band sections of the indicated localized regions of the spectral densities describes dissipative losses of energy under the action of resistive forces of soil due to the vibrational movement of the aggregate. Since it is worthwhile to adopt low-frequency regular components of processes $\sum_j \int_\omega S_{st\,j}(\omega)$ as a useful signal (of working resistance, power consumption, etc.), it is absolutely natural to determine power and dynamic effectiveness in the form of a certain coefficient, which is similar in the physical sense to the efficiency (of a working member or the aggregate as a whole):

$$\eta_\omega = \frac{\sum_j \int_\omega S_{st\,j}(\omega)\, d\omega}{\sum_{ji} \left[\int_\omega S_{st\,j}(\omega)\, d\omega + \int_\omega S_{di}(\omega)\, d\omega \right]},$$

where $\sum_i \int_\omega S_{di}(\omega)\, d\omega$ is the spectral density of the dissipative and other components of nonuseful power consumption.

Calculations show that for the most passive and active members, the coefficient η_ω does not exceed 0.3-0.5. A similar level of values of this coefficient for working members of soil-processing machines is also obtained by other authors, but under different initial conditions [38]. For aggregate as a whole, the maximum level of nonuseful consumption and the minimum value of η_ω exist for trailer (tractor-drawn) machines.

The manifestation of dynamic properties of working members, machine aggregates and the properties of materials under processing is the basic trend of the formalized description of operational resistances. As regards the investigation method of the fluctuation of operational resistances, the consideration of the processes in the form of stationary growths provides acceptable results.

26. MODELING OF POWER CHARACTERISTICS AND THE PREDICTION OF RATIONAL CORRELATION AMONG BASIC PARAMETERS OF MACHINE AGGREGATES

The power and operational indexes of machine aggregates and their basic

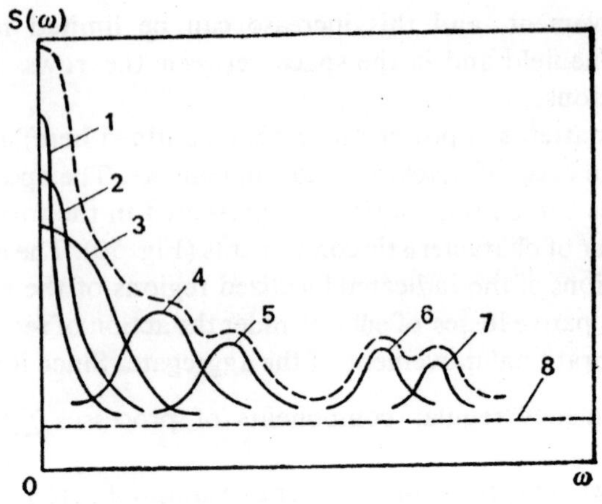

Fig. 50. Schematic representation of the spectral structure of power processes:
1-3—quasi-static components of proper working resistance, change in relief and displacement, respectively; 4-5—dynamic components, conditioned by angular (longitudinal and transverse) vibrations of the supporting structure (or tractor) and by vertical vibrations of the suspension points of the working members, respectively; 6-7—dynamic components, which are conditioned by the dynamic properties of the working members or the machine; 8—component similar to noise.

parameters collectively determine technical-operational characteristics. For example, an operational characteristic such as productivity depends considerably on the width of the claw, the power consumption of an aggregate and the power of the engine mounted on it. Hence, modeling and the prediction of power characteristics of the aggregates follow the aim of exposing rational correlations among their basic parameters, which can be taken as the width of claw B, mass M_a, and the power of engine N_{en} [58-60].

The general relationship for the construction of a static mathematical model is of the form

$$W = \frac{N_{en}}{N_{abv}} \eta_e v_m \tau,$$

where N_{abv} is the combined expense of power on the unit width of the claw of the aggregate in the function of the movement speed; η_e—the load factor for the engine.

The initial data in tabulated or graphical forms are compiled on the basis of accumulated statistical material, the organization of special experiments and if necessary—the determination of the values of power consumption of machines and working parts [60].

On the basis of the experimental data, the coefficients of regression equations are determined, for example, in the form:

$$P_t = a_0 + a_1B + a_2v + a_{12}Bv + a_{11}B^2 + a_{22}v^2;$$
$$M_t = b_0 + b_1B + b_2v + b_1Bv + b_{11}B^2 + b_{22}v^2;$$
$$M_t = c_0 + c_1v + c_{11}v^2,$$

where P_t is the tractive resistance; M_r—the moment of machine resistance; and M_t—the moment of resistance to the movement of the tractor.

Combined power, necessary for the working of the aggregate, is determined as follows:

$$N = \frac{1}{75}(P_tv + M_r\omega_1 + M_t\omega_2).$$

The *productivity of the aggregate* for various values of the width of a claw and movement speed is predicted on the models which are described in the previous paragraphs.

On the basis of the results of such modeling, it is possible to establish the significance of various methods and the possible intensification limits of mechanized processes on such machine parameters as the width of the claw and the working speed, and then to determine the necessary power source.

The distribution of all possible combinations of these parameters can be represented [58] by lines of equal productivity (w = const). Such graphs for beet harvesting machines are shown in Fig. 51.

If one deviates from the possible limitations of a technological and technical nature, then the direction of the maximum increase of productivity

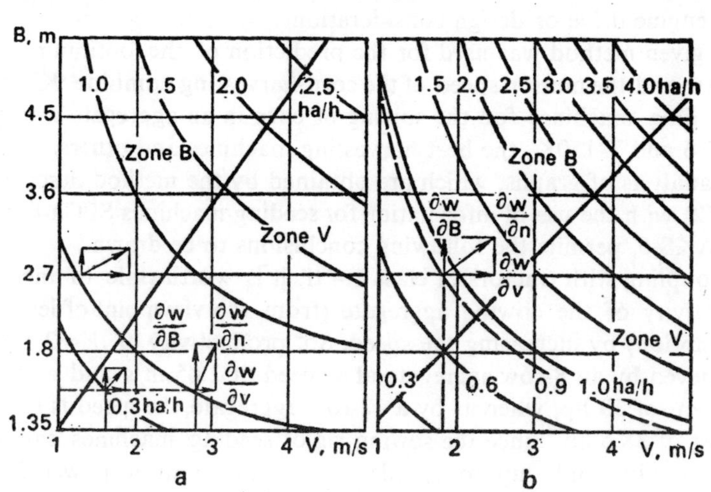

Fig. 51. Field of possible combinations of the width of the claw and the movement speed of beet-harvesting aggregates: a—for beet combines; b—for root-harvesting machines.

can be considered as the rational change in direction of the parameters of the aggregate in time, i.e., direction of gradient n:

$$\frac{\overrightarrow{dw}}{dn}=\frac{\overrightarrow{dw}}{dB}+\frac{\overrightarrow{dw}}{dv_m}.$$

Investigation of the partial derivatives $\delta w/\delta v_m$ ha/km and $\delta w/\delta B$ ha/h m, which characterize functions of limiting effectiveness of given factors (see Fig. 51), shows that in the region of high working speeds, the dominating effect on the productivity rise comes from component $\delta w/\delta v_m$, on limiting the width of the claw to its mean values; and on the contrary, in the region of low working speeds, the higher productivity rise is ensured by the width of the claw.

Such analyses help in determining *zones with a dominating effect of the width of the claw* (zone B) and the movement speed (zone v) on the productivity. This is extremely important for testing power saturated aggregates for high speeds and aggregates with wide claws (Fig. 51).

Further on the graph of $N=f_1(B, v)$, by using relationship $W=f_2(B, v)$, lines of equal productivity can be drawn. By analyzing the obtained graph along with the necessary engine power, the transmission losses and the power supply for starting or for overcoming unevenness, it is possible to determine the optimum motion speed and the width of the claw from the criterion of maximum productivity.

In the analysis, limitations on speed arise because of the quality of work, those in speed and the width of the claw because of the tractive properties of the engine drive or design considerations.

The given method was used for the prediction of the optimum width of the claw and the motion speed of the corn-harvesting combine 'Khersonets', the seeding machines for the sowing of corn in an aggregate with tractors YuMZ-6 and T-150K, the beet-harvesting machines and others.

An analysis of graphs, which are obtained by the method described here (Fig. 52) with the use of information for seeding machines SPCh-6, SUPN-8 and SAKS-6, permits the following conclusions to be drawn.

In coupling with tractors of class 1.4 tf, it is worthwhile to increase the productivity of the sowing aggregate (from the viewpoint of lesser power consumption) by increasing the speed. So, productivity of $W=2.5$ ha/h can be achieved by an 8-row aggregate at a speed of 2.05 m/s and a power requirement of 33 hp, whereas by a 12-row aggregate, the speed is 1.4 m/s and a power of 38.5 hp. Since the sowing rig of seeding machines can provide sowing quality only up to speeds of 2.2 m/s with a power limitation (YuMZ-6), the optimum is an 8-row aggregate that can only provide productivity up to 2.6 ha with a power requirement of 37 hp (14.2 hp h/ha). The tractor will work with a certain underloading in this case.

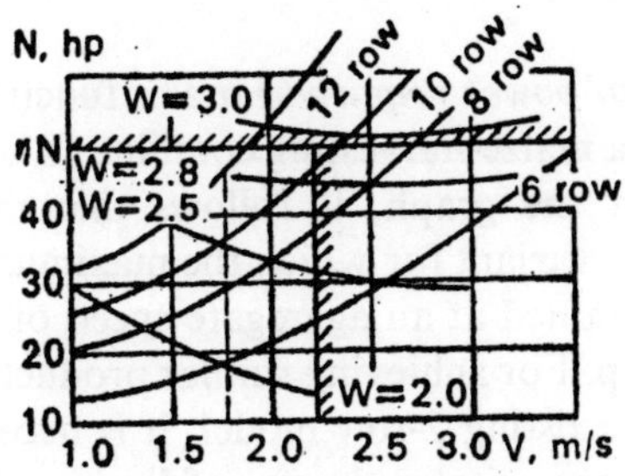

Fig. 52. Forecast of the power requirement for optimum speed and the width of the claw of aggregates for corn sowing with tractors of class 1.4 tf.

Higher productivity (2.8-3 ha/h) with the use of a 12-row seeding machine can be achieved by some increase in the working speed, but the specific power consumption will be more—16.4-18 hp h/ha at a power requirement of 54 hp.

For machines with a width of a claw more than eight rows on coupling with tractors of type T-150K, the increase in productivity by an increase of speed is connected more with higher power consumption than by an increase of the width of the claw. In the aggregate with tractor T-150, an 18-row seeding machine gives the optimum results, the productivity of which can be 4.4 ha/h at a power of 134 hp h/ha; (30.4 hp h/ha) whereas, a 12-row aggregate with tractor T-150K can provide productivity of only 3.2 ha/h. The power of tractor T-150K in the last case will not be fully utilized (Fig. 53).

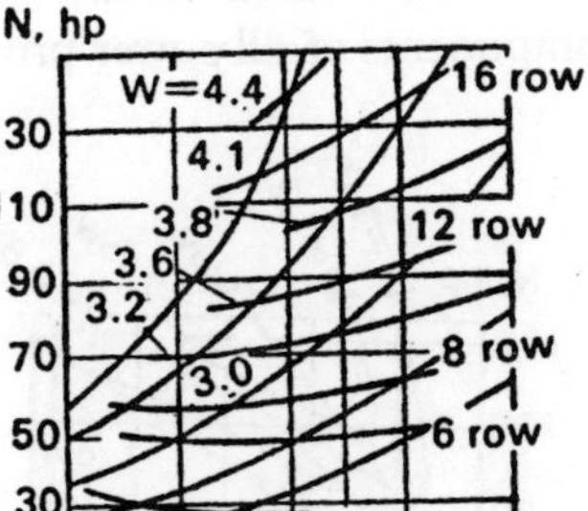

Fig. 53. Forecast of the power requirement, optimum speed and width of the claw of aggregates for sowing maize with tractor T-150.

Prediction results of power requirement as a function of motion speed and width of the claw of a maize-harvesting combine 'Khersonets' are represented in Fig. 54. From the graph, it follows that the most rational is the machine with a 3-row variant for which the maximum operational productivity of 1.3 ha/h is obtained at an aggregate speed of 2.5 m/s and a power requirement of 120 hp. For achieving similar productivity by a corn-harvesting combine with the existing 2-row model, it is necessary to raise the speed up to 3.7 m/s, which necessitates the use of limiting the tractor's power.

In conclusion, an example on the use of a *polynomial power model* (see Table 16) for the synthesis of power N_i as a function of M_a and v_m [58] on the basic criteria of the specific power consumption of beet-harvesting combine, while considering its unit mass (hp/t), carrying capacity (hp s/kg) and turnout (hp h/ha) will be considered along

$$N_{i\Omega} = a_{111}v_m^3 + (a_{11} + a_{112}M_a)v_m^2 + (a_1 + a_{221}M_a^2)v_m + a_{22}M_a^2 + a_a M_a + a_0$$

with similar models on the basis of quadratic functions for other elements of power consumption (torques and traction resistance). Skidding and fuel consumption are described by linear relationships on estimated parameters:

$$\delta_\Omega = \exp\left[b_1 \frac{P_t}{M_a} + b_{11}\left(\frac{P_t}{M_a}\right)^2 + b_2 P^0 + b_{22}(P^0)^2 + b_0\right],$$

where M_a is the mass of the tractor of the self-propelled machine.

The coefficients of the model for beet combines and root-harvesting machines (Tables 15, 16) are calculated on the basis of Maksimchuk's investigations, using the method of least squares. It is natural that the static power characteristics of harvesting machines and aggregates are substantially nonlinear (except for individual items, which allow linearization). Here, low-frequency components of all power processes, which are obtained

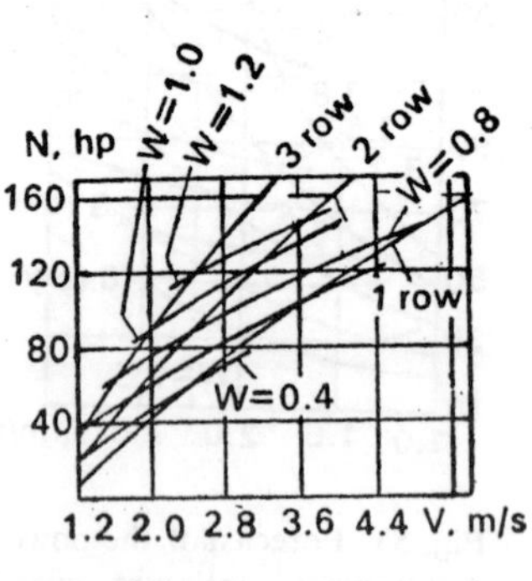

Fig. 54. Prediction of optimum speed and width of claw of corn-harvesting combine with tractor T-150K.

Table 15

Parameter	v_m, m/s	v_m^2	P^0, kgf/cm²	P^0v_m	w^0, %	v_mw^0	B, m	B^2	Bv_m	Q, qtl/ha	Qv_m	QB	P^0w^0	b_0
Force on hook, P_h, kgf	154.99	14.37	84.54	—	38.74	−8.28	−3817.43	2072	86.94	—	—	—	—	809.46
Force on self-movement, P_{sm}, kgf	321.40	60.95	208.60	−6.22	189.88	−24.6	928.17	—	103.85	—	—	—	10.89	—
Moment on shaft of machine, M_{PTO}, kgf m	−108.26	7.71	—	—	0.06	—	90.01	61.43	36.79	0.243	0.138	−0.271	−72.71	—

Table 16

Parameter	Mv_m, m/s	v_m^2, m²	P^0, kgf/cm²	P^0v_m	w^0, %	v_mw^0	B, m	B^2	Bv_m	Q, qtl/ha	Qv_m	QB	P^0w^0	b_0
Traction resistance, P_{tr}, kgf	−476.81	24.04	−35.10	—	−65.62	13.28	34.60	139.67	242.36	—	—	—	—	1980.97
Force on self-movement, P_{sm}, kgf	−293.62	80.58	−1.62	6.8	3.68	4.20	150.16	—	27.63	—	—	—	−0.04	22.55
Moment on shaft of machine, M_{PTO}, kgf m	−28.02	5.71	—	—	−0.24	—	7.23	3.72	8.65	0.1	0.02	−0.06	—	15.82

by averaging over suitable intervals of the path, follow sufficiently closely the normal law of distribution. This confirms the narrow-zone and dominating influence of the linear part of the system (Table 17).

The investigations of polynomial models were accomplished on a computer by the numerical method. The obtained results (Fig. 55) show that the optimum width of the claw of beet combines for the existing energy budget structure on the just criterion of specific energy consumption (hp/t), consists of 0.9 m at $v \geqslant 2.0$ m/s and on root-harvesting machines, it shifts in the region of the width of the claw of 1.35-1.81 m. This confirms the fact that the accepted mechanical-technological principles of beet harvesting turnout are extremely uneconomical in terms of power consumption when harvesting is realized with the use of technology.

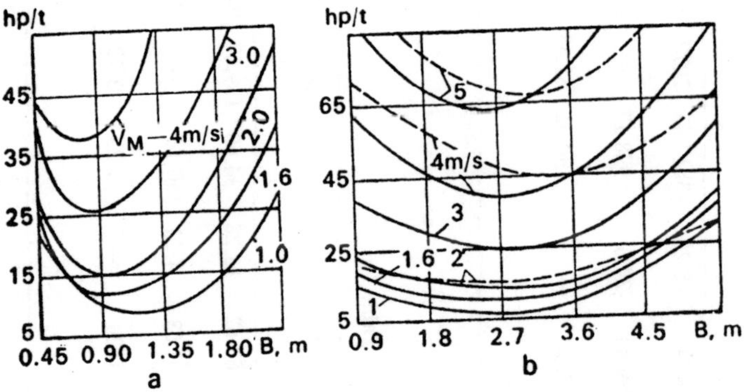

Fig. 55. Optimization of parameters of (a) beet combines and (b) root-harvesting machines with respect to the criterion of specific power consumption (hp/t), using disk and rotor-fork type (dotted) digging elements.

A more general presentation of the *specific power consumption of beet-harvesting aggregates*, for a unit mass under processing, is provided by data accumulated over many years. Beet-harvesting combines with a large widths of claw (1.35-1.8 m) have advantages in respect of this index. Combines of the pulling-type KST-3 consume somewhat higher power because of the high power-consuming digging elements.

Machines for two-stage harvesting show less specific power consumption as compared to combines. With the increase of beet rows and root-harvesting aggregates, their power consumption on the indicated criterion considerably decreases, while this pattern is more characteristic of self-propelled aggregates. This is caused by higher movement stability and a lower level of dissipative forces. In connection with this, the optimum speeds of low power-consuming machines are raised up to 7.2-8 km/h.

Table 17

Parameter	Combines						Root-harvesting machines								
	KST-3			SKD-3			UKS-3			KS-6			'Parma'		
	Speed, m/s														
	1.5	1.7	2.6	1.68	2.6	3.1	1.55	2.1	2.64	1.55	1.94	2.12	1.84	2.3	2.8
Traction resistance, kgf:															
P_{cr}	1010	1100	1720	1200	1600	1880	1080	1280	1680	—	—	—	—	—	—
$\sigma_{P_{cr}}$	460	515	683	430	556	650	364	520	580	—	—	—	—	—	—
Moment on axle-shaft, kgf m:															
M_{ax}	—	—	—	—	—	—	—	—	—	870	1063	1180	1010	1250	1560
$\sigma_{M_{xa}}$ (for self-propelled machines)	—	—	—	—	—	—	—	—	—	210	295	335	310	358	426
Power on drive of working parts, hp:															
N	23.0	25.0	27.3	24.5	32.2	38.6	18.2	22.5	28	—	53.5	61.3	25.1	32.5	34.6
σ_N	5.2	5.65	6.68	4.6	7.8	10.5	2.65	3.6	—	—	16.8	17.6	5.3	6.9	7.2
Probability of coincidence with normal law of distribution according to criterion χ^2:															
P_{cr}	0.81	0.40	0.74	0.10	0.58	0.89	0.12	0.10	—	—	—	—	—	—	—
M_{ax}	—	—	—	—	—	—	—	—	—	0.10	0.15	0.23	0.16	0.28	0.41
N	0.99	0.52	0.27	0.17	0.21	0.21	0.11	0.35	—	0.22	0.38	0.56	0.17	0.59	0.45

For the existing structure of a power budget, the optimum power consumption on the unit supply (hp/kg) and turnout (hp h/ha) is achieved when the width of the claw of the combines for root-harvesting machines is 1.35-1.8 m ($Mv_{m\Omega}=7$ to 7.2 km/h) and $B\approx 2.7$ m ($Mv_{m\Omega}=7.5$ to 8.5 km/h), but for different working speeds. For the first criterion, the optimum is obtained at the upper level of the motion speed. Beyond the limits of indicated speeds, the increase rate in power consumption exceeds the productivity increase.

With a change in the structure of the power budget due to a 1.5-2 times decrease in traction resistance, the optimum width of the claw of root-harvesting machines shifts into the range of 9-12 rows.

The obtained results, together with a generalized analysis of the parameters for various types of beet-harvesting machines under development, lead to the conclusion that established power conditions for optimality largely hold good in the case of self-propelled beet-harvesting and root-harvesting machines with claws of widths 2.7-5.4 and 2.7 m, respectively, and power saturation 10-12 and 18-20 hp/t and for light and medium conditions, 13-15 hp/t. The optimum power saturation of the 3-row combine aggregates is 12-14 hp/t.

In the estimation and selection of the type of engine, it is necessary to keep in view [58] that for beet-harvesting and similar aggregates with a considerable value of quasi-static load (60-70%) and undesirable parametric interferences, which are connected with aggregate vibrations, the characteristic of an engine with a large linear range of variation of moment is preferable, i.e., soft characteristic with high storage for the torque moment (SMD-17, AM-03, etc.). On the intensifying of the working processes, when the active working parts with dynamic loads dominate, it is more effective to use engines with a more rigid mechanical characteristic and a large power storage (SMD-18K, SMD-60-64 and others) in order to decrease the fluctuations in the power processes. It is for this reason that machines such as SKD-3 and KST-3, which possess a large traction resistance and a high level of dispersion of traction processes, do not couple well with tractor T-150, while the root-harvesting machine KS-6 differs in that it has an increased fluctuation of working speed and a much lower efficient use of engine power.

CHAPTER 9

Operational and Technological Evaluation*

27. GENERAL

Operational and technological evaluation is carried out in actual conditions of machine operation and is directed at determining comparative operational (productivity, fuel consumption, lubricants and other materials) and basic technological indexes.

The concept of operational and technological evaluation was comparatively recently formulated in connection with attempts to tie-up operational and technological indexes regarding the working quality of machines with the economic effectiveness of their application. The use of agrotechnological indexes, which are determined while taking zonal conditions into consideration, permits a more objective estimation of the potentialities and the effectiveness estimation of the potentialities and the effectiveness of a technique during its operation over a wide range of service conditions.

Comparative agrotechnological testings (agroestimation) on specially selected plots of limited dimensions, conducted in the traditional way, cannot always provide a basis for the bringing in of corrective amendments in economic analysis, which takes into account the losses and damages of the main product, the extent of utilization of the technical possibilities of the machines, etc.

For the *operational and technological evaluation of agricultural technological processes*, the following parameters and indexes are determined: working regimes (speed, actual width of claw, processing depth, sowing norm, feed, etc.); agrotechnical indexes (product quality losses, plant damage, etc.); operational and technological indexes (operating time for the test period); productivity calculated for one hour (net, technological, shift, operational time); fuel consumption and the use of primary and auxiliary materials; operational coefficients (for working strokes, technological servicing, technological process reliability, the use of technological, shift and operational time); operating time on technical and technological breakdowns; number of personnel for servicing; machine universality (capability of machines to

*V.A. Yasinetskii has assisted in the preparation of Chapter 9.

work on other crops, different types of jobs to simultaneously fulfill several technological operations, etc.); potentialities and effectiveness of the working of machines in a technological complex.

Indexes of operational and technological evaluation serve as the initital data for a comparative economic estimation of machines.

The procedure of conducting an operational and technological estimation of various machines in state testings on the machine-testing stations of *Goskomsel'khoztekhnika* has been standardized. It consists in the determination of indexes which characterize the working conditions, operational and technological coefficients and indexes of effectiveness.

It is recommended that the comparison of machines for operational and technological evaluation be carried out in maximum comparable external conditions.

The determination of external condition is based on data concerning their effect on the working quality of machine. In order to select and justify the set of indexes of external conditions, which can be subjected to control, it is first necessary to conduct special investigations. Generally the results of test condition are related on the basis of *a priori* information and accumulated earlier experience. A scientific approach toward the estimation of conditions can be based on ranking the effect of factors as well as on the quantitative estimation of the degree of their effect. Naturally, factor analysis should comprise a considerable part of the general test theory.

According to state standards for harvesting machines, it is possible to determine the following characteristics of conditions: variety and yield of seeds and biological masses and their variation along the row or area; number of plants or mass on 1 m or m^2, their dimensions; physical characteristics of soil and relief, etc.

Along with the typical indexes characteristic of a crop, a series of indexes characterizing test conditions appear to be sufficiently common.

Soil moisture and moisture in the material under processing, weed infestation of soil and crop, stubble and stones, etc. can be included among the widely accepted indexes. Methods of determination of these indexes are regulated by the corresponding industry standards. The number and dimension of the areas can be determined from the recommendations given in Table 18, depending on the machines purpose or the technological method.

For testing, the preceding crops are considered similar (equivalent) if the difference in the individual characteristic factors which affects the working of the machine (soil moisture harvested culture, crop capacity, weed infestation, length of the run, soil density, etc.) does not exceed 20%.

A more objective approach can be provided on the basis of the comparison of experimentally obtained data with the normalized value of the function of the individual characteristics of conditions and their statistical estimation in a zonal aspect.

Table 18

Purpose (type of machine or technological method)	Accounting areas	
	number, pieces	dimension
Soil-processing implements and machines for all-round inter-row processing, plows, cultivators, flat-cutters, stubble plow, narrow and rotary hoe, ridgers, rippers	5-6	Width equal to the claw width of the implement or machine, length 0.5 m
Technological methods for tilling of cereals, rice, peas, etc —all-round or with space less than 25 cm between rows	20	0.5×0.5 m
Technological methods for tilling of cultivated crops: corn, cotton, sunflowers, potatoes	20	Width equal to two spaces between two rows, length 10 m
Machines for harvesting cereals, rice, peas, etc.—all-round or with space less than 25 cm between rows	10	0.5×0.5 m
Machines for harvesting of cultivated crops: beets, corn, hemp, vegetable-cultures, cotton, castor-oil plants, sunflowers, groundnuts, etc.	5	Width equal to two spaces between two rows, length 10 m
Machines for harvesting beets, potatoes	5	1×1 m

In the existing methodical manuals, methods for determination of the effect of operational zonal conditions on the indexes of the working of machines under test are not sufficiently developed. However, such a determination of information would permit a more thorough analysis of the data obtained from laboratory, field and technological experiments as well as disclose the factorial structure and obtain quantitative estimations for each of the parameters in the systems framework. Designing the zonal model can also serve as the basis for a solution of this problem.

Photographic timekeeping is utilized for operational evaluation, since it combines pictorial study of the working day along with operational timekeeping.

The order of photographic timekeeping and the development of observation files is determined by industry standards.

Individual operational timekeeping is used in the case of frequently repeated operations. In this case, 10 to 50 and, if necessary, more measurements of the time consumed on each operation are made (Fig. 56), and the mean value and dispersion (Table 19) are calculated.

The advantages of timekeeping for each operation are its simplicity and the possibility of subsequent mathematical modeling of the operational coefficients in dependence to the working conditions and operation regimes of the machine.

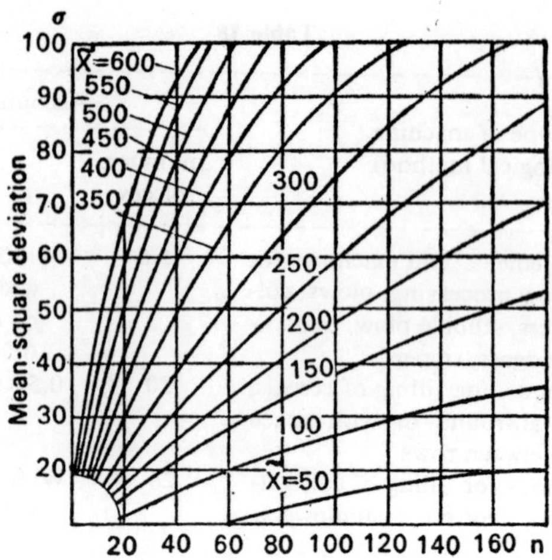

Fig. 56. Nomograms and table for determination of the necessary number of operation cycle measurements.

Table 19

$\tilde{x}$	σ											
	10	20	30	40	50	60	70	80	90	100	110	120
50	61	246	553	983								
100	15	61	138	246	384	533	753	983				
150	7	27	61	109	171	246	355	437	553	683	826	983
200	4	15	35	61	96	138	198	247	311	384	465	553
250	2	10	22	39	61	98	120	153	199	246	297	354
300	2	7	15	27	43	61	84	109	158	171	207	240
350	1	5	11	20	31	45	61	80	102	125	152	181
400		4	9	15	24	35	47	61	72	96	116	138
450		3	7	12	19	27	37	49	61	76	92	109
500		2	6	10	15	22	33	39	50	61	74	88
550			5	8	13	18	25	33	41	51	61	73
600			4	7	11	15	21	27	35	43	52	61

In the practice of agricultural machine testing, it is necessary to determine the values of operational coefficients at various working conditions and modes, including the extremal. In accordance with the general rule, it is necessary to conduct at least three control sessions for each preceding crop and to detemine the operational coefficients on them. However, such a method of operational testing is labor consuming and does not permit time for the machine to be checked for various preceding crops and operational modes over a short agrotechnological period. This can be compensated by the use of *mathematical modeling on the basis of time-keeping for each operation.* For example for the determination of the coefficient of working strokes,

$$K_0 = \frac{1}{1 + \frac{t_{turn} v_m}{3.6L}},$$

(where t_{turn} is the time for one turn, s; v_m—the working speed m/s; L—the length of drive, m) or coefficient of technological servicing (at the unloading of the product or the loading of technological material)

$$K_1 = \frac{1}{1 + \frac{Q_p W_h t_t}{3600 g_b}},$$

where W_h is the machine productivity for one hour of network, ha; Q_p—the crop produce or introduction norm of the technological material, quintal/ha; t_t—the time for one technological operation, s; g_b—the container capacity (bunker, box, cart, etc.), in quintals; it is sufficient to give the corresponding values of the variable characteristics for different regions of the given zone and to calculate their values on the computer.

The synthesizing coefficients are calculated on the basis of partial coefficients. For example, the coefficient of technological time utilization is determined by the coefficients of the working strokes and technological servicing. In determining the coefficient of shift operational time utilization, the coefficients of the working strokes, technological and technical servicing, reliability of the technological process, availability K_r and normalized rest time are taken into account. While obtaining the combined coefficient of availability for the aggregate or the complex, the corresponding coefficients for each machine, included in the aggregate, and the nature of construction of the process (parallel or sequential connection of machines and aggregates) are taken into account:

$$\sum K_r = \Pi K_{seq}; \qquad \sum K_r = \sum K_{par}.$$

The combined coefficient is generally obtained by formula [12]:

$$K_{comb} = \frac{1}{\sum 1/K_i - (n-1)},$$

where K_i, n are the values and number of the partial coefficients.

In processing the results of photographic time-keeping and the determination of the mean values of the temporal elements of the production process, the sharply distinguished values are discarded by statistical methods.

From the observation data, the time budget of the machine operation shift or of the worker (when the observed process is manual) is compiled.

Mean operational and technological indexes for the whole testing period for a given type of job can be determined as the weighted mean for the zone's values in terms of the percentage ratio of the preceding crops in the test zone. This percentage ratio is obtained on the basis of statistical data by formula:

$$C_{av} = \frac{\sum_{i=1}^{m} G_i \alpha_i}{100},$$

where C_{av} is the mean-weighted operational and technological index of the compared machine; C_i—the operational and technological index of each ith preceding crop; α_i—the percentage ratio of the ith preceding crop in the zone; m—the number of ith preceding crop in the zone.

Mean-weighted operation and technological indexes for productivity, fuel consumption, electric power and materials are used in economic calculations.

The ability of the machine to accomplish the given technological process on works of different types and on various preceding crops is characterized by the universality of the machine. This universality is estimated by the potentiality of various types of job accomplishments; the quantity of different technological schemes; the simultaneous accomplishment of some technological operations; the values of limiting indexes of different factors and backgrounds and the values of limiting work regimes.

28. CHARACTERISTICS OF REAL-INDUSTRIAL CONDITIONS

The changeability and diversity of actual-industrial agricultural production conditions decisively influence the operational and technological indexes of machine operation and aggregates and, to considerable extent, determine the structure of industrial processes and the ultimate economic efficiency.

One of the important conditions for the solution of the problems on the optimum planning of operational testing is the availability of reliable information concerning actual industrial conditions for the given region or zone. Here, the use of modern mathematical methods for the analysis of arising situations, predictions of generalized data about technological and operational indexes of the aggregate for large regions and even zones and finally, the operational plan, rational construction and scientific organization of industrial processes all have an importance.

In order to attain the aforementioned objectives, it is necessary to make assumptions convenient for the numerical methods regarding the basic characteristics of regions or zones, i.e. the zonal model. An especially important value of the zonal model appears in the modeling of the technological and operational indexes of the aggregate operation and the analysis and synthesis of the optimum structure of industrial processes.

As actual-industrial conditions possess a probabilistic nature, the development of their models must necessarily be based on probabilistic-statistical concepts.

The *physical state of the soil* and biological masses is determined in the scales of the soil-climatic zone by the mateorological conditions and complex energy processes, occurring on the earth's surface and in the atmosphere, involving a large number of factors (cumulative, compensating, etc.). These factors are related in a complex cause-and-effect manner. Since the probability of most of the factors acting in one direction is small, and their large deviations are rarely encountered; therefore, it can be assumed that the space-time distribution of the state of the characteristics of external conditions and the results of the functioning of machines are equivalent in a statistical sense. This is practically confirmed by the known stability over many years as well as by the finiteness of the characteristics of meteorological, rheological, agrotechnical and other cycles.

From the viewpoint of applied investigations and testing, it is necessary to have exhaustive statistical data about the physical parameters which characterize the state and change of the characteristics of soil and agricultural plants over a large period of time during the accomplishment of works, vegetation and harvesting.

A statistical analysis of the foregoing factors can be carried out for the characteristic regions and the zonal activity spheres special investigative results and the collection of data over many years of MTS on the basis of a special investigation. Due to a large volume of the required data, the number and place of determination of various factors can be planned by the method of random testings with a coverage of typical conditions and their extremal values.

Statistical characteristics of a number of factors under investigation are generalized and distribution laws established as a result of the collections of regional statistical material. For example for the purpose of testing, the basic indexes of zones and subzones are of interest (Table 20).

Meteorological conditions can be described mainly by the periodicity and the quantity of precipitation, which should be taken into account particularly during the periods of crop harvesting.

While considering energy problems and the dynamics of machines and aggregates, it is extremely important to have the surface characteristics of the fields determined [24]. However at present, the distribution laws of these characteristics can only be approximately indicated: the normal law for the height of unevenness and the exponential law for area distribution on the angles of the slopes.

In order to study the megastructural characteristics of field surfaces on typical massifs, the angles of the slopes are noted by a gyroscopic instrument with a 'level' base, equal to 1.2 m. The accuracy of measurements

Table 20

Indexes	Zones							
	Woodlands	Forest-steppe	Northern and central steppe	Southern steppe	Hilly zone of the Crimea*		Hilly zone of the Carpathia**	
					foot-hills	hilly areas	foot-hills	hilly areas
Arable area, %	51	61	72	76	39	—	61/30	33/36
Preferential inclination, degrees	3.1	2.3	1.2	1.0	7	17	6	14
Mean-weighted length of drives, m	401	746	1160	1230	630	180	550	190
Specific resistance in processing of soil, kgf/cm	0.40	0.43	0.55	0.61	0.47	—	0.67	—
Preferential dimensions of fields for crop-rotation, ha	16	70	90	110	14	—	16	—
Annual precipitation, mm	590	540	450	360	515	700/560	716/850	900/950
Duration of period without precipitation, days	160	165	180	190	—	200/238	264/283	226/259
Mean yearly temperature, °C	10.1	10.8	11.2	11.7	12.4/9.8	7/7.5	10/12	6/8

*In numerator indexes for southern slopes of a hilly zone of the Crimea are given, while in denominator—indexes for northern slopes.
**In numerator indexes for eastern slopes of a hilly zone of the Carpathia are given, while in denominator—indexes for western slopes.

increases because of leveling, and the conditions in which they are carried out approach the conditions of the working of the aggregates. On the obtained realizations (length about 5 km), the spectral content of low-frequency components, which have two typical ranges of frequencies—depending on the prevalent wave of macrorelief—is determined (Fig. 57).

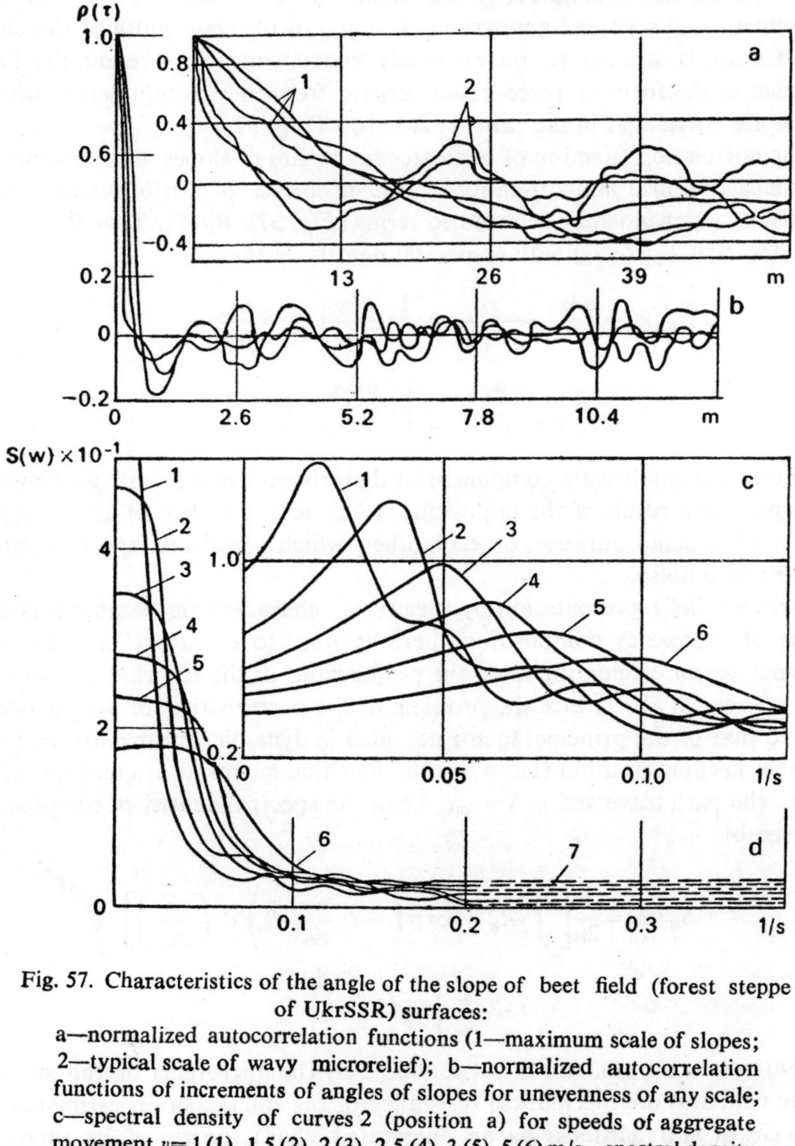

Fig. 57. Characteristics of the angle of the slope of beet field (forest steppe of UkrSSR) surfaces:
a—normalized autocorrelation functions (1—maximum scale of slopes; 2—typical scale of wavy microrelief); b—normalized autocorrelation functions of increments of angles of slopes for unevenness of any scale; c—spectral density of curves 2 (position a) for speeds of aggregate movement $v = 1$ (1), 1.5 (2), 2 (3), 2.5 (4), 3 (5), 5 (6) m/s—spectral density of curves 1 (position a) for the same speed regimes (7—uniform frequency distribution of intensity—Gaussian noise).

When the machine moves at a speed of 1-4 m/s, the lower frequencies of change in the angles of macrorelief slope are concentrated in a narrow interval of 0-0.20 l/s, beyond which a Gaussian distribution of noise alone acts with weakly-expressed fluctuations. This is confirmed by the type of spectral densities and the nature of the correlation functions of the increments of the angles of the slopes (Fig. 57, b). The significant correlation function of the increments had a spread approximately equal to the claw width of the plow (1.3-1.8 m). It seemed to be extremely convenient to represent the field surfaces in the form of three characteristic frequency components, whose scales are correlated in the ratio 1 : 2-3 : 100 (Fig. 57).

The correlation function of ordinates for angles of slopes, as a function of the megastructural noncorrelation of the processes of various scales, will consist of independently determined terms (Fig. 57), $R_\alpha(T_1, T_2) = R_{\alpha_1}(t_1, t_2) + R_{\alpha_2}(t_1, t_2) + R_{\alpha_3}(t_1, t_2)$ with a spectral density of the type:

$$S_\alpha(w) = \frac{\sigma^2\alpha_2}{\pi} \frac{\alpha_1}{w^2 + \alpha_1^2} + \frac{1}{2} \frac{\sigma^2\alpha_2}{\pi}\left[\frac{\alpha_2}{(w-\beta)^2 + \alpha_2^2} + \frac{\alpha_2}{(w+\beta)^2 + \alpha_2^2}\right] + \frac{\sigma^2\alpha_3}{\pi\alpha_3}.$$

The tiniest small-scale component of disturbances due to soil unevenness is formed as a result of the imposition of a large number of independent pulses of a small duration on each other, which permits its approximation to the white noise.

Surface relief representation by means of characteristic components (in terms of frequency composition) permits one 'to construct' any type of external action originating from the peculiarities of the soil-climatic zone.

It is necessary to reduce the problem of the perturbation of soil unevenness to that of the principal factor accepted in dynamical problems—namely to time, keeping in mind that when the machine moves at a speed of $v_m =$ const, the path traversed is $S = v_m t$. Then the spectral density of the process is given by:

$$S_\alpha(\omega) = \frac{1}{2\pi}\left[\int_{-\infty}^{\infty} R_\alpha(\tau) \exp\left(-i \frac{\omega}{v_m} S\right) d\left(\frac{S}{v_m}\right)\right] = \frac{1}{v_m}\left[S_\alpha\left(\frac{\omega}{v_m}\right)\right].$$

Taking into account the effect rendered by the microrelief of the surface of the field and the mechanical resistance of the soil (generally estimated on the basis of soil hardness as an integrated characteristic of its physico-mechanical properties), attempts are made to develop suitable technical means for their estimation [55].

As mobile agricultural aggregates with adequate accuracy for the purposes of testing are generally represented as linear models, for the analysis of such models as the input factors, which are disturbances in the form of the totality of operating conditions, the statistical independence of these factors is a desirable condition. The presence of a linear connection or correlation between the input processes leads to unnecessary complications of the model and the requirement of additional information. In order to reveal the closeness of the linear relationships between the factors which characterize the test conditions as continuous processes, the function of coherence $\gamma(\omega)$ was used, which is determined by the following expression:

$$\gamma(w) = \frac{|S_{12}(\omega)|^2}{S_1(\omega)\,S_2(\omega)},$$

where $|S_{12}(\omega)|^2$ is the module of the mutual spectral density of processes, which characterizes conditions of machine operation, and $S_1(\omega)$ and $S_2(\omega)$ —the autospectral densities.

The value of the coherence (Fig. 58) between the profile and the soil hardness from the field section is considerably less than unity in the entire range of existing frequencies. Thus one can consider these processes to be statistically independent.

After the selection of preceding crops and a description of the test conditions, operational and technological evaluation and power rating can be carried out.

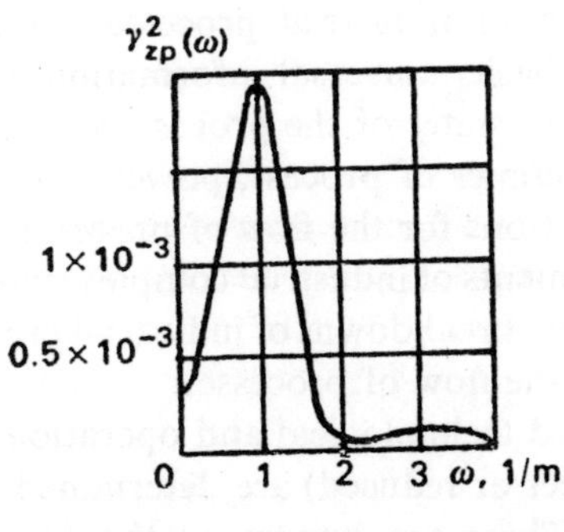

Fig. 58. Coherence for function processes, which characterize field profile and soil hardness (from Dubrovskii's data).

29. MODELING AND FORECASTING OF OPERATIONAL EFFICIENCY INDEXES

General considerations. The operational and technological testing of agricultural machine aggregates and complexes is coupled with the study of the processes of their functioning or the so-called industrial processes. An important stage in solving the problems of analysis, synthesis and optimization of industrial processes lies in developing mathematical modeling, used widely in the said area of study as well as in many other related fields [13, 31, 55 59, 72].

The basic problem of the mathematical *modeling of agricultural industrial processes consists in establishing operational relationships* over a wide range of possible actual-industrial conditions for various machine parameters and process schemes as also in the search for the optimum alternatives of an industrial process. Modeling is particularly effective in designing new types of industrial processes, when the characteristics of the individual elements alone are known, but in reality, their combined functioning was never observed. Investigation characteristics, beyond the limits of exhaustively studied values of machine parameters and processes and on the basis of accumulated information on the existing and experimental techniques of machines and industrial processes forms the essence of prediction for machine parameters and production processes.

In order to obtain a solution for concrete types of processes, it is necessary to design models of machine operation and engines in terms of an algorithm for different values of their parameters, to select suitable limitations and to optimize parameters of subject under study or to the estimate of the degree of their optimality.

The sequence of operations, performed in industrial process modeling, remains the same as has been considered for the general case.

Certain properties have also to be simultaneously accounted for.

In the formalization of industrial processes, it is necessary to take into account the following basic, universal information: time-intervals regulating the duration of different states of the process; kinematic states of machine movement or the dynamics of process; power modes of process operation, actual-industrial conditions for the flow of processes; conditions for worker occupation and the elements of industrial complexes; reliability of the elements of industrial complexes; breakdowns of industrial processes and their reasons; and characteristics of the flow of processes.

The above-mentioned technological and operational indexes, productivity and expenditures (direct or reduced) are determined as the *basic indexes for industrial processes.* These are known as the aim function and conserve generality in solving the optimization problems.

Investigations show that the distribution of different elements of time (working motions, etc.) for agricultural industrial processes carries an asymmetric character. This is due to the predominating influence of a wide range of factors (different distances for transportation of materials from the machines to the place of storage, labor consumption for removal of technological and technical breakdowns, etc.).

The nature of asymmetric distribution of the temporal elements can be approximated with a certain accuracy by the gamma distribution, the

Weibull distribution and the logarithmic normal distribution. However, the processes under investigation differ in that they show long 'tails' to the right. Therefore, it is preferable to use logarithmic normal distribution in a number of cases. As may be seen from the data of Table 21, statistical characteristics for the standstill regimes are satisfactorily described by the normal logarithmic distribution in a series of machines. The time spent on turnings also follows the logarithmic normal law. In this, problems can be easily solved with the help of logarithmic transformations.

The comparative analysis of time intervals on the above-mentioned double-parameter families of distribution shows that for specific values of mathematical expectation and dispersion, the normal logarithmic distribution is distinguished by its longest 'tail'. Closely following it is the Weibull distribution (for the coefficient of variation more than unity); then comes the gamma distribution. These distribution properties show a good correspondence between the investigated empirical relationship and the normal logarithmic law [55].

The 'length' of the shift of mobile machines lies within the purview of two basic laws: the normal and the normal logarithmic. The duration of the shift, the productivity for one hour of net and shift time for the combines, the density of idle time distribution, which often bears an asymmetric nature because of technological and technical reasons, conform well with the latter distribution law. The productivity of machines whose working process is more stable (for example, seeding machines and cultivators), satisfactorily follows the normal law of distribution (Table 22).

The kinematic states of the stable movement of machines, which characterize the movement speed, will depend on a large number of factors, including the above-considered time-characteristics of the cycle; the hypothesis about their asymptotic normal distribution can be accepted [58]. However, experimental checking of different movement speeds showed that speed distribution does not always follow the normal law. When the loading of the engine approaches the maximum, the normal, law is violated, and the asymmetry and length of the 'tail' to the right increases. This may be explained by the fact that the formation of the law of velocity distribution is connected with the movement stability and results when the law of 'masses and velocities' is applicable.

Before, going into a discussion of the general description of industrial process modeling certain arguments about operational indexes should be established, limited to the description of relationships which relate aim functions with kinematic states of movement and the generalized characteristics of the random nature of time elements [39]:

$$W = 0.1 B v_{\mathrm{w}} \tau_{\mathrm{mov}} \tau_{\mathrm{cyc}} = W_{\mathrm{cyc}} \tau_{\mathrm{cyc}},$$

$$E_{\mathrm{d}} = \frac{10 \sum_{i=1}^{n} C_i}{B v_{\mathrm{w}} \tau_{\mathrm{mov}} \tau_{\mathrm{cyc}}},$$

Table 21

Model, machine	Law of distribution	m, min	σ	A	k	v	Probability	Number of experiments
Beet-harvesting combine SKD-2	Normal logarithmic	7.01	11.57	9.43	330.54	1.64	0.539	154
-do- SKD-3	-do-	5.54	8.83	8.84	278.30	1.59	0.662	140
-do- KST-3	-do-	10.96	150.09	2605.0	0.27	13.68	0.660	157
Root-harvesting machine 'Parma' (USA)	-do-	4.71	74.50	19.48	305.35	15.67	0.998	146
Beet-seeding machine 2STSN-6A	-do-	5.74	9.40	9.28	316.72	1.63	0.543	150

Table 22

Aggregate content	Law of distribution	M	σ	A	k	v	Probability of accordance	Number of experiments
Beet-harvesting combine	Logarithmic-normal	0.27	0.39	7.15	161.01	1.42	0.429	150
SKD-2 with tractor T-54S	Normal	0.27	0.041	−0.29	−0.88	15.14	0.001	150
Beet-harvesting combine	Logarithmic-normal	0.35	0.51	7.17	161.86	1.42	0.934	310
SKD-3 with tractor DT-75	Normal	0.33	0.054	−0.98	1.21	15.31	0.001	310
Root-harvesting machine UKS-3	Normal	0.46	0.12	−0.71	−0.41	26.71	0.001	250
Grain-harvesting machine SK-4	Logarithmic-normal	1.22	1.92	8.59	258.00	1.57	0.628	100
Seeding machine 2STSN-6A	Logarithmic-normal	1.41	2.04	7.38	174.69	1.45	0.220	320
(sowing)	Normal	1.43	0.37	0.048	0.67	25.85	0.415	320
Seeding machine—cultivator	Logarithmic-normal	1.81	2.60	7.32	170.60	1.44	0.002	260
2STSN-6N (feeding)	Normal	1.46	0.26	−0.16	−0.72	18.33	0.12	260
Cultivator KRN-4.2 (cultivation)	Logarithmic-normal	2.98	4.31	7.38	174.24	1.44	0.466	300
	Normal	2.42	0.50	−0.40	−0.38	20.70	0.676	300

where B is the claw width of the aggregate, m; v_w—the working speed km/h; τ_{mov}—the coefficient of utilization of time of movement in working cycles, τ_{cyc}—the coefficient of the utilzation of the cyclic time of the shift; W_{cyc}—the productivity of working cycles; $\sum_{i=1}^{n} C_i$—the elements of deductions and expenses on unit productivity, generally a linear function of W and E_d—the direct expenses.

Generally asymptotic values W_{cyc} and τ_{mov} are considered

$$\tau_{mov} = \frac{T_w}{T_w + T_i},$$

where T_w is the cumulative working time (movement) and T_i—the cumulative idle time (nonproductive time expenses).

It is extremely convenient to express τ_{mov} in terms of specific values of time elements of the flow calculated for the unit path traversed:

$$\tau_{mov} = \frac{1}{1 + v_w\left(\frac{t_t}{L_2} + t_{T(L)} + t_{y(L)}\right)} = \frac{1}{1 + v_w \sum_{i=1}^{n} t_{i(L)}},$$

where $t_{T(L)}$, $t_{y(L)}$ are the specific elements of time per unit path (h/km), which are spent on technological and technical servicing as a function of the aggregate's parameters; t_t—the time consumed in turning $\sum_{i=1}^{n} t_{i(L)}$—the cumulative nonproductive time.

Then the expression:

$$W_{cyc} = 0.1B \frac{v_w}{1 + v_w \sum_{i=1}^{n} t_{i(L)}}$$

will represent a random variate with a mathematical expectation given by:

$$M_{W_{cyc}} = 0.1B \int_{v_1}^{v_2} \int_{t(L)_1}^{t(L)_2} \frac{v_w}{1 + v_w \sum_{i=1}^{n} t_{i(L)}} P(v_w, t_L) dv\, dt_{(L)}$$

and the dispersion by:

$$D_{W_{cyc}} = \int_{v_1}^{v_2} \int_{t(L)_1}^{t(L)_2} \left[0.1B \frac{v_w}{1 + v_w \sum_{i=1}^{n} t_{i(L)}} - M_{W_{cyc}} \right] P(v_w, t_{(L)})\, dv\, dt,$$

where $P(v_w, t_{(L)})$ is the combined distribution function.

The corresponding probability models can be developed on the basis of static characteristics of productivity with the help of power relationship [60, 72]:

$$W = 0.1 \frac{N_e}{\sum N_i} \eta_e v_w \tau_{sh} = \frac{0.1 N_e v_w \tau_{sh}}{N_{T_{vB}} + N_{C_{vB}}},$$

where $\sum N_i$ is the specific power consumed for the unit claw width; η_e—the coefficient of loading of the engine; $N_{T_{vB}}$, $N_{C_{vB}}$—the power consumed in the speed function and the claw width on the traction of the machine and the self-movement of the engine.

Components of a power budget are generally given in tabular form and are obtained by generalizing data collected over many years or with the help of calculated values.

The general scheme of agricultural-industrial process, considering the already described ideas, is presented in Fig. 59. Generalization and study experience provide the basis for concluding that the Monte Carlo statistical test method is most important for mathematical modeling as it permits the use of random variates of a different kind.

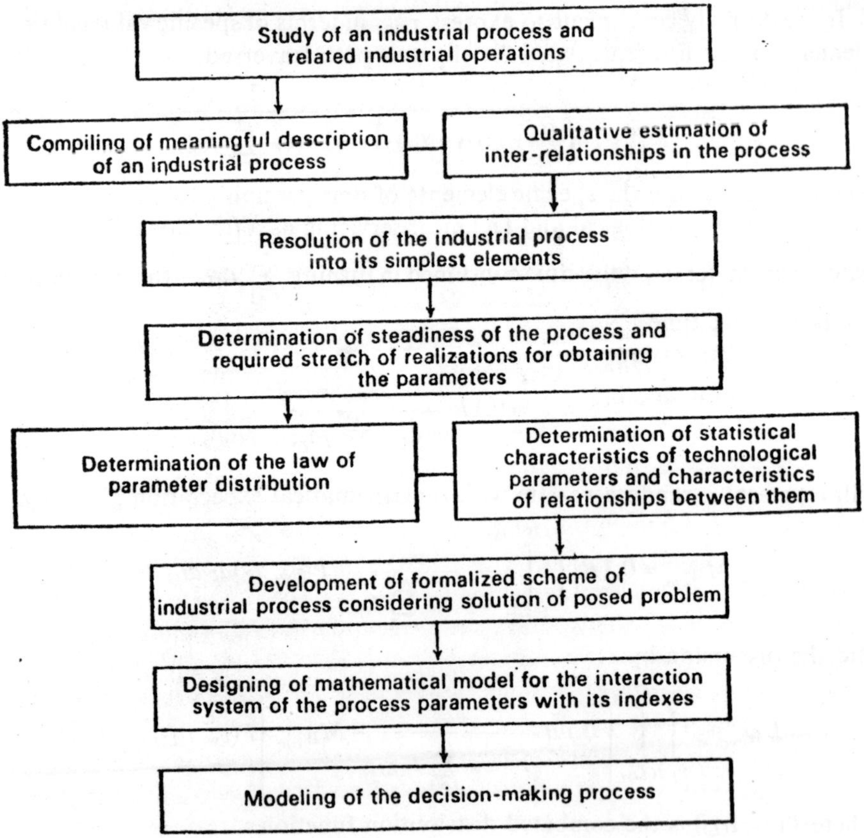

Fig. 59. General scheme for formalization of industrial processes.

Modeling and forecasting of operational indexes. As experience shows in the practice of testing, it is worthwhile to use models which by methods of statistical analysis on a computer, can predict the productivity and economic indicators of sowing and harvesting aggregates as functions of their parameters, regimes and operating conditions [39, 59].

In these cases, statistical models in the form of analytical expressions related to the indexes of effectiveness are preferably used. Models of a polynomial type are seldom used.

Well-developed modeling algorithms for sowing and harvesting aggregates are designed while considering the relationships which reflect the like between productivity W and reduced (direct) expenses of funds U and the following characteristics of the aggregate:

(1) B is the claw width in meters; k—the number of plowshares; v—the movement speed in m/s; k_r, k_{rtp}, k_t—the coefficients of availability and the reliability of the technological process and technical maintenance; α_7—the time norm on the technological servicing of a tractor; v_{tr}—the transportation speed, km/h; and Q_1, Q_2—the capacity of bunkers for seeds and fertilizers, kg, respectively;

(2) for technological and industrial processes: g_1, g_2 is the sowing norm (seeds/m) and introduction of fertilizers (kg/ha); μ—the mass of 1000 seeds and t_t, t_s, t_f—the time for the turning and loading of seeds and fertilizers, s, respectively; and

(3) for external conditions: L—the length of the traverse, m; T—the duration of the working shift, h; S—the field area, ha; and l_x—the distance of transport passage, km.

In accordance with the methodological assumptions of state standards, direct expenses depend on productivity (shift and operational):

$$U=\frac{T_1}{W_{sh}}+\frac{T_2}{W_{op}}+C,$$

where $T=\sum_{i=1}^{n} C_i$—the sum of expenses and C—the cost of fuel and other inputs which is calculated, as a rule, on unit turnout.

The algorithm of modeling takes into account a situation when the capacity of bunker is not sufficient for a run to the other end of the field at a given sowing norm (crop capacity), and it is necessary to accomplish the loading of seeds and fertilizers (unloading of crop) on two sides of the field (or along the traverse). In this case, coefficients T_1 and T_2, which take into account the labor expenditure and the loading (unloading) means of technological material from two sides are used in the modeling formula.

Using the statistical characteristics of operational coefficients, which include the claw width, the movement speed, the yield capacity of the technological bunkers and transportation means, etc. it is possible to reliably determine the corresponding arbitrary or absolute characteristics for general-zonal or specific conditions by the statistical test method.

Some very interesting examples will be considered.

For the *modeling of the productivity of sowing aggregates*, the following relationships are used, having been obtained on the basis of the formalization of the operational sequence, accomplished during sowing and which takes into account the methodological specifications on the determination of operational and economic indexes provided in the state standards:

operational productivity:

$$W_{op}=\frac{0.36Bv}{\sum_{i=0}^{3}\alpha_i+k_{9\cdot 2}+\frac{1-k_r}{k_r}},$$

shift productivity:

$$W_{sh}=\frac{0.36Bv}{\sum_{i=0}^{3}\alpha_i+k_{9.2}},$$

where

$$\alpha_0=\frac{0.36Bv}{W_0};\qquad \alpha_1=\frac{1-k_{rtp}}{k_{rtp}};$$

$$\alpha_2=\frac{1-k_r}{k_r};\qquad \alpha_3=k_7+k_{21}+\frac{k_8}{60};$$

$$k_{9.2}=\frac{0.72LBv}{T_{du}v_{tr}W_0};\ W_{0l}=\frac{0.36\,Bv}{\sum_{i=0}^{3}\alpha_i};$$

$$k_{9.1}=\frac{l_x\,0.36Bv}{v_{tr}S};\qquad k_7=\left(1+t_t\frac{v}{L}\right)\alpha_7;$$

$k_{9\cdot 2}=\frac{0.72LBv}{T_{du}}$ is the technological productivity, which is determined in terms of correlation between the number of traverses for a single dressing by seeds

$$n_1=\frac{1000Q_1}{q_1k\mu L}\ \text{and fertilizers}\ n_2=\frac{10000Q_2}{q_2LB};$$

$$\Delta t=\frac{t_s}{n_1}+\frac{t_f}{n_2},\ \text{if}\ \left.\begin{matrix}n_1\geqslant 2\\ n_2=1\end{matrix}\right\}\ \text{or}\ \left.\begin{matrix}n_1=1\\ n_2\geqslant 2\end{matrix}\right\};$$

$$\Delta t=\frac{t_s}{n_1}+\frac{t_f}{n_1},\qquad \text{if}\ n_1\geqslant 2,\ n_2\geqslant 2,\ n_1\leqslant n_2;$$

$$\Delta t=\frac{t_s}{n_1}+\frac{t_f}{n_2},\qquad \text{if}\ n_1\geqslant n_2,\ n_2\geqslant 2,\ n_1>n_2.$$

The weightage of the field area in estimation of the mean productivity of the aggregate, considering the condition of the zone, is evaluated from the coefficient $k=(10000S)/BL$, which characterizes the number of traverses for the processing of the field with area S.

Figure 60 presents the flow chart of the program which realizes the reduced model for the case, when characteristics t_s, t_t, t_f, v are considered as

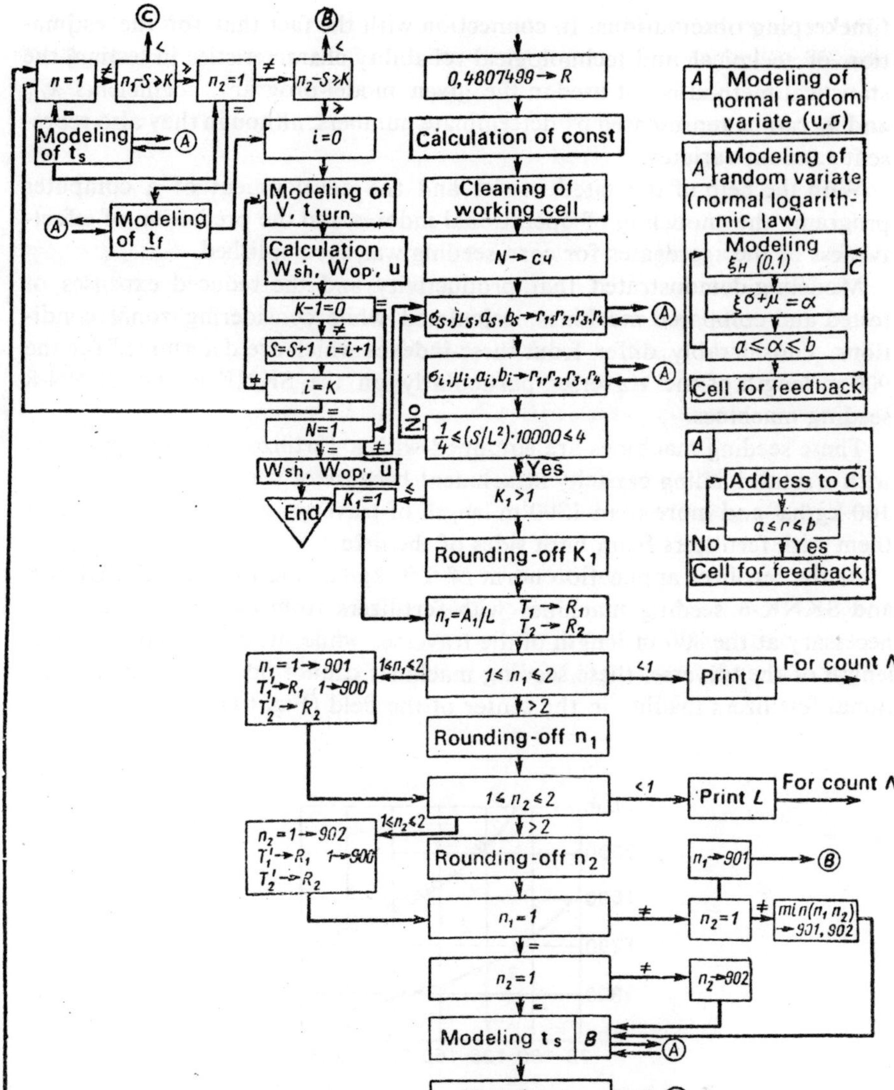

Fig. 60. Flow chart of the program for operational forecast and the economic indexes of sowing machines considering zonal conditions.

random variates, distributed with respect to truncated normal laws, and S and L—to a normal logarithmic law. The parameters of the laws of distribution and characteristics of the aggregate are evaluated from the results of timekeeping observations. In connection with the fact that for the estimation of technical and technological reliability characteristics in testing, the statistical method is not used in the given model-program; coefficients k_{rtp} and k_r can be represented by determinate numbers, although they also represent random variates.

With the help of the cited model and the development of a computer program, the modeling of operational indexes and the prediction of effectiveness of the aggregates for corn seeding was accomplished.

Modeling demonstrated that productivity and the reduced expenses of tested and compared machines, calculated while considering zonal conditions, considerably differ from these indexes which are determined for the 900 m length of the traverse, particularly on the SKNK-6 and SUPN-8 seeding machines.

These seeding machines are equipped with a fertilizer-sowing apparatus with a small loading capacity attachment because of which, at the norm of 100 kg/ha and more than 1300 m length of traverse, it is necessary to load them with fertilizers from both sides of the field.

At the fertilizer application norm of 150 kg/ha, the loading of SUPN-8 and SKNK-6 seeding machines with fertilizers from both sides is already necessary at the 800 m length of the traverse, while at more than 1700 m length of the traverse, these seeding machines cannot operate without additional fertilizers loading in the center of the field (Fig. 61).

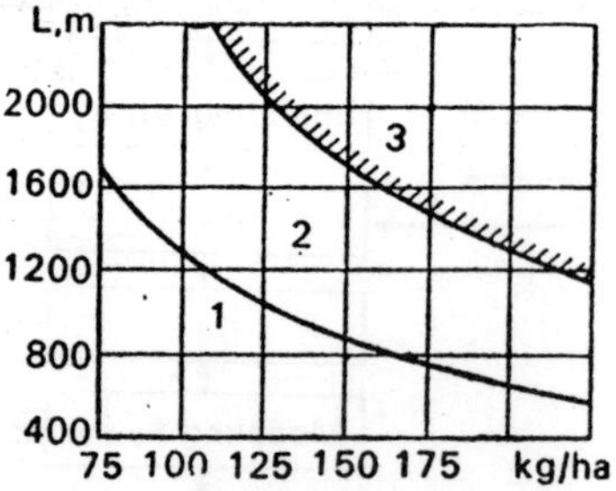

Fig. 61. Prediction of limitations caused by the bunker capacity of SKNK-6 and SUPN-8 seeding machines as a function of the length of the traverse and fertilizer application norms:
1—loading zone from one side of the field; 2—loading zone from both sides; 3—zone of additional loading in the central part of the field.

In this connection, seeding machines equipped with a fertilizer-seeding apparatus with an increased bunker loading capacity do not have limitations on the length of the traverse. They demand less time for fertilizers loading in the overall shift time.

In this way, the prediction of effectiveness of the tested machines showed that a 6-row seeding machine with an increased bunker capacity has higher technoeconomic indexes in the forest-steppe zone conditions as compared to the SUPM-8 8-row seeding machine. The 6-8 row seeding machine is also inferior to the 2SUPP-6 12-row seeding machine in terms of productivity, but excels it in deduced costs.

The problem of estimating the beet seeding-machines parameters was solved by the statistical testing method on the basis of the engine's model. The coefficient of the utilization of the shift time was formed by a special subprogram, considering the statistical data of the dimensional characteristics of fields and elements of shift time structure.

Modeling results of the productivity for estimating beet-seeding machines of different claw widths show that the nature of change of the mathematical expectation of productivity as a function of parameters of aggregate (Fig. 62) depends considerably on the correlation of the basic parameters of the aggregate. At a lower degree of engine loading for tractors (50-70%), the productivity of the aggregates monotonically increases—more intensely on aggregates with a large claw width. With the increase of the degree of engine loading (more than 85-90%), the nonuniformity of the speed regime and the variation of the cyclic coefficient of the utilization of shift time begins to influence the mathematical expectation of productivity more noticeably.

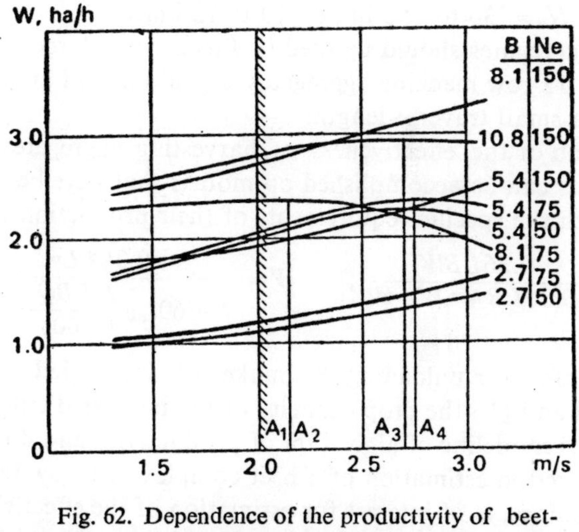

Fig. 62. Dependence of the productivity of beet-harvesting machines on motion speed, claw width and engine power.

This involves the appearance of a bend in the flow of the corresponding curves of productivity and their gradual downward slope.

The obtained results have made it possible to draw the important conclusion that in the range of an agrotechnically allowable time speed (up to 6.6-7.2 km/h), at present aggregates with a large claw width possess a considerable advantage. The optimum aggregate parameter for these conditions $B = 8.1$ m (18-row); $N_e = 80$ hp, $v = 7.2$-8 km/h. However, by further increasing the working speeds, the advantage of an 18-row aggregate on the productivity is preserved only up to speeds of 8.5-9.5 km/h, and after that the productivity advantages belong to a machines with a claw width 5.4 m (12-row), for which the optimum working speed is 9.5-10 km/h. The same advantage at the power saturation of 80-90 hp tractors is preserved up to speeds of 10-12 km/h.

A further rise in the aggregate productivity can be achieved only due to an increase of engine power up to 130-150 hp (see Fig. 62). Here, the optimum claw width is 8.1 m, working speeds up to 12 km/h are allowed and only agrotechnical criteria are to be limited because with the speed increases, the sowing quality considerably decreases—especially the distribution of seeds in a row.

Aggregates with a claw width of 10.8 m (24-row), because of the increase of the technological idle time, do not have a considerable advantage as compared to the 18-row seeding machines—even at extremely high power saturation (150 hp); and for speeds greater than 8 km/h, they are inferior to the 18-row seeding machines.

All the foregoing arguments only confirm the expediency of combining aggregates of two basic types: $B = 5.4$ m, $N_e = 80$ to 90 hp, $v = 8$ to 10 km/h, and $B = 8.1$ m, $N_e = 130$ to 150 hp, $v = 10$ to 12 km/h.

Wide claw aggregates should be used in farms with large sowing areas, while standard 12-row machine aggregates should be used in small farms or on fields with a small traverse length.

The prediction of the effectiveness of harvesting aggregates considering zonal conditions can be accomplished on models that have been modified in accordance with the specific requirements of their production process:

$$W_0 = \frac{0.36LBW_1}{0.36LB + t_t W_1 \cdot 60}; \qquad W = \frac{0.36BLv}{L + 60t_u v\left(\dfrac{LBQ}{100q} - 1\right)},$$

where t_u is the time for unloading the bunker of size q filled with the harvested material and Q—the crop capacity of the harvested culture, Qtl/ha.

The statistical modeling of algorithm of productivity and direct expenses, which was realized in estimation of a beet complex of 6-row beet and root-harvesting machines, is of interest for estimation of the effectiveness of connected working machines and prediction of the best variants of their use.

The essence of the algorithm lies in the fact that the productivity of the root and beet-harvesting machines is modeled first either with the use of models or according to the law that sufficiently reliably describes the real distributions of the actual values of productivity of each aggregate separately. Then, depending on the variants of the beet and root-harvesting machines, the corresponding productivity of the complex is selected: in dependent work, it is less; in the work of the beet-harvesting machine independent of root-harvesting—the equal productivity of root or beet-harvesting machine taking into account the cumulative positive difference $W_{1i} - W_{2i}$.

From the results of the mathematical modeling of the productivity indexes, it is established (Table 23) that productivity characteristics, averaged over a wide range of conditions that are characteristic for the forest-steppe of the Ukrainian Republic, undergo a considerable variation, i.e. the dispersion of productivity reaches values of 0.15-0.20 (ha/h)^2. The main contribution in the productivity dispersion is brought about by variation of the length of the traverse and yield, and also by the dispersion of the movement speed [58].

Table 23

Type of machine and mechanized processes	$M v_{m1}$, m/s	σv_{m1}, m/s	MW_{Ω}, ha/h	$\sigma_{W\Omega}$, ha/h
One-phase harvesting				
3-row beet combines	2.0	0.04	0.58	0.12
Two-phase harvesting				
Beet-harvesting machine:				
6-row	1.8	0.30	0.95	0.22
12-row	1.9	0.35	1.80	0.48
Root-harvesting machine:				
3-row (sequential)	2.1	0.40	0.6	0.10
3-row with bunker-accumulator	2.2	0.45	0.65	0.11
6-row	2.0	0.40	1.20	0.24
Three-phase harvesting				
Root-harvesting machine:				
6-row	2.5	0.50	1.34	0.23
9-row	2.0	0.35	1.71	0.30
12-row	1.9	0.35	2.20	0.43

Comparison of the obtained statistical characteristics of operational-economic indexes with their calculated magnitude on the mean values of factors shows (Table 24) that by taking account of these working peculiarities, it is possible to obtain a more objective estimation of the indexes of effectiveness. The indexes undergo a shift in their mean values and a high dispersion, caused by the nonlinear nature of the dependence of productivity

Table 24

MW, ha/h	σW, ha/h	Calculation of mean values U_{av}, rubles/ha	MU_r, rubles/ha	σU_r, rubles/ha	Absolute value $MU_r - U_{av}$, rubles/ha
0.50	0.09	62.94	64.65	10.72	1.71
0.55	0.33	62.94	76.89	53.41	13.95
0.66	0.36	62.94	81.56	83.83	19.62
1.00	0.09	38.01	38.79	2.52	0.78
1.10	0.18	38.01	39.28	5.33	1.27
1.04	0.37	38.01	45.09	27.60	7.07
1.50	0.09	29.71	30.46	0.98	0.75
1.50	0.18	28.71	35.36	1.25	5.65

on the movement speed and the operational characteristics of the working cycles (whereas reduced expenses experience the same on productivity) and also by the considerable variation of the zonal conditions and the difference in the distribution laws of many factors from the normal distribution.

It is for this reason that in real conditions, the productivity decrease of machines in field-crop cultivation reaches 15-20%, and the mathematical expectations of labor expenses and reduced expenses for similar reasons get shifted to the increasing side by 5-20%, particularly with the rise of the productivity dispersion.

If the BM-6 top-harvesting machine could work independently of the KS-6 root-harvesting machine, then for the obtained values of shift productivity, the reduced expenses would have been 71.59 as against 84.55 rubles/ha.

Calculations have shown that the sum of reduced expenses on separate beet harvesting can be reduced by firstly increasing the steadiness of the working of the beet-harvesting machine.

If the mean productivity of the BM-8 remains equal to 0.68 ha/h, but the mean-square deviation is decreased to 0.15 ha/h (coefficient of variation 22%), then the sum of reduced expenses would also decrease from 84.55 to 76 rubles/ha.

At the mean value of productivity of BM-6 equal to 0.95 ha/h and a variation coefficient of 20%, the sum of reduced expenses decreases from 84.55 to 67.58 rubles/ha. If the productivity of KS-6 is higher than the stable value (for average value of 0.93 ha/h and the coefficient of variation 19.3%), then the sum of reduced expenses can be decreased to 49.57 rubles/ha.

In Fig. 63, a, b, the dependence of productivity for one working hour of BM-6 and KS-6 machine on the movement speed, the length of traverse and the crop capacity is shown. The curves show that the BM-6 beet top-harvesting machine is inferior to the KS-6 root-harvesting machine due to its

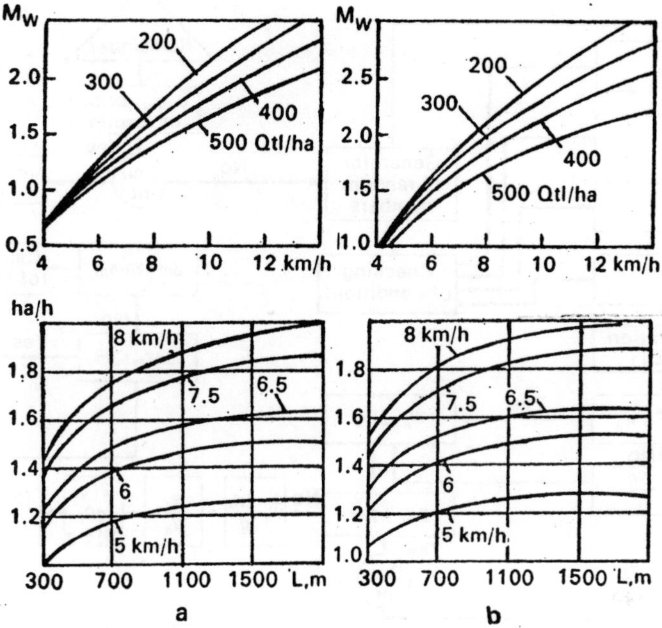

Fig. 63. Dependence of predicted productivity for one working hour in the strip, of (a) BM-6 and (b) KS-6 beet-harvesting machines on movement speed length of traverse and beet-yield.

trailer replacement halts on the above-indicated index. In the working of the complex on sections with length of traverse of less than 900 m, a sharp decrease in productivity for one working hour in the strip is observed.

The effectiveness of the line-harvesting of an agricultural crop, depending on the provision of autotransport, the distance to delivery points, and the crop capacity is predicted by the statistical test method with the use of a model, which includes an algorithm for modeling the productivity of the harvesting aggregate, depending on the quantity of autotransport and the time of the cycle trip as well as on the combined harvesting expenses and use of autotransport (Fig. 64). The modeling sequence and calculation procedures in solving the problem are given below.

Random values are modeled on the selected distribution laws: $W = QW_c$ is the productivity of the aggregate; Q—the crop capacity of the harvested crop; v_t—the speed of the autotransport; and t_u—the time for the unloading of the transport at the delivery point.

Then the cycle trip time is determined as follows:

$$T_{ij} = \frac{g}{W_{ij}} + \frac{L_t}{v_{tij}} + \frac{L_t}{v_{t(i+1)j}} + t_{aij},$$

where g is the carrying capacity of the transport and L_t—the distance to the load's delivery place.

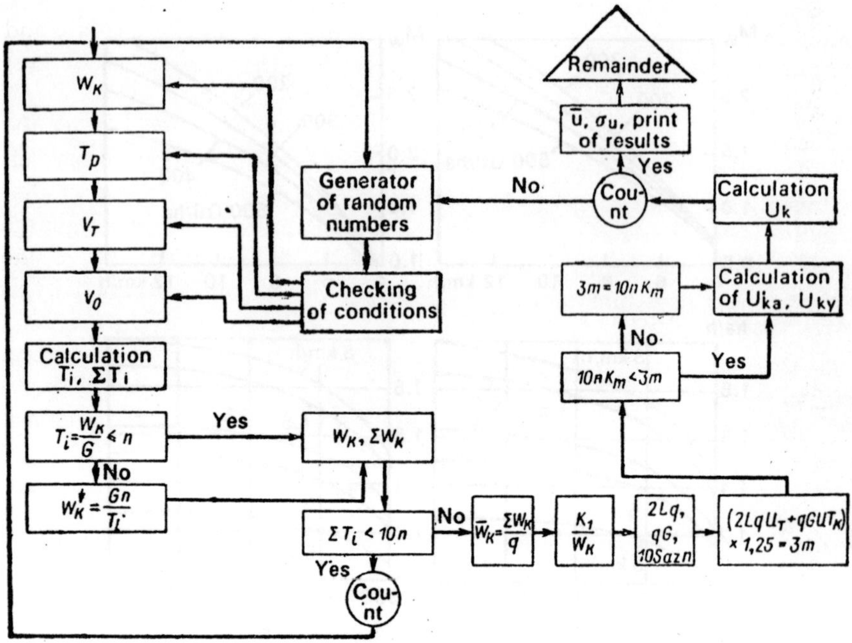

Fig. 64. Flow chart for transport problem modeling.

If $T_{ij}\,\dfrac{W_{ij}}{g} > n$ (insufficient automobiles), then the productivity of aggregate $W_{ij} = g_n/T_{ij}$.

When condition
$$\sum_{j=1}^{k} T_{ij} \geqslant T_{du}n$$
(T_{du} being the shift duration) is fulfilled, calculations are made for the mean productivity of the aggregate $\overline{W} = \sum_{j=1}^{k} W_{ij}/k$; direct expenses $U_{1i} = C/W_j$; and expenses on autotransport

$$U_{2i} = \frac{2L_t k\xi_l - T_{du}\xi_a n + k(g\alpha_1 + L\alpha_2)}{\sum W_{ij}},$$

where ξ_l, ξ_a are the expenses for one kilometer of traverse and on one working hour respectively; α_1, α_2—the coefficients which take account of expenses for one trip; and $U_i = U_{1i} + U_{2i}$—combined expenses.

In accordance with the statistical test methods, the listed operations are performed a sufficiently large number of times, after which the mean value and dispersion of the total expenses are calculated from formulas of mathematical statistics.

The prediction of the effectiveness of a complex for separate beet harvesting with the use of described models permitted the more reliable estimation of its parameters, the identification reserves for productivity increase and the provision of sound recommendations for further improvement of beet top and root-harvesting aggregates.

The sum of reduced expenses obtained as a result of prediction on the complex as a whole, differs by 13% from the sum of reduced expenses, calculated by the actual method. This difference is explained by the fact that the shift productivity of the aggregates is interconnected and varies in large limits (in calculation this is not considered by the method), while the dependence of the sum of reduced expenses on the shift productivity is nonlinear.

Dependence of deduced costs (rubles/m) in the root harvesting of beet by BM-6 and KS-6 machines on the quantity of automobiles with the carrying-capacity of three tons for distances of transportation 6 and 20 km is shown in Fig. 65.

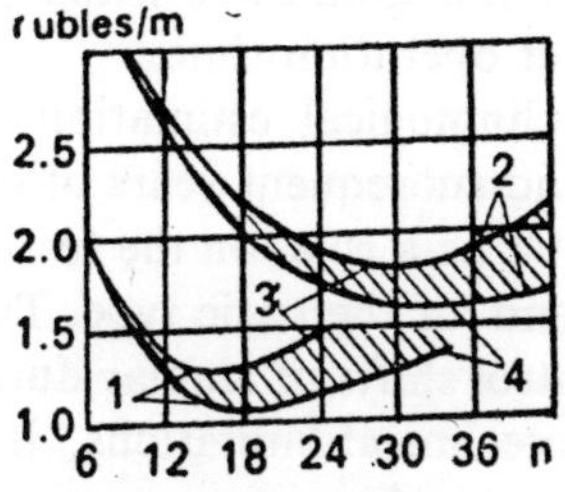

Fig. 65. Forecast of deduced costs on harvesting with a KS-6 root-harvesting machine and transportation of beet-roots at a crop capacity of 350 Qtl/ha by an automobile:
1—L_t=6 km; 2—L_t=20 km; 3, 4—movement speed of KS-6 machine 5.5 and 7.5 km/h respectively.

As follows from the graph, at the movement speed of a KS-6 machine equal to 5.5 km/h (actual working speed on farm test results) and a yield of 350 Qtl/ha, the minimum of deduced costs are provided by 15 automobiles for a distance of 6 km and by 27 automobiles for a distance of 20 km. With an increase in the movement speed of the root-harvesting machine up to 7.5 km, deduced costs are further reduced using the previous assigned number of automobiles, while the minimum is achieved for 18 (distance 6 km) and 32 automobiles (distance 20 km).

Both examples of modeling and the prediction of operational and technological indexes of the working of machines could be continued; however, the number of examples is less important than the principal understanding of the procedure for model design and algorithm realization on a computer .

30. TECHNOLOGICAL EFFECTIVENESS IN OPERATIONAL CONDITIONS

The indexes of agrotechnological effectiveness, determined generally in certain idealized laboratory conditions, characterize the comparative potentialities of the machines under test. In an ordinary operation, which the operational (farm) testing approaches, the agrotechnological indexes change (decrease) to a known degree and generally depend on external operational conditions as well as on the variation of operational indexes. Simultaneous determination of operational and agrotechnological indexes appears as the basis for the technological estimation of machines.

All the controlled shifts on a particular agrotechnical background, in which machines had worked in strictly comparable conditions, are picked out from the photochronometration data for the determination of operational and technological indexes. It is accepted that the number of control shifts for this purpose should not be less than three with a general duration, as a rule, of not less than 15 h of net operation time.

The operational and technological estimation of production samples of machines of the second and subsequent years of output (which do not serve as an etalon), is carried out, as a rule, on the test results in the zone characteristic for the background of the basic type. This testing volume consists of not less than three control shifts of similar duration.

Before formalizing agrotechnical limitations, it is necessary to establish the background parameters and operation regimes on which the indexes of effectiveness depend. This can be demonstrated by the example of grain and beet-harvesting combine.

While testing combine grain-harvester, the harvester as well as thresher losses are individually accounted for. Generally, the former depend mainly on the movement speed, whereas the latter on the feed of the grain-mass into the threshing machine. These arguments of the function of losses can be reduced to productivity for one hour of network of the combine.

The above-mentioned relationships for harvesters and thresher grain losses are different. The dependence of harvester grain losses on productivity is satisfactorily represented by a linear function [12] $X_{\text{hl}} = aW_2 + b$.

The dependence of the thresher grain losses on productivity is satisfactorily approximated by the exponential curve of the type $X_{\text{tl}} + ae^{BW}$.

In both equations, *a* and *b* are the coefficients to be determined (generally on a computer) from the results of experimental data.

A great deal of attention is given to the determination of the thresher losses in the testing of grain-harvesting combine. Vast data, accumulated over many years of testing of different designs of combines in different zones of the country, are used to obtain these relationships. These relationships on the harvesting of winter wheat for different threshers of a combine grain-harvester according to the data of KUBNIITiMa are represented in Fig. 66.

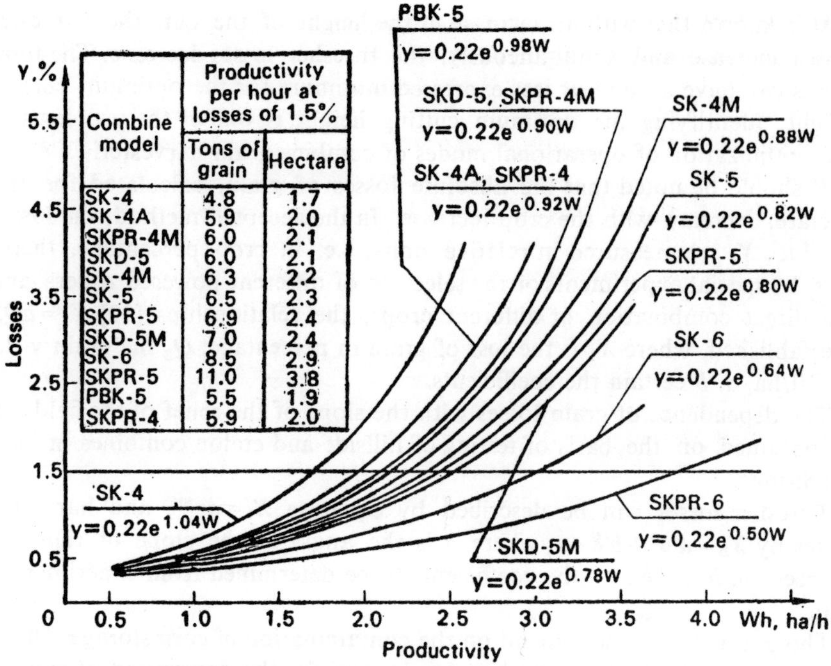

Fig. 66. Thresher losses for the different combines on the selection of swaths of winter grain 'Bezostaya-1' with a crop capacity of 28.7 Qtl/ha.

Depending on the harvesting duration, the harvesting machine losses can be described by a polynomial of the second order (Fig. 67):

$$X_1 = an^2 + bn + c.$$

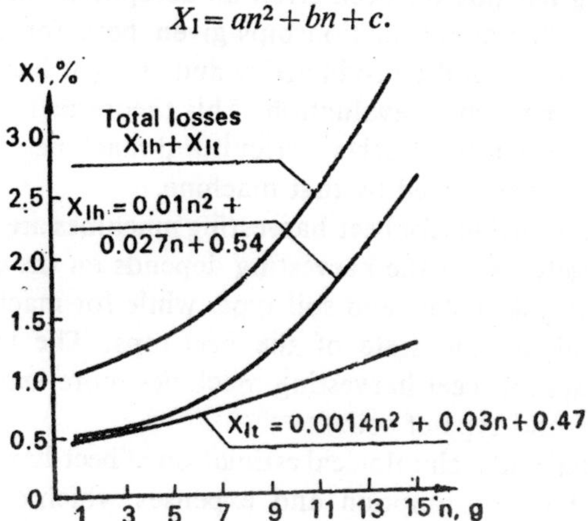

Fig. 67. Grain losses for combine SK-5 as a function of the duration of harvesting of winter wheat 'Bezostaya-1' (Q=50 Qtl/ha, W_4=1.6 ha/h, h_{av}=0.2 h; X_{hl}—harvester losses; X_{tl}—thresher losses).

It is known that with an increase in the height of the cut, the harvester losses increase and simultaneously, the thresher losses decrease. The total losses can have a more or less expressed minimum for the optimum cutting height. Identifying the optimum cutting height is one of the problems of the optimization of operational modes of combine grain-harvester.

It should be noted that the absolute losses of grain, calculated for one hectare, increase with the crop increase. In the accepted method, the losses produce X_1 are measured in relative units, i.e. in crop percentage, therefore, by special experiments on the selection of different powered rollers and the direct combination of different crops, the relationships [12] $X_1 = aQ_g^b$ is established, where X_1 is the loss of grain in percentage; Q_g the grain yield in Qtl/ha, a, b certain the coefficients.

The dependence of grain losses with the slope of the relief of the field can be obtained on the basis of testing of hillside and etalon combines in hilly conditions.

Thresher losses can be described by equation $X_t = ae^{Bb}$, and harvester losses by $X_h = a'\delta^2 + b'\delta + c$, where δ is the angle of the slope of relief in degrees, a, b, a', b', c—the coefficients to be determined from experimental data.

The grain losses also depend on the contamination of corn storage. Often the polynomial of the second order is used for the description of experimental data:

for thresher losses, $X_t = a\mu^2 + b\mu + c$, for harvester losses $X_h = a'\mu + b'$, where μ is the contamination of corn storage in terms of mass, %.

Integrated function of losses for harvesting machines for the whole duration of harvesting has not yet been given an acceptable form.

The various mathematical relationships given here for the collection of produce (grain loss) with the productivity and state of backgrounds are the theoretical basis for machine evaluation. This theoretical principle can also be used for the evaluation of other agricultural machines, independently of the operation to be performed by that machine.

The various relationships for beet-harvesting machines are shown in Fig. 68. In this, the completeness of the harvesting depends on the crop capacity as well as on the physical state and soil type, while for machines of the pull-type it also depends on the state of the beet-tops. The technological and operational indexes of beet-harvesting machines more explicitly depend on absolute moisture and type of soil.

In the operational and technological estimation of beet-harvesting machines, the additional labor consumption and expenses which occur due to considerable deviations of agrotechnological indexes of mechanized processes from agro-requirements [58] and established standards (depending on the cutting quality) or the cluttering of the heap by beet tops and soil (Z_{T_1} man-hour/ha), are $MZ_{T\Omega 1} = -0.41(P_1 + P_2)_\Omega^2 - 272.40(P_1 + P_2)_\Omega + 337.16$,

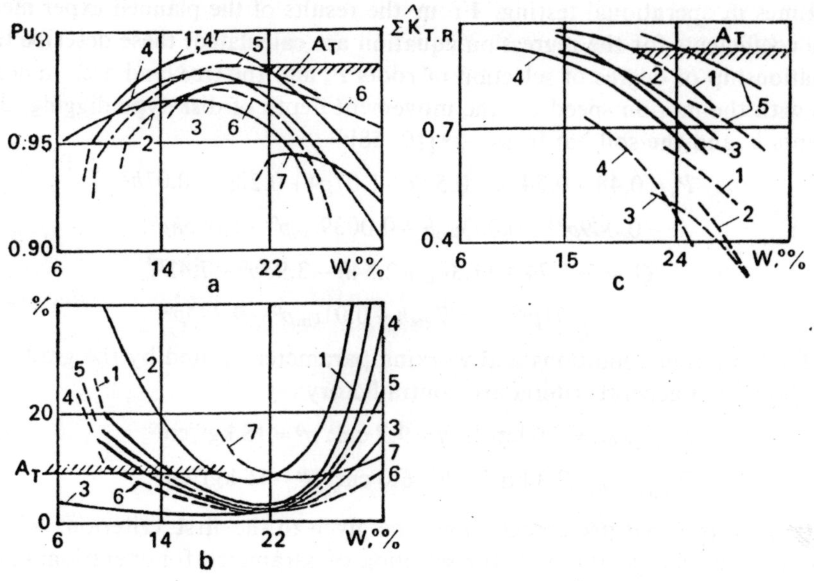

Fig. 68. Relationships of completeness of (a) harvesting, (b) clogging of heap by soil and (c) coefficient of technological reliability of machines on absolute soil moisture for different working parts and designs of beet-harvesting machines: a, b—1—disk-type digging elements of 2-3 row combines with an automated control; 2—share-type digging elements; 3—pulling apparatus at different speeds of changing and a frontal aggregate scheme; 4—disk-type digging elements of 6-row root-harvesting machines with manual and automated (4′) controls; 5—rotor-fork type digging elements of 6-row root-harvesting machine with manual control; 6—rotor-type digging elements of firm 'Vikon'; 7—digging elements of the oscillating-stroking type made by Volvo company; c—1—KS-6 root-harvesting machine; 2—beet-root combines of the SKD-2 type; 3—RKS-6 root-harvesting machine; 4—KST-3A beet combine; 5—single-pass root-digger of the Jean Moreau Company (France) on medium loamy soil—continuous line or hard-stroke line; A_t—agrotechnical limitations.

depending on the value of recurring losses (Z_{T_2})

$$MZ_{T\Omega_2} = 6.25 - (1 - P_{u\Omega})^2 - 12.50(1 - P_{u\Omega}) + 1.31.$$

Similar losses of manual labor on the elimination of technological rejects and the reductions of the cost of the harvested crop are estimated, depending on the value of nonrecurrent losses and the beet wastage at the lower cutting.

Accumulation and generalization of data about the relationships of technological and operational indexes of different machines, in perspective permits to accomplish the modeling of technological indexes. Here, it is not possible to go about without the application of designed experimental methods, which is especially, necessary for the determination of optimum working regimes in operational testing. From the results of the planned experiment the coefficients for the regression equation are calculated; these describe the relationship of degree of selection of roots P_s and the free land area in heap Q_3 with the motion speed v_m, the movement depth of disk-type digging elements h, and the soil hardness—p^0 [10, 58]:

$$P_s = 0.48 - 0.34v_m + 0.52h + 1.07p^0 + 0.22v^2 + 0.07h^2$$
$$- 0.049p^{02} + 0.053v_m h + 0.0039v_m p^0 - 1.095hp^0;$$
$$Q_3 = -1.74 + 11.3v_m + 2.54h - 3.95p^0 - 7.42v_m^2$$
$$+ 2.71p^{02} + 1.87v_m h - 0.01v_m p^0 - 9.77hp^0.$$

The optimum conditions and working parameters found by the gradient method from general criteria are contradictory:

$$Ps_{max}: v_m = 7.0 \text{ km/h}; h = 9.2 \text{ cm}; p^0 = 18 \text{ kgf/cm}^2;$$
$$Q3_{min}: v_m = 7.3 \text{ km/h}; h = 6.0 \text{ cm}; p^0 = 16 \text{ kgf/cm}^2.$$

It is natural that preference should be given to the first criterion, which should also serve as the basis for selection of parameters for operational and technological estimation.

The determination of coefficients of the technological efficiency of the working processes has considerable importance for the estimation of the operational possibilities and the agrotechnological indexes in different conditions. Unfortunately, this requires conducting tedious and labor-consuming experiments.

The generalized results which are obtained on the basis of results of many years of indirect experiments (see Fig. 68), permit the forecast of indexes of technological and operational effectiveness for different conditions, including vast regions of a given zone.

Such data should be accumulated and, on its basis a standard card index should be created for the indexes of operational-technological effectiveness of agricultural machines and aggregates.

Hence, the important problems for the development of operational and technical estimation of machines lie in the following:

1. Establishment of a interrelationship between the values of factors which determine external conditions and of variables which characterize the working quality of machines. On this basis, standard recommendations for the rational selection of indexes which characterize the working conditions of machines is developed.
2. Automation of chronometric observations with data processing from timekeeping on the testing day.
3. Creation of equipment for obtaining first hand information and systems for transformation and processing. New trends in this direction are characterized by the use of systems which provide the accumulation of information

on carriers. This is useful as computer input. The systems are also useful for retrieval of information on the day of testing and at the particular testing place.

Systems AIIS-EI and 'Urozhai' can be considered as examples of systems which provide operative transformation and processing of the experimental data.

System AIIS-EI permits realization of the transformation from the old principle of obtaining information to the principle of processing information directly on the day of testing. The system includes [55]: sensors for measuring the work of aggregate determination of fuel consumption, the traversed path and the frequency of engine rotation, block of internal storage of the telemetric transformation device, ratio-channel of communication with the cyclic call for the user, block for collecting information in the computer center at the MTS. After processing information on the computer, it is possible to obtain up to 30 operational indexes for the aggregate, depending on the given algorithm.

In the 'Urozhai' systems, portable magnetic-recording carriers and a centralized system for the collection and storage of information are used as inputs for the computer.

CHAPTER 10

Methods for the Estimation of the Reliability of Agricultural Machines

31. BASIC CONCEPTS

Since the problem of estimation of the reliability of agricultural machines is extensive and involves many aspects, some basic methodological and practical considerations will be briefly dealt with in this chapter.

It is important that information about reliability is obtained in the early stages of the experimental testing and the finishing of a machine. Generally, reliability parameters can be predicted during the process of the product's design and are controlled after its manufacture.

The testing of experimental samples appears to be the most complicated problem in the theory of reliability testing. As is well known, difficulties are caused primarily by the small number of samples which are subjected to testing. Moreover, each of them has its own peculiarities.

In spite of the fact that a considerable number of assemblies and parts are generally aggregated, it is very important that in the test stage of the experimental samples, the various weak spots in the machines are located so as to reduce the time spent in their necessary modification. The general theoretical base of the thoroughly studied problems and state standards can serve as a basis for solving reliability problems.

The terminology and methods for testing reliability are standardized so as to provide compatibility in the results obtained in different conditions and at different times by many investigators.

The *preliminary estimation* of reliability indexes and the determination of their readiness for batch or mass production are envisaged in the methods used in the state testing of experimental samples. The correspondence of reliability indexes with normatives, which are stipulated by technical objective designing and by the extent of the design development of machines, is taken into account. In this connection, serialized and modernized machines are subjected to control testing for the control of quality and reliability parameters during production.

It is well known that the reliability of any product is expressed mainly by its flawlessness, repair accessibility, nonimpairment qualities and durability. In agricultural technological process testing, the indexes of flawlessness and repair accessibility are only partially determined. At present, it is difficult to fully estimate durability because of the insufficient development of complex rapid test methods.

The methods for testing *reliability* envisage determination of the following basic indexes: full operating time (of each complexity group), specific labor consumption for planned technical servicing, a search for and elimination of rejects, service life, gamma-percentage of service life and the availability and technical utilization of machines.

As indicated earlier, the reliability of tractors, agricultural machines and aggregates is estimated by comparing the actual with the normative indexes. In the absence of normatives, a comparative estimation by means of comparison of the actual with the reliability indexes of the machines' analogs is permissible.

Indexes of flawlessness, suitability to technical servicing and repair accessibility are determined for the duration of the proposed volume of work.

The reliability of agricultural machines could be characterized to a large extent by a guaranteed operating time before the given service life is over. If the service life is not given, then it is worthwhile estimating reliability when the machines have achieved operating time which corresponds to the depreciation period. The normative values of reliability indexes for machines should be appropriately established for the corresponding operating periods. However, these questions have not yet been fully solved for agricultural machines.

For calculation of reliability indexes, it is necessary to measure the operating time for tractors and the self-propelled chassis in motor hours and for agricultural machines—in net operating time. The reliability indexes are quantitatively characterized by the consumption of time, labor-technical servicing resources, recovery time and the repair of agricultural techniques.

The methods of collection and processing of testing information are determined by industry standards. Calculation of operating time is made by the method of continuous recording of the agricultural machines' net operating time. When the machines are used for more than one month in a year, the net operating time can be determined by calculating the operating time in physical units and the productivity for one hour of net operation in a control shift. The recovery time of machine restoration is realized by keeping time of the process of searching and removing failures, including the duration and labor consumption for these normative processes.

The time consumed in carrying out the planned technical servicing is determined by the chronometration of the control technical services.

The basic method for testing the reliability of agricultural machines is by testing in actual farm conditions. However, quick test methods are important since the limited agrotechnical periods and the seasonal prevalence of most types of work do not always permit the estimation of machine reliability in a short period of time.

Rapid testing is subdivided into rig testing, ground testing and simulation testing. Their type is determined by the availability of technical resources, mathematical tools as well as by the duration and required accuracy of the obtained results.

Rig testing is conducted on stationary installations with special loading devices that can simulate frequent loading [55].

The loading devices of rigs simulate the influence of the soil-road backgrounds and technological materials, the traction resistances (for self-propelled machines) or traction forces (for tractor-drawn machines) and special factors (climatic, vibrational, etc.). For machine testing, the rigs are equipped with several types of loading devices.

Ground testing of mobile machines is accomplished on special grounds with simulators for the reproduction of influences of soil-road backgrounds and traction resistances.

Simulation testing is generally accomplished in a real time scale with a physical limitation of the technological process and the working medium.

Certain properties which a particular product should possess in the operation process, are controlled for that product in the form of technical conditions. Any deviation in the product's properties from the technical conditions is considered a fault. The state of fault is denoted by the term 'failure'. Failures can be of different types according to their operational significance. Product failures can be classified according to their various indicators.

Design defects (failures) appear as the result of design imperfection.

Technological defects appear because of the deviation from the accepted manufacturing technology of units and parts. As a whole, the quality of individual parts, units and products, has an unavoidable random variation. The variations of quality, maintained within sufficiently narrow tolerance limits, do not at all affect the reliability of the system. For sharp variations of quality, the reliability of one specimen set may appear to be considerably below the reliability of the others. Therefore, technological defects lead to the reduced reliability of a part of the specimen from the general totality of the manufactured machines.

Operational defects generally appear because of violations of operational instructions, which lead to premature failures. Generally, such violations relate to only a part of the machine under operation and should be accounted for during designing.

Aging and wear of the system appear in the operation and holding process. In metals, plastics and other materials, irreversible changes build up, which affect the strength, coordination and interaction between the parts and finally lead to failures.

Thus variations in the faultless working time evolve from variations in the manufacturing quality, operating conditions and from the processes of wear and aging. The following schemes for origin of failure can be pointed out in view of this.

The theory of instantaneous defects assumes that the failure of an element occurs at a random moment of time when the load exceeds a certain level S_d.

It is characteristic that the failure of an element occurs independently of the state or the duration for which the element may have been in operation.

The theory of accumulating changes explains the situation when a failure is formed because of the gradual accumulation of defects (gradual aging or wear). For a number of working parameters of the element, the tolerance limits are established beforehand, and anything beyond them is qualified as failure. The time elapsed before the parameter exceeds the tolerance limit, is the time of flawless operation.

The theory of relaxation is characterized by the gradual accumulation of defects that may appear not as a direct but as an indirect reason for failure. So, the wear of an engine crosshead and guides leads to an increased gap in their conjugation. This leads to jamming in unfavorable external conditions in contrast to the theory of accumulating defects. Here, the tolerance limit for the working parameter (clearance) is not fixed. The accumulation of defects only leads to an increased probability of failure. The combination of a gradual accumulation of defects with an uneven change of the subject's state is characteristic of this example. Here one can cite the case of the destruction of a statistically indetermined system under the action of cyclic loading or the failure of a duplicated electronic system for when the reserved elements fail, the load on the working elements increases.

The theory regarding the influence of certain independent reasons is most typical and is determined by the situation when several reasons for failures act simultaneously. Often it so happens that among the set of reasons that cause the failures, there are one or two predominant ones. The effect of the remaining factors will be so weak that any failure due to them will be practically nonobservable. Therefore in the investigation of the reliability of the subject, the predominant reason for failure is first identified and then, only if it is necessary, the effect of the remaining reasons is considered.

32. BRIEF INFORMATION ON THE METHOD OF RELIABILITY TESTS ON TRACTORS AND AGRICULTURAL MACHINES IN OPERATIONAL CONDITIONS

Failure is defined as an event comprising the breakdown in the working ability of a machine part or in the working parameters exceeding the limits and specified by normative technical documentation.

In testing, the failures of a machine product are classified according to the groups of complexity established in accordance with the industry normative-technical documentation.

The service life of the component or the unit, in which the appearance of a defect in the initial stage does not practically affect the working ability and, up to the moment of replacement or repair, does not violate their functioning, is counted as operating time after which the component or unit is replaced (repaired).

The following operations are performed in the search for and elimination of failures: disassembling and assembling of the unit with the detected component; removal of the detected component and mounting of a new or restored component; adjustment of the mechanisms of the unit in connection with elimination of failure; restoration of the working ability of the component (unit) without its replacement; organizational elements such as delivery of the detected component (unit) to the workshop and delivery of the restored or new component (unit not included in the individual set of spare parts, waiting for mobile repairing rigs, mobile repair workshop, welding unit, lifting crane, etc.).

For the calculation of numerical indexes of reliability on observation results, the following indexes are recommended by the standards [55].

The gamma-percentage service life R_γ is found from the variation series.

For this, the member of the series is established; the serial number of which $N(R_\gamma)$ is determined by equation:

$$N(R_\gamma) = \frac{N\gamma}{100} + 0.5,$$

where γ is the given stipulated probability in percent.

The service life of the product, which has the serial number $N(R_\gamma)$, appears as the gamma-percentage service life for the batch of the product.

If the calculation result from the equation is a fractional number, then the gamma-percentage service life is calculated by the following equation:

$$R_\gamma = R_i - (R_i - R_{i+1})\,[(N(R_\gamma) - N(R_i)]N(R_{i+1}) > N(R_\gamma) > N(R_i),$$

where $N(R_i)$, $N(R_{i+1})$ are the nearest small and large serial members respectively of the variational series, in relation to the calculated expression, R_i—the service life of the product corresponding to the serial member of the variational series, R_{i+1}—the service life of the product corresponding to the serial member N_{i+1} of the variational series.

The use of the method of maximum likelihood is allowed for finding the gamma-percentage service life in repeated truncated testing, when the operating time of the objects, eliminated from testing, can be less than the operating time up to the tolerance level, and it is not possible to construct, a variation series. It is considered that the distribution of service life follows the Weibull law, with parameters b and t_0 and the probability density given by:

$$f(t)=\frac{bt^{b-1}}{t_0}\exp\left(-\frac{t^b}{t_0}\right).$$

If in the testing of N objects, the operating times before failures are denoted by $R_1, \ldots, R_n$ and the operating times without failure of objects by $T_{n+1}, \ldots, T_N$, then the estimation of the parameter can be found from the equation:

$$\frac{\sum_{i=1}^{n} R_i^b \ln R_i + \sum_{i=n+1}^{N} T_i^b \ln T_i}{\sum_{i=1}^{n} R_i^b + \sum_{i=n+1}^{N} T_i^b} - \frac{1}{b} - \frac{1}{n}\sum_{i=1}^{n} \ln R_i = 0.$$

The estimation of the parameter t_0 is found from the equation:

$$t_0=\frac{\sum_{i=1}^{n} R_i^b + \sum_{i=n+1}^{N} T_i^b}{n}.$$

Then, the gamma-percentage service life R_γ is obtained from the following formula:

$$R_\gamma=\left(-\ln\frac{\gamma}{100}\right)^{1/b} t_0^{1/b}.$$

If in service-life testing, a small number of service-life failures are observed (for example, less than hour), then the prediction of the gamma-percentage service life can be made by the method intended for a small number of service-life failures. In this, it is considered that the distribution of service life follows the law of Weibull with the coefficient of variation V or parameter b, which are known either from previous information or are established from expertise.

If N objects were subjected to test and $R_i, \ldots, R_n (n \leqslant 3)$ are the operating times before failures, $T_{n+1}, \ldots, T_N$ are the operating times of the remaining objects which are removed from testing, then the gamma-percentage service life can be calculated by the following formula:

$$R_\gamma=\left(-\ln\frac{\gamma}{100}\right)^{1/b}\left(\frac{\sum_{i=1}^{n} R_i^b + \sum_{i=n+1}^{N} T_i^b}{n}\right)^{1/b}.$$

For operating time T (motor hours, hours, physical units), the number of failures is determined from the equation:

$$m=\sum_{j=1}^{r}\frac{1}{N_j}\sum_{i=1}^{N_j} m_{ij},$$

where m_{ij} is the number of failures of the ith product in the jth interval; N_j—the arbitrary number of the product operating in the jth interval; and r—the number of intervals of the proposed operating time.

The number of failures of the first m_{I}, second m_{II} and third m_{III} group of complexity for operating time T is determined in Eq. (59). In this case, failures on the first, second or third group of complexity are considered separately.

The determination of operating time on failure after achieving the assigned operating time T (for tractors—in motor hours, for agricultural machines—in hours of network) will be considered.

The operating time on failure T_0 is found from equation $T_0 = T/\sum m$.

The operating time on failure of the first (T_0^{I}), second (T_0^{II}) and third (T_0^{III}) complexity groups for operating time T is determined in a similar way.

Availability is determined by the following equation:

$$K_{\text{a}} = \frac{aT_0}{aT_0 + T_{\text{r}}},$$

where a is the coefficient for converting motor hours into hours, it is found from the normatives for the tractor, for agricultural machines $a = 1$.

The mean time for restoration is given by the following:

$$T_{\text{r}} = \frac{\sum\limits_{i=1}^{N} t_{\text{gr}\,i}^{\text{sub}}}{\sum\limits_{i=1}^{N} m_i'},$$

where $t_{\text{gr}\,i}^{\text{sub}}$ is the total time for search and elimination of failures of the ith product; and m_i'—the number of failures and faults of the ith product.

The total time for search and elimination of the ith product is found from the following equation:

$$t_{\text{gr}\,i}^{\text{sub}} = \sum_{j=1}^{m_j} (t_{ij}^{\text{sub}} + t_{\text{gr}\,ij}^{\text{n}}),$$

where t_{ij}^{sub} is the time for search and elimination of the jth failure and $t_{\text{gr}\,ij}^{\text{n}}$—the time for normalized elements for elimination of the jth failure.

The coefficient of technical utilization is found from equation:

$$K_{\text{t}} = \frac{\sum\limits_{i=1}^{N} aT_i}{\sum\limits_{i=1}^{N} aT_i + \sum\limits_{i=1}^{N} t_{\text{gr}\,i}^{\text{sub}} + \sum\limits_{i=1}^{N} t_{\text{gs}\,i}^{\text{sub}}},$$

where T_i is the operating time of the ith product in motor hours for tractors and in network hours for agricultural machines and $t_{\text{gs}\,i}^{\text{sub}}$—the duration of planned technical servicing (technical care) for the working period in accordance with the industry standard.

Specific labor consumption in the search for and elimination of failures is determined from the equation

$$S_1^{\text{sub}} = \frac{\sum_{i=1}^{N} S_{1\,i}^{\text{sub}}}{\sum_{i=1}^{N} T_i},$$

where $S_{1\,i}^{\text{sub}}$ is the total labor consumed in the search for and elimination of failures of the ith product,

$$S_{1\,i}^{\text{sub}} = \sum_{j=1}^{m} (S_{\text{gr}\,ij}^{\text{sub}} + S_{\text{gr}\,ij}^{\text{n}}),$$

where $S_{\text{gr}\,ij}^{\text{sub}}$ is the labor consumption in the search for and elimination of the jth failure, $S_{\text{gr}\,ij}^{\text{n}}$—the labor consumption for normalized elements in the elimination of the jth failure.

The specific cost in the search for and elimination of failures is determined from the equation:

$$\bar{C}_{\text{gr}}^{\text{sub}} = \frac{\sum_{i=1}^{N} C_{\text{gr}\,i}^{\text{sub}}}{\sum_{i=1}^{N} T_i},$$

where $C_{\text{gr}\,i}^{\text{sub}}$ is the cost of the search for an elimination of failures of the ith product, which includes wages (with extra charges) of workers engaged in eliminating the faults, T_i—the budgeted cost of the replaced components (units) except for components which are used from the separate banks of spare parts.

The specific cost of planned technical servicing for search and elimination of failures:

$$\bar{C}^{\text{sub}} = \frac{\sum_{i=1}^{N} C_{\text{gr}\,i}^{\text{sub}} + \sum_{i=1}^{N} C_{\text{gs}\,i}^{\text{sub}}}{\sum_{i=1}^{N} T_i},$$

where $C_{\text{gs}\,i}^{\text{sub}}$ is the total cost of the planned technical servicing of the ith product for the period of operating time. This cost is determined from the industry standards.

Labor consumption and the cost of planned capital repair is found in accordance with the acting normative-technical documentation.

33. BASIC PRINCIPLES AND METHODS FOR RAPID TESTING OF AGRICULTURAL TECHNOLOGICAL PROCESSES*

In the present practice of testing, the data about machine failures in real operative conditions serves as basic information about machine reliability. Taking into account the fact that a limited number of machines are subjected

*Section 33 is prepared with the assistance of G.V. Makushin.

to test, the primary task is to improve the system of centralized collection of information, the method for the estimation and prediction of the indexes of durability and faultlessness of machine aggregates, on the basis of data on their failures during the test period as well as in general farm conditions.

Use of current information from testing on all the testing stations and its centralized processing on computers permit the achievement of a more effective estimation stage and the prediction of reliability indexes of machines under test [55].

In connection with the aforesaid, the role of rapid testing, with its main objective that of acquiring information on product reliability within a period of testing, is much less than the guaranteed durability.

In such testing, model conditions are sought for establishing the volumetric and fatigue strengths of the elements of constructions, their vibrational endurance and wear resistance. Rapid testing is generally conducted on fields, tracks and rigs.

The known methods of testing on grounds and rigs can be divided into high-speed tests and those conducted under normal modes according to the rigidity of used parameters.

In the development of different methods of rapid testing, certain results have been achieved, and grounds for strength testing have been constructed and are functioning in the ONIS NATI[1], KubNIITiM. The soil grounds for rapid testing are available at various machine-testing stations; different rigs for rapid testing are devised and introduced. In specific cases, methodological prerequisites have been designated for carrying out effective service-life testing. On test grounds, for example, the testing can be done two to five times faster by maintaining operational regimes, and up to 30 times faster on narrowing the test regimes. The scientific basis for fatigue testing [28] has been evolved, and the simulation testing of the technological part of machines under test are widely used in practice.

In the general case, the frequent reproduction of operational loads forms the basis of rapid testing. On test grounds and tracks (Fig. 69), this is achieved by movement over an artificial unevenness. In other cases, lugs are mounted on the wheels of the machines (Fig. 70).

At present, autotest grounds are being widely used for the rapid testing of the carrier-system of automobiles, tractors, trailors and other machines. Such test grounds are provided with vast stretches of tracks with different types of covering; sometimes they are equipped with special types of obstacles which make it possible to select test parameters close to the operational. The circular test grounds of a small diameter (20-50 m) are also used.

[1]NATI—Gosudarstvennyi soyuznyi nauchnoissledovatel'skii traktornyi institut (State All-Union Scientific Research Institute of Tractors)—Translator.

ONIS—Oblastnyi nauchno-issledovatel'skii Sovet (Oblast' Scientific Research Council) —Translator.

Fig. 69. Testing on test areas: a—for tractor and b—for combine.

On the test grounds of KubNIITiM [55], for example, it is foreseen that the trailor under test moves on the track of the ground provided with a profile of cemented combs whereas the K-700 tractor-prime mover moves on the side of an even road. This method is complicated, expensive and besides builds excessive loads in the traction-hook equipment. This does not permit an accurate reproduction of the picture of failures that corresponds to the conditions of normal operation. Moreover on test grounds, it is fairly difficult to reproduce the complete spectrum of operational loads, because on any specific machine, the stationary obstacles act differently depending on the chassis, wheel width, rigidity of suspensions and tires. The method for

Fig. 70. Track grousers, mounted on the wheels of a 2PTS-4 tractor-trailer.

selection of obstacles for each mode and machine is still not sufficiently developed. It is based on the rough identity of spectral densities of the load (Fig. 71).

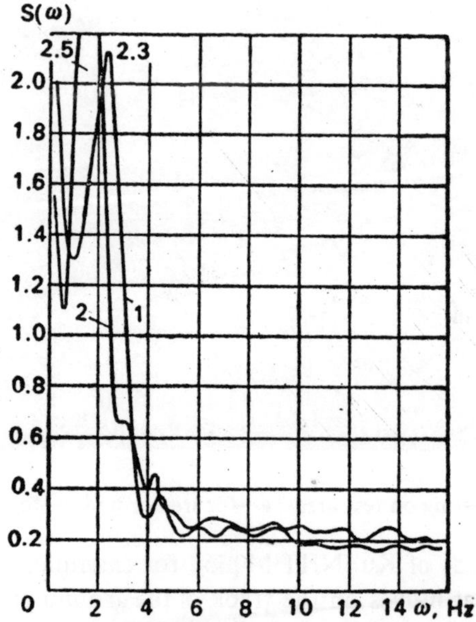

Fig. 71. Spectral density of a load on a trailer frame, while working in natural conditions or with the use of track grousers.

In field testing, resonance occurs because of the effect of elastic elements of the suspension and the tires. Thus, for example, for most machines it becomes necessary to calculate the resonance speed:

$$v_{res} \approx 0.06\omega_{1,2,3},l,$$

where ω_1, ω_2, ω_3, are the natural frequencies of the front and rear axle and spring-loaded and l—the pitch of the grouser in meters.

On the calculation results of the resonance speeds, the working speeds are selected so that they differ by $\pm$ 10% from the calculated resonance speeds.

However, in spite of these precautionary measures, resonance appears, nevertheless, as a result of the change of the elastic properties of suspension and the carrier, which can lead to distortion of the actual loads.

In the test process, the elastic suspension properties and the elements can change up to four times. Besides that, elastic elements (tires, springs, rubber shock absorbers, etc.) considerably filter the load.

In high-speed operation, the increase of the obstacle height basically leads to energy dissipation in the elastic elements. The latter results in more frequent failures of elastic elements (springs, rubber shock absorbers), a fact that should be considered while estimating the construction.

The specific power capacity of the elastic elements varies within wide limits. In handbooks prepared by the SAE (American Society of Automobile Engineers), the values of the specific power capacity for different types of elastic elements are given (Table 25).

Table 25

Elastic element	Specific power capacity per unit mass, kg·cm/kg
Laminated spring	760-1150
Spiral-buffer spring	1270-2540
Coil spring	1780-2800
Torsion bar	2540-3800
Rubber producing shear	5080-10160

Boosting the loading cycle with a change in speed is possible only within fairly narrow speed intervals. The different loading types correspond to different movement speeds on an uneven track. At low speeds twisting of the frame takes place, and at high speeds there is bending. Very sharp changes of speed can lead to a change in the test conditions, i.e. a transition from the twisting of the frame to its bending and, naturally, to a change in the nature of the catastrophic failures.

It thus follows that for field testing indicated limitations do not bring about an increase in the acceleration coefficient more than 30 times, thereby making it difficult to conduct rapid service-life testing of carrier systems.

In spite of the drawbacks mentioned here, and also considering the high cost of erection and operation of test grounds, the ground test method is suitable for the approximate estimation of the strength of basic units in the initial stage of building the machines with small accelerated testing coefficients.

Hence in recent years, rigs for rapid testing and processing on the reliability of frames, carcass, housings and other carrier constructions are getting a wide application in this country as well as in other countries. The rigs for the fatigue testing of frame constructions, using different methods for the excitation of loads, have also obtained a similar dissemination.

Rigs for generating loads with the help of a crank mechanism are well known. They provide bending and twist for the testing of frames. These rigs are mounted on massive force fields that can accept dynamic loads. The regulator systems provide the possibility of changing the frequency of the cyclic loading. Introduction of equipment that controls the eccentricity of the crank makes it possible to change the level of the dynamic load.

Rigs for generating the dynamic load by means of a hydraulic system are used in two modifications: with the use of either hydropulsation or hydroelectric aggregates. Both the aggregates mentioned here possess force-pump systems, i.e. pumps which create a fluctuating pressure of the fluid in the hydraulic cylinders and transfer the load to the frame under test or the moving apparatus.

Rigs with an inertial cyclic load excitation have found a wide application in the study of full-scale designs. Their basic merit lies in their universality and simple construction (Fig. 72), while demerits include the impossibility of a direct measurement of the loading cycle and the necessity for the use of powerful vibrators in the testing of less rigid designs of large dimensions. The possibilities of an inertial loading method are limited because of the quadratic dependence of excited loads on the frequency of loading and the complexity in maintaining the vibrational regime.

In recent years, leading undertakings of the autotractor industry of the USSR and also certain foreign firms dealing with the fatigue testing of

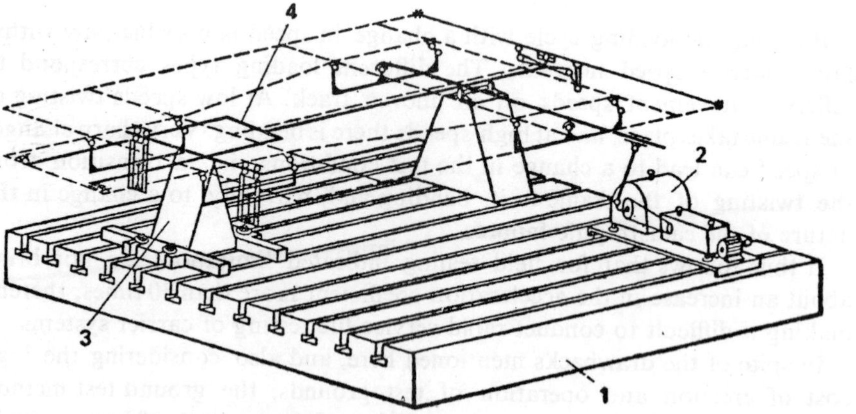

Fig. 72. Diagram of the rig for fatigue testing of a mobile machine carrier: 1—force floor; 2—generator; 3—supporting components; 4—component frame under test.

different carrier systems have used elementwise methods of carrier system rig testing, thus making it possible to considerably accelerate the conducting of testing, to reproduce the spectrum of loadings more completely and to control load values more accurately.

Besides the fact that by the introduction of elementwise test methods of the carrier systems, the effect of the suspension and the tires, which cause resonance, is eliminated and the possibility of an increase in the loading frequency appears. This permits a considerable acceleration of the test process.

The method of rapid fatigue testing used at the ONIS NATI, permits the testing of frames of caterpillar tractors to be carried out about 480 times faster. An approximately similar value of the acceleration coefficient, proposed by G.G. Balovnev (MIISP), is obtained by rigs for the testing of tractor-trailers. Here an increase in the loading frequency, comparable to operational conditions, is allowed so that the fatigue distortional characteristics are not considerable, and the effect of side phenomena, such as heating up, additional inertial loads, etc. (which appear at increased frequencies) can be accounted for with sufficient accuracy.

For ferrous metals, an increase in the loading frequency is allowed within the limits of the first and second orders.

For aluminium alloys, which are more sensitive to the increase of loading frequency, the distortion of test results is observed in the range of loading frequencies between 100 and 1000 cycles/min.

It is necessary to mention the achievements that have been made in the sphere of testing equipment and devices in foreign countries. Considerable success in this direction was achieved by the two firms: Shenk in the Federal Republic of Germany and MTS in the US. The 'Gidropul's' servohydraulic testing system is manufactured on the aggregate principle and is intended for various types of material, component, unit and machine testing as a whole. Individual components of the 'Gidropul's' system (Fig. 73) from the servohydraulic regulating closed circuit. They can be compiled for the most varied test problems. The application range of the system covers varied testings, including small samples for the classical testing of material as well as the testing of machine prototypes as for example, automobiles.

While the resonance machine accomplish only the sinusoidal vibrations, the servohydraulic testing devices can imitate almost any type of loading. Hence, with the help of these devices, it is possible to carry out static or quasistatic and dynamic testing with any nature of loading, including testing vibrational endurance. The loading pattern is changed by an electronic controlling device with the use of a recording of the actual loading on a magnetic tape. The correspondence between the given and reproduced loading regimes is shown in Fig. 74. A similar principle of testing lies at the basis for testing devices of the MTS firm.

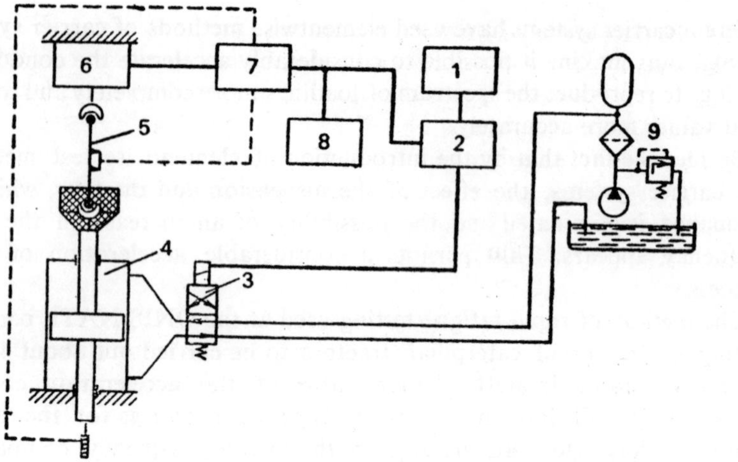

Fig. 73. Diagram of the 'Gidropul's' system made by the Shenk firm: 1—pickup for the given objective; 2—regulating amplifier; 3—servo-valve; 4—cylinder; 5—sample; 6—dynamometer; 7—measuring amplifier; 8—oscillograph; 9—high pressure pumping aggregate.

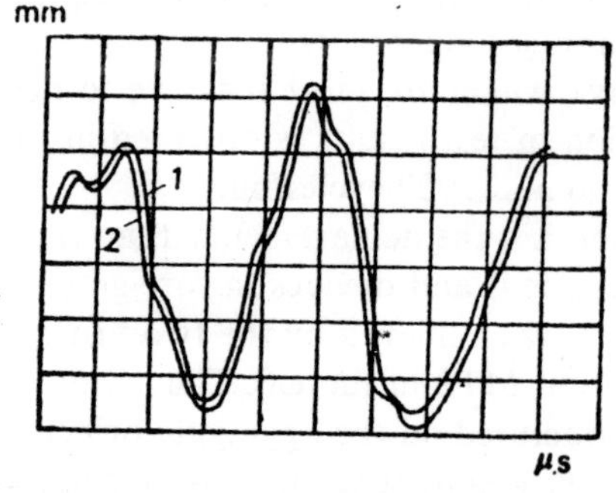

Fig. 74. Variation plots of (1) given and (2) reproduced loading cycles on the 'Gidropul's' rig.

Rig testing is based on patterns of fatigue limits. Experimental investigations of the fatigue strength of materials and constructions are generally conducted for variable stresses, which are represented by a periodic (cyclic) function. The cycle of variable stresses, i.e. the set of all values of stress, corresponding to the period of its change, is characterized by maximum and minimum stresses of the cycle, τ_{max} (σ_{max}) and τ_{min} (σ_{min}), by the mean (const) stress of the cycle, τ_m (σ_m), by the amplitude (variable) of the stress of the cycle, $\tau_a(\sigma_a)$ and by its mode and frequency. Fatigue strength (resistance to fatigue) is characterized by dependence of the number of loading cycles N until the fracture on amplitude of the cycle $\tau_a(\sigma_a)$ for $\tau_m = \text{const}$ or on the maximum stress of the cycle τ_{max} (σ_{max}) at τ_{min} (σ_{min}) = const.

The absolute value of the maximum stress, which does not give rise to a fracture at any number of cycles, is known as the fatigue limit. Many factors considerably affect the fatigue strength: the type of the stressed state, the value and sign of the mean stress of the cycle, the activity of the medium and its temperature, the modification, the order to load alternation of different values, the structure of the material, metal contamination by nonmetallic inclusions, the distribution uniformity of the alloying elements, the surface quality, the dimensions and shapes of the components, the technology of their manufacturing, the distribution of stresses within the component, the residual stresses, etc.

The hypothesis of the power analog of fatigue fracture and the melting of metals has been proposed at the Institute of Metallurgy of the USSR Academy of Sciences. On the basis of this hypothesis, the process of the fatigue fracture of the metal is considered as consisting of three sequential stages, which can be represented in the form of a generalized fatigue diagram (Fig. 75). Here *ABCD* is the fatigue curve, *A'B'C'D'*—the line for the beginning of the formation of the submicroscopic cracks and *A'C*—the line for the beginning of the formation of microcracks.

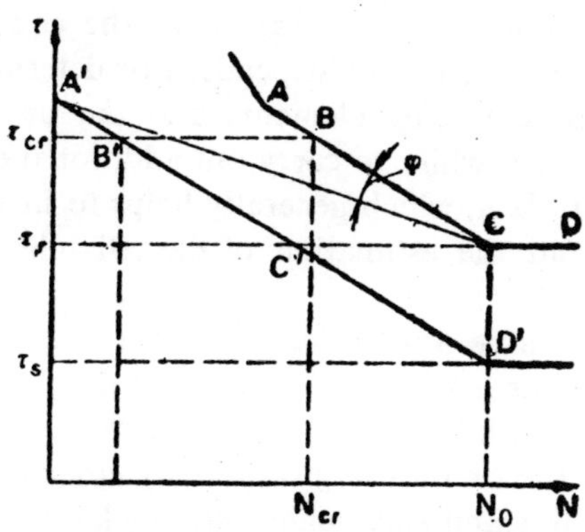

Fig. 75. Fatigue diagram for metal.

On the basis of a generalized diagram, the following criteria for fatigue are concluded: N_{cr} is the critical number of cycles (corresponding to point *C* of the fatigue curve), on achieving the critical number of cycles at the stress equal to the endurance limit τ_r; irreversible distortions of the crystal lattice and submicroscopic cracks start appearing; τ_{cr}—the critical fatigue stress (corresponding to point *B* of the fatigue curve), under the action of critical stress, the fracture comes at N_{cr} cycles, N_0—the limiting number of cycles (corresponding to point *C* on the fatigue curve). The number of cycles determines that minimum stress at which a fracture can still appear, τ_s—the cyclic limit of elasticity (this is the stress below which irreversible distortions of the crystal lattice do not appear howsoever large the number of loading cycles may be). α is the cyclic constant of the fracture (corresponding to section $B'C'$ of the diagram), which is numerically equal to the difference between the critical stress and the endurance limit when the machine's frame has undergone twist, $\alpha = \tau_{cr} - \tau_r$.

According to the technical principles accepted in machine building, the reliability testing of complicated machines, as a rule, should be built on the component-unit-aggregate prototype scheme. It is most worthwhile to preferably carry out the reliability testing of components, units and aggregates on rigs, which maintain the given test conditions with a higher accuracy and carry out testing at a considerably faster pace.

In organizing the rapid testing of mobile agricultural machines, the machine tester comes across the difficulty of totally reproducing all the types and varied loads met within the operation of the machine. The coefficients of the rapid testing of individual units and components are found to be different; in this some components and units appear to be insufficiently loaded, while others may be overloaded. Therefore, in the testing of complicated mobile agricultural machines, one of the important questions which requires a first priority solution is the determination of the service life of the base unit-carrier system or the frame, on the condition of which the service life of the machine generally depends. For determination of the reliability indexes of the remaining elements and the units of the machine, in most cases, it is worthwhile to carry out additional elementwise testing or to collect statistical data, which generally helps to increase the authenticity of the testing and the estimation of the reliability parameters of the machine.

The most simple, reliable and accessible for the carrying out of rapid testing of carrier systems is the set of equipment developed by the Institute of Mechanics of the Ukrainian Academy of Sciences.

The basic units on the rig (housing station, dynamic generator movement force floor, block for stabilizing loads and block for programming loads) are universal and can be used for the testing of different design frames. The device for the clamping of construction (under test) for each of its sizes is separately made. The design under test is clamped on a reinforced concrete force floor, which has in-plan dimensions of 6 × 6 m and a mass of about 120 t (see Fig. 72).

Clamping of the device to the power floor and the exciter of dynamic movements should satisfy the following requirements: the rigidity of clamping device should be very much higher than the rigidity of the device under test, the clamping devices should provide degrees of freedom, similar to operational conditions, and should not generate forces and reactions in the supports which are not characteristic to normal operational conditions.

It should be noted that the correspondence of rig regimes to the operation mostly depends on how meticulously these requirements are observed. The scheme of clamping the frame of the KTU-10 fodder-distributor to the power floor is shown in Fig. 72. With the help of a rigid imitator of springs, clamped to the spring bracket of the frame, the design under test is suspended over flat springs, which permit the frame to swing along the longitudinal axis of the fodder distributor. Besides that, the flat spring is fixed on a stationary bracket on the top through the bear housing thus enabling the beam to freely turn around the transverse axis of the fodder-distributor while twisting the frame; however, the turning of the beam around the longitudinal axis is firmly held by clamping.

The rapid test method on strength foresees the tensiometric study of the subject under test, the construction of a formalized load block (Fig. 76), the development of a loading scheme with calculation of time and the number of cycles on each stage of loading and, particularly, the carrying out of testing up to the time of fracture. The adequacy of the obtained fractures to real conditions is estimated from generalized data, collected in a real operation. In the absence of these, preference is given to the product which holds the maximum number of loading cycles.

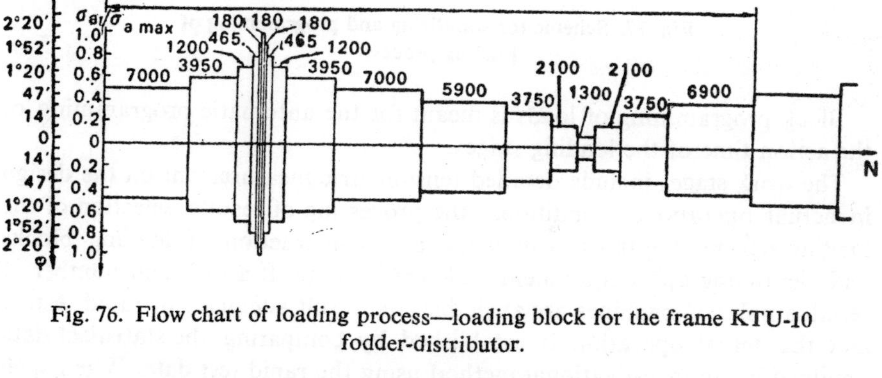

Fig. 76. Flow chart of loading process—loading block for the frame KTU-10 fodder-distributor.

The vibrational amplitude of the deforming beam can be changed in motion with the help of a programmed remote control regulator of the radius of the crank of the exciter for dynamic movement within the limits 0-28 mm (Fig. 77). The maximum force in connecting the shaft of the exciter is 500 kgf. The exciter is put in motion through the linked regulator with the help of an electric motor of 3 kW power. A lower power consumption by the drive and small forces in the connecting shaft are explained on the ground that testing is performed in the mode close to the resonance.

The block for load stabilization has 12 stages of loading, each of which can be regulated within wide limits. The frequency of stabilized loads varies from 2 Hz to 1 kHz. The load is controlled and regulated with the help of direct and feedback links through a polarized relay. A direct link is accomplished through the traction electromagnets of the exciter for dynamic displacements, which is used for switching on the transmission box in order to increase or decrease the load; feedback is accomplished with the help of an inertial sensor for displacements, which measures the frames angle of twist.

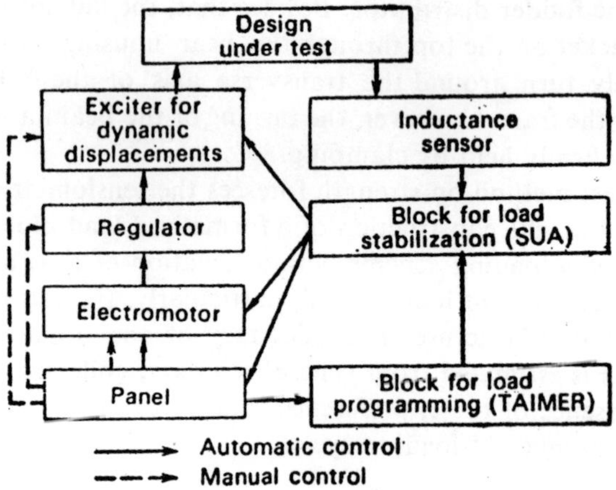

Fig. 77. Scheme for stabilizing and programming of loading process.

Block programming of loads is meant for the automatic programming of the action time of the loading stages.

The work stages include detailed tensiometric measurement on the design in actual operational conditions, the processing of results, selection of the loading scheme, the basis of its values and construction of loading blocks and the tuning up of equipment and rapid testing. If a sufficient number of serial samples of machines are available, the result adequacy of rapid testing and the actual operation is established by comparing the statistical data obtained by an investigational method using the rapid test data. When such an appropriate methodological approach is adopted, the results possess a sufficient degree of coincidence.

The experience obtained from the application of the proposed method of testing enables the process of testing to be hastened by one or two orders as compared to ground testing or with experimental operation.

The mounting complex of branched, kinematic chains, which drive working organs into motion on the carrier system, is the characteristic peculiarity for most of the agricultural machines. The effect of frame deformation, particularly when they lose their initial rigidity with the appearance of additional loads, should necessarily be accounted for in the estimation of the reliability of transmission elements and in carrying out rapid testing (report). In this, it should be borne in mind that around 40-60% of the power consumed by agricultural machines is spent by the technological part of machines on nontechnological requirements, which appear due to structural and manufacturing defects.

The statistical indeterminability of mechanisms, which is manifested in the fact that the mechanisms are unique and base deformations appear as a reason for the emergence of considerable additional loads and have nothing in common with the technological loads. The problems of these peculiarities are related to the sphere of optimum synthesis of structure of mechanisms. In principle, their solution is not difficult and does not lead to any considerable rise in the price of the machine.

However, while designing the machines, these problems are still not solved in the required details. This reduces the reliability of the machines and increases the duration for modification of new agricultural techniques.

The study of the effect of frame deformations of the KTU-10 fodder-distributor on transmission loads has shown that the loads increase by 1.8-2.4 times.

Hence, for an increase of reliability and a reduction of power consumption of machine drives, it is necessary to take this important factor into account.

Imitation testing generally does not give considerable acceleration. For example, in the testing of grain-harvesting combines with the application of different substitutes (leather wastes, rubber products) or load selectors, the coefficient of acceleration is close to unity. While using substitutes for the loading of technological communications, it is necessary to take into account the reduction in loads on the working members of the machines with the wear and change in their properties (Fig. 78) [55].

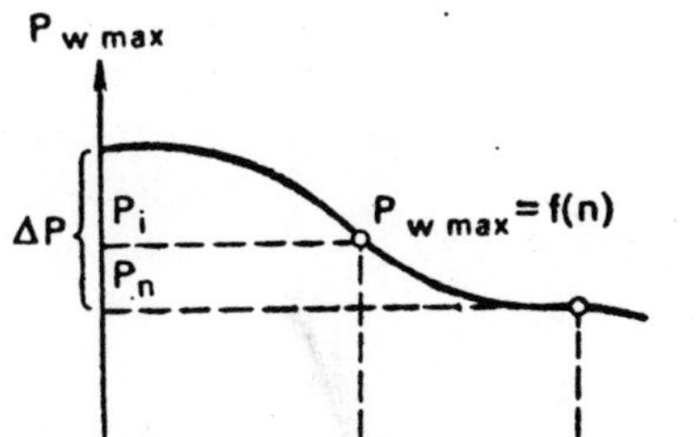

Fig. 78. Dependence of the load connecting shaft of the load selector on the accumulated number of cycles.

In view of the foregoing discussion, the material used for imitation requires constant renewal or replacement.

In systems approach, the problem of the reliability of agricultural machines and the development of rapid test methods as an effective means for establishing the reliability parameters requires for its solution the consideration of a large number of factors. It can only be successfully accomplished with the application of the findings on the strength and wear-resistance of materials, the concepts of the mechanism and machines theory and, particularly, its important branch, i.e. dynamics.

The question of reliability acquires an extremely high significance in large systems of machine complexes for field-crop cultivation and farm equipment, which are functionally related to the structure of industrial processes.

CHAPTER 11

Economic Estimation of Agricultural Technological Processes and Decision-making on Test Results*

34. SIGNIFICANCE AND PROBLEMS OF ECONOMIC ESTIMATION IN STATE TESTS

Economic efficiency is the most important index for decision-making in regard to proposing new machine designs for production, since indexes of economic effectiveness synthesize all the results of operational and technical tests.

The fundamental concepts and indexes (meant for utilization in estimation), which are used for determining the comparative economic effectiveness of new and, in particular, agricultural technological processes, have been standardized.

Initial information for an economic estimation of machines comprises of the normative reference materials (material cost, tariff tables, norms of depreciation, coefficient of capital investment effectiveness, etc.) along with test results.

Sometimes technical and technological reliability indexes and also qualitative indexes of the working of the machines appear to be low due to imperfection of the experimental sample. At the same time, it is known that the technology accomplished by a particular machine offers the perspective for further development. In such cases, the indexes of economic effectiveness (Fig. 79) under prediction can be determined, so that it is possible to estimate the potential of the subject under test.

Economic indexes characterizing the machine are: labor consumption on a single unit or on the annual volume of work, direct operational expenditure on a single unit or on the annual bulk of work, capital investment on a single unit or the annual volume of work, deduced costs on a single unit or the annual bulk of work.

*This chapter has been prepared in collaboration with V.A. Yasinetskii.

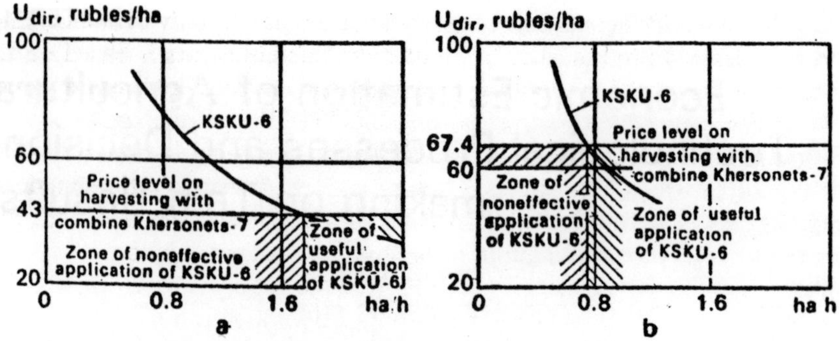

Fig. 79. Plot of deduced costs against the operational productivity of a KSKU-6 corn harvesting machine for an annual load of 170 h:
a—harvesting of corn with removal of corncobs; b—harvesting of corn with threshing of corncobs.

Apart from these indexes, it is recommended that the labor consumption in maintenance and periodical technical servicing and also the total labor consumption in the operation of machines calculated for a single unit or the annual bulk of work be furnished.

It is supposed that the machine has been tested and the above-mentioned indexes have been for the machine under test as well as for the standard machine. Decision-making is now required about the worthiness of the machine against which a comparison was made.

The initial data for decision-making appears to be the economy in labor required for the operation of the machine, the release of the labor force determined from the peak period of its requirement, the economies in the operational expenditure and the annual economic efficiency, considering the changes in the quantity and quality of produce and economy in basic inputs. This last index is the basic indicator for estimating economic effectiveness.

For decision-making regarding machine production, it is necessary, apart from merely comparing the numerical data, to take into account other indicators, such as improvement in the working condition of the machine-operator, the possibility of releasing a large number of workers, competitiveness among the workers, etc.

Along with the development of a general methodology of testing, methods of the economic estimation of a machine were also developed and modified. These methods can be divided into two major groups:

(1) the economic estimation of specialized machines intended for accomplishing a single agricultural operation (plowing, cultivation, seeding, etc.). For such a group of machines, it is assumed that their introduction into production cannot substantially influence the economic effectiveness of carrying out other agricultural operations that may or may not be directly linked with process under study; and

(2) the economic estimation of tractors, transports, universal machines and complexes. Here the calculation of economic indicators is based on the assumption that the technological processes being introduced have a considerable bearing on the effectiveness of many machine complexes.

Indexes regarding the use in agricultural practice of an optimum fleet of machines (consisting of a series of machines) are compared with the indexes regarding the use of a fleet in which machines under test are included. The theory of model economy is utilized to study such a comparison.

35. METHOD FOR DETERMINATION OF ECONOMIC INDICATORS

Labor consumption for the unit operation of a machine (hectare, ton, etc.) is determined as the ratio of quantity of the consumed labor to the hourly shift productivity of the aggregate. Naturally, the unit of this index is expressed in the man-hours required for unit work.

In general, the formula is of the form:

$$Z_{1e} = \frac{L}{W_h K_{sh}},$$

where L is the number of workers servicing the aggregate, W_h—the productivity of the aggregate in one hour of net operating time and K_{sh}—the coefficient of shift time utilization.

In individual cases, for machine (equipment) servicing, apart from the main group of technical personnel, some auxiliary workers are involved. For example, on the hop-picking machine, 15-20 persons are engaged in addition to the operator, whereas on the transplanting-planting machine, 8 additional persons are required. With the objective of ensuring a proper operational organization, technical personnel carry out a monthly technical servicing of the machine until the arrival of auxiliary workers on the shift.

That is why labor consumption on unit work is determined on the actual working time of the auxiliary workers:

$$T'_{sh} = T_{sh} - T_5 + T_{coinc},$$

where T'_{sh} is the working time of auxiliary personnel in h, T_{sh}—the shift time in h, T_5—the time for technical servicing of the machine in h, and T_{coinc}—the working time of auxiliary personnel, which coincides with the time for technical servicing of the machine and elimination of failures in h.

The labor consumption on unit work is determined by the formula given below on the basis of main and auxiliary personnel:

$$Z_{1e} = \frac{L_{m.w}}{W_h K_{sh}} + \frac{L_{aux.w}}{W_h K'_{sh}},$$

where $L_{m \cdot w}$ and $L_{aux.w}$ are the numbers of main (technical) and auxiliary servicing personnel, $K'_{sh} = T_h/T'_{sh}$ (T_h is the net working time of the machine in h).

Labor consumption is determined from the accomplishment of the basic process as well as from the associated operations. The machines being compared exert a definite influence on the labor consumption in these operations.

In view of the fact that the transport running the unit is not utilized in the preparatory period of the working unit for operation before the beginning of the shift (technical servicing is carried out by foreman and other personnel) and lies idle during elimination of the failures of the main aggregate during the shift, labor consumption on associated transport operations is determined by the following formula:

$$Z_{1a} = \frac{L_{tr} t_{ct} m}{60 \gamma_{ac} K'_{sh}},$$

where L_{tr} is the number of servicing personnel on the transport aggregate, t_{ct}—the time for a complete cycle in min, m—the mass of material carried by the transport, calculated over the unit working time of the main machine (placement norm for one hectare, crop-yield from one hectare, etc.) in t/ha, γ_{ac}—the actual load-carrying capacity of the transport in t, and K'_{sh}—the coefficient of utilization of the working time of the machine serviced by the transport, which is determined as follows:

$$K'_{sh} = \frac{T_h}{T_{sh} - T_5 + T_{coinc}},$$

where T_h is the net working time of the main aggregate is h, T_{sh}—the shift time of the working of the main aggregate in h, T_5—the time for technical servicing of the aggregate in h and T_{coinc}—the time of forced idling of the transport, comprising of technical servicing and the elimination of failures of the main aggregate during the shift in hours.

The total labor consumption on the operation of compared machines, calculated for unit working, include (apart from the above-mentioned) the labor consumption on periodical technical servicing and on repairs (search for an elimination of failures):

$$Z_g = Z_{1e} + Z_{tr} + Z_{ts} + Z_{ta},$$

where Z_{tr} is the actual labor consumption on repair and Z_{ts}—the labor consumption on periodical technical servicing.

The total labor consumption on tested and compared machines does not provide a basis for specific conclusions because the farm requirement of the labor force at the working time of the compared machines is not known. That machine is considered more effective which releases the labor force in peak periods of an agricultural cycle. That is why a calculation of the labor consumption is recommended to be accompanied by an analysis of the graph of the working force requirement of a typical farm in the zone.

The most important economic indicator is provided by the direct operational cost values on the unit work. Direct expenses consist of the following elements: salaries of servicing personnel, cost of consumed materials (fuel and lubricants, etc.), expenditure on technical servicing and repairs, depreciation.

Salaries of the main technical personnel are calculated according to the following formula:

$$C_{s\,m.w}=\frac{\sum r_{r\,m.w}}{W_h K_{sh}},$$

where $\sum r_{r\,m.w}$ is the sum of the hourly wage tariff of the main technical personnel in rubles/h, W_h—the productivity for one hour of net work, hectare (ton, ton-km, heads), K_{sh}—the coefficient of shift time utilization.

Wages of the auxiliary workers for unit work is determined by the following formula:

$$C_{s\,m\cdot w}=\frac{\sum r_{r\,a.w}}{W_h K'_{sh}},$$

where $\sum r_{r\,a.w}$ is the sum of the hourly wage tariff for auxiliary workers in rubles/h.

Expenditure on fuel, lubricants and other materials are calculated by formula:

$$C_m=\sum_{i=1}^{i} q_{mi} P_{mi},$$

where q_{mi} is the consumption of the ith material (or power) on the unit work, P_{mi}—the price of the ith material in rubles/g.

Material consumption is determined from the technological estimation of the machines.

Expenditure on technical servicing is determined from the actual salaries and wages, the spare parts and the materials and deductions for tool wear. Specific expenditure of technical servicing is given as follows:

$$C_t=\frac{\sum C_{pp}}{T_{n.1} W_h K_{op}},$$

where $\sum C_{pp}$ is the total annual expenditure, $T_{n.1}$—the normative annual loading of the machine, K_{op}—the coefficient of operational time utilization.

Expenditure on repairs should be determined by the methods described in industry standards. The total expenses are calculated for the operational period equal to the service life of the machine. Specific expenditure is calculated by the following formula:

$$C_r=\frac{\sum C_{pp}}{T_p W_h K_{op}},$$

where $\sum C_{pp}$ is the total expenditure on repairs during the operational period equal to the service life of machine and T_p—the service life of the machine in hours.

Apart from the method described here, presently expenses on the machine's repair are determined by the following formula:

$$C_r = \frac{BE_r}{T_{n.1} W_h K_{op}},$$

where B is the price of the machine in rubles and E_r—the normative coefficient of deductions on repair for one year of operation.

This method is simpler and is often used in economic estimation.

When the machine is unjustifiably high priced or when, for the obtained operational indexes and proposed price, the machine does not provide the necessary yearly economic effectiveness, its cost can be conveniently determined graphically (Fig. 80). The quantum of yearly economic effectiveness (losses) for two or three values of the machine's price serve as the initial data for plotting the graph.

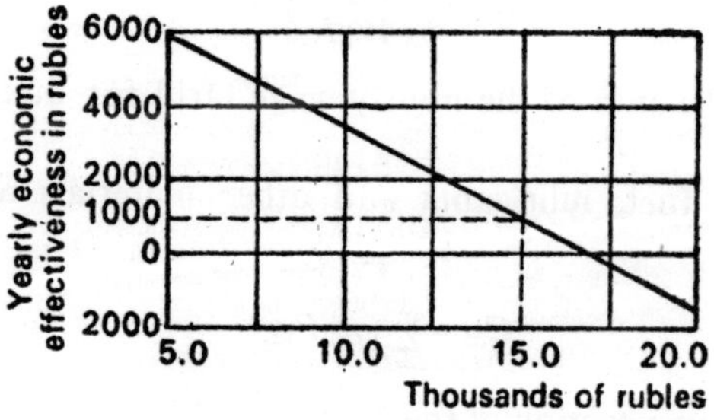

Fig. 80. Determination of the possible book price of hop-picking machine ChMP-1,6.

The determination of the maximum permissible price for the obtained operational indexes is shown in Fig. 80.

Deduction for renovation imply the annual expenditure incurred on replacement by a similar machine*. Specific deductions on renovation are determined by the following formula:

$$C_r = \frac{BE_r}{100 T_{op} W_h K_{op} (n + n_0)},$$

where E_r is the normative coefficient of depreciation, T_{op}—the duration of the use of the machine in a day in hours; n—the number of days of machine

*Deductions for renovation—These refer to the annual allowances from the profit margin being put aside with the intention of replacement of existing machines on completion of service life—Translator.

use for the main (given) form of work during a year and n_0—the duration of machine use for other forms of work (apart from the main job) during a year.

If the technological operational processes of the compared machines are not similar and if certain additional operations (preparatory, associated or sequential) are required for bringing the field or the production to a comparable level, then, in such cases, economic indicators are determined for each operation in terms of the unit operation of the machine as described earlier. These indicators, however, do not consider the associated transportation to be accomplished by vehicles.

The operational expenditure on associated transport operations, accomplished by vehicles with respect to the unit working of the main machine or aggregate, is described by

$$C_{tr}=\frac{\eta_{tr}\left[P_h+v_{op}(P_{km}+\frac{1}{2}\gamma_{ac}P_{tk}+\frac{P_t}{2l}\right]}{100W_hK'_{sh}},$$

where η_{tr} is the number of vehicles servicing the aggregate, P_h—the expenditure on one hour of the vehicle's working in kopecks/h, v_{op}—the operational speed of the vehicle in km/h (determined in accordance with OST 70.2.17-13), P_{km}—the expenditure on one km of travel kopecks/km, γ_{ac}—the actual load-carrying capacity of the transport in tons, P_{tk}—the expenditure on one ton kilometer, kopecks/t km, P_t—the expenditure on one trip, kopecks/trip, l—the distance covered by vehicles in km, W_h – the productivity for one hour of net time of the aggregate under servicing, unit of work/h, K'_{sh}—the coefficient of shift time utilization of the machine being served by the vehicle.

Expenditure on one hour of automobile operation includes the salary of the driver and the expenditure on the renovation of the rolling stock calculated for one hour of operation. While determining expenditure on renovating the rolling stock, the amount of work by a particular vehicle is taken to be 1800 hours.

In the absence of normative initial data, it is assumed that the operational expenditure on associated transport operations can be determined from the following formula:

$$C_{tr}=\frac{\eta_{tr}\Phi_c}{W_hK'_{sh}},$$

where Φ_c is the normative cost of one automobile hour in rubles.

The number of transport necessary for the uninterrupted servicing of the aggregate is given by:

$$\eta_{tr}=\frac{W_hK_smt_{tr}}{60\gamma_{ac}},$$

where W_h is the productivity of the serviced aggregate for one hour of net work, units of work in hours, K_s—the coefficient of the working runs of the aggregate under service and m—the seeding norm or yield calculated per unit area in t/ha.

Specific capital investments on associated transport operations:

$$K_{sp\ tr}=\frac{B_{tr}m}{T_{zon\cdot tr}W_{tr}K'_{sh}},$$

where B_{tr} is the book price of the transport in rubles, $T_{zon\cdot tr}$—the zonal loading time of the transport in hours, W_{tr}—the productivity of the transport in t/h, K'_{sh}—the coefficient of shift time utilization for machines being compared and served by the transport.

The transport productivity, which serves the working aggregates, is defined by:

$$W_{tr}=\frac{60\gamma_{ac}}{t_{ct}}.$$

After determining individual elements, total specific direct expenditure is calculated

$$U_{dir}=C_s+C_m+C_{tr}+C_r+C_{ta}.$$

Deduced costs consist of direct costs, specific capital investments and losses due to waste of the produce and deterioration of its quality.

Specific capital investments are calculated by formula:

$$C_{coeff}=\frac{BE_{coeff}}{T_{op}W_hK_{op}(n+n_0)},$$

where E_{coeff} is the normative coefficient of the effectiveness of capital investments (0.15).

Losses due to produce waste, a composite part of the deduced costs, are determined by:

$$C_w=\frac{X_wQ_w}{100}(P_r-C_b),$$

where X_w is the produce loss from the main yield in %, Q_w—the main yield of produce (crop, productivity of one animal) in quintals, P_r—the realization price of the produce in rubles/quintal, C_b—the direct expenditure for preparing one quintal of produce up to trade standards.

Losses due to the deterioration in quality of the produce or profit because of its improvement (also a composite part of the deduced costs) are determined as follows:

$$C_{lg}=\frac{X_kQ_p}{100}(P_r-P_{rn}),$$

where X_k is the quantum of produce which does not satisfy trade standards (either substandard or above standard) in %, Q_p—the main yield of produce in quintals, P_r—the realization price of standard (high class) produce in rubles/quintal, P_{rn}—the realization price of substandard (low class) produce in rubles/quintal.

Besides the indicated components of deduced costs, additional expenditure on preparatory, associated and concluding (sequential) operations are taken into account. This expenditure arises only in those cases where the operations of the machines under comparison are different. If the preparatory, associated or concluding operations of the machines under comparison do not have differences in expenditures, then these expenditures are not included in the total deduced costs.

The annual profit (economic effect) for machines used in plant growth is calculated from the formula:

$$D_r = \sum_{i=1}^{i} C_i W_{h\,i} K_{op\,i} T_{c\,i} n_i - \sum_{j=1}^{i} C_j W_{h\,j} K_{op\,j} T_{c\,j} n_j,$$

where C_j is the specific deduced cost of the machine (new) under test on the jth operation, C_i—the specific deduced cost of the machine under comparison on the ith operation, $W_{h\,j}$, $W_{h\,i}$—the productivity for one hour of net machine operation of test and comparison machines, $K_{op\,j}$, $K_{op\,i}$—the coefficients of utilization of the operational time of the machines under test and comparison, $T_{c\,j}$, $T_{c\,i}$—the duration of the operation of machines under test and comparison over a day, n_j, n_i—the number of days of operation of the machine under test and comparison for the jth and ith operations.

If the machine is intended for accomplishing a single operation, the annual economic effect is determined from the difference of the specific deduced cost (of machines under test and comparison), multiplied by the annual work load accomplished by the machine under test.

It is necessary to focus attention on the peculiarities of the economic estimation of machines and equipment for the mechanization of plant growing and stock-breeding processes when the annual operating time remains constant, for example, the loader hay column, the stationary transporters for agricultural material, manure removal from the premises, etc.

An increase in the productivity of these machines leads to a proportional decrease of the annual work load in hours: $W_{op} T_p = \text{const}$. With the increase of productivity from one side, the employment of servicing personnel is reduced and both the power consumption on process accomplishment and machine cost (in view of the increased metal consumption) is increased.

Hence, a portion of the total expenditure, consisting of energy consumption and deductions on renovation, repair and technical servicing, increases with an increase in productivity, whereas the expenditure on payment for live labor is reduced.

From the calculation results (applicable to hay columns), a graph is plotted (Fig. 81) from which it can be concluded that the combined operational expenditure on unit output manifests as an initial decrease, and then shows as an increase.

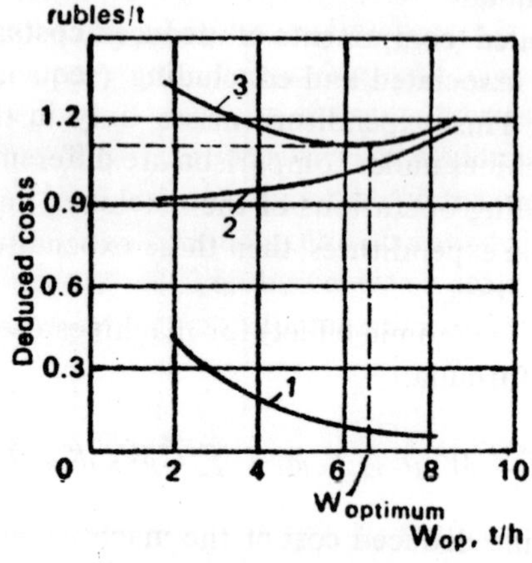

Fig. 81. Relationship between deduced costs on column unloading and its productivity: 1—wages; 2—depreciation and power charges; 3—total expenditure.

Taking the value of $W_{op} = 6$ to 7 t/h as a guide for RBV-6, the optimum solution for deduced costs (for accepted labor wages) is obtained.

With the wage increase for line labor, the optimum solution will shift to the side of higher values of W_{op}.

36. MODELING AND PREDICTION OF ECONOMIC EFFECTIVENESS INDICATORS

The formulation of the economic effectiveness indicators of machines under test as well as their complexes and the industrial processes in the conclusive stage of systems analysis (from the results of the estimation of agricultural technological, power and operational characteristics of machines) has proven to be worthwhile. Optimization problems in tests, i.e. a search for the optimum conditions of carrying out operations, the basic parameter values of the machines and aggregates, the rational composition of machine complexes, etc. are, as a rule, still not solved in spite of their urgency.

It is clearly seen from the example in Fig. 82 that the efficiency of the aircraft's refueller of toxic chemicals will ensure an optimum expenditure on complex operations only when the pump efficiency is doubled.

With the aim of optimization and the prediction of machine parameters within the limits of some integrated curve, which describes the economic effectiveness of a given type of machine whose shape is not completely known, it would be worthwhile to estimate the possible character change in the economic effectiveness indicators under consideration. For this purpose, it is most rational to use modeling methods of operational-economic characteristics for various potentially possible aggregates, conditions and operational modes. In these cases, the use of extrapolation methods is not permissible as it can lead to miscalculations.

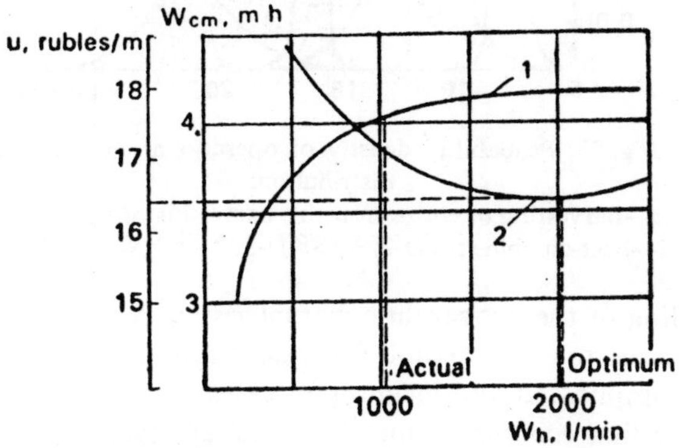

Fig. 82. Dependence of the operational efficiency of (1) aircraft and (2) direct expenditure on the application of toxic chemicals on the efficiency of a refueller.

In accordance with the systems approach, it is worthwhile to determine the economic effectiveness indicators for all possible states of operating conditions, relative to some probability distribution of external factors, by using the method of statistical description of indicators and system of model algorithms (see Fig. 89).

Using the characteristics of operating conditions, it is possible with the statistical test method to estimate arbitrary and absolute operational-economic characteristics over a wide range of the basic parameters of the beet-harvesting machines which accomplishes various mechanized processes.

A variation of the arbitrary statistical characteristics and the operational-technological indexes, which reflect machine operation results in the given region or zone, on the average leads to a considerable dispersion (Table 23 and Fig. 83) of the economic effectiveness indicators. Characteristics of deduced costs, obtained by statistical test and by the earlier described numerical methods for determined values of factors (as shown in Table 24), confirm the increase in deduced costs with the increase in productivity dispersion. It becomes obvious that standard methods can only be used without large errors for the estimation of the economic effectiveness for machine operation in certain specific conditions with a low dispersion of operational indexes. In the reverse cases, it is possible to incur serious errors in the economic estimation of machines.

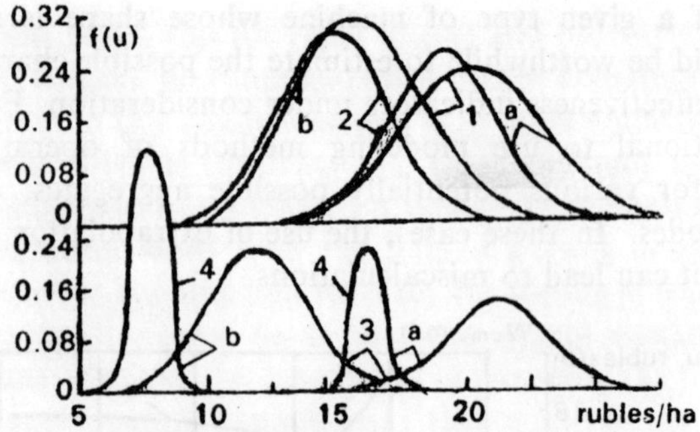

Fig. 83. Probability density of operational expenditure distribution:
a—harvesting of entire crop; b—harvesting of beet roots;
1—beet-combine SKD-2; 2—SKD-3; 3—UKS-3; 4—KS-6.

The modeling of the probability characteristics describing economic indicators is the most reliable method for their estimation and should be used in the most complicated and meaningful cases.

Considering that economic indicators and an overwhelming number of techno-operational indexes of the aggregates (specific metal and power consumption, fuel consumption, etc.) are formed as quotient $Q^i = \frac{\sum C_i}{W}$ ($\sum C_i$ are the elements of operational expenditure on labor or resources), the hourly and shiftwise productivities of most machines, as shown by investigators [39, 55], closely follow the normal law of distribution. The probability density form of distribution of the indicated criteria and indexes given below can be used in modeling. It is, then, possible to determine their statistical characteristics (MQ_i, DQ_i) by numerical methods:

$$P_{Q_j} = \frac{2}{\sqrt{2\pi}} \frac{MW\sigma^2_{\Sigma C_i} + M\sum C_i \sigma_W Q_i}{(\sigma^2_{\Sigma C_i} + Q_i^2 \sigma_W^2)} \exp\left[-\frac{(M\sum C_i - Q^i MW)^2}{2(\sigma^2_{\Sigma C_i} + Q^2 \sigma_W^2)}\right].$$

A mechanical interaction of the distribution density for direct operational expenditure (Fig. 83) shows that linear density is symmetric with respect to the mathematical expectation. A more compact distribution is obtained for root-picking machines. A combination of the complicated processes of root and beet tops harvesting in one aggregate involve a considerable expansion of the variation range in operating costs.

For beet-picking machines under consideration, the possible level of economic effectiveness of the mechanized process and its relation to the magnitude of capital investments for harvesting aggregates, having various basic parameters can be easily estimated. This is calculated from the prediction of the ratio between the deduced costs on a KST-3A combine and the possible investments on new types of machines, in the function for the degree of productivity increase (as obtained from the generalized numerical model [58]):

$$MU_{dir} = \frac{0.169 \times 30.74 K_b + 7.84 + W\,13.89}{19.52 K_w}\, U_{ae},$$

where K_b, K_w are, respectively, the degree of book price increase, and the productivity of the complexes of the machines under prediction, and W—the productivity in ha/h.

Investigations have shown that the already described drop pattern in the increase of economic effectiveness with the increase of K_w and K_{sp} are, on the whole, peculiar to the trajectories of the prediction curves of economic effectiveness (Fig. 84).

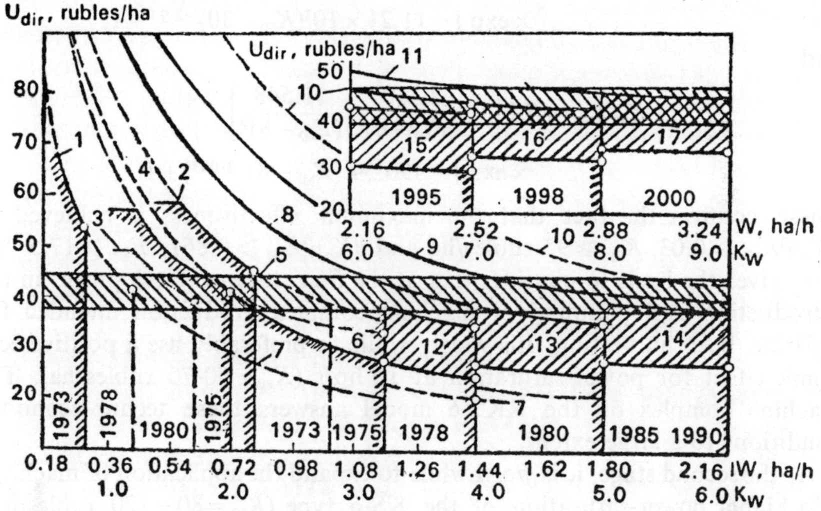

Fig. 84. Prediction of the possible values of deduced costs from productivity I_p and the degree of its escalation K_w for new types of beet-picking machines: single-phase (combine) harvesting at speeds of 7-7.2 and 9 km/h, $K_w = 1.2$ (1.3), separate two-phase harvesting by machine complexes on the basis of model KS-6 (2.5), K_w (3.3) and RKS-6 (6), $K_w = 2$, combined two to three-phase harvesting by unified machine complexes on the basis of KS-6 and RKS-6 models, $K_w = 3.5$ (4), three-phase harvesting with a formation of thick rows, $K_w = 1.6$ (7), a range of expected effectiveness of universal machine complexes (at $K_w = 4.9$, 7.2, 10, 15, respectively) predicted for the period 1990-2000 (12-17).

In the long-range taking the decided significance of increasing economic effectiveness into consideration in estimating the parameter formulation of beet-picking and other types of machines, it is necessary to adhere to the strategy that ensures correspondence among the proportions of deduced costs, the degree of productivity increase and escalation in capital investment. Capital investments usually tend to escalate in view of the increase of power-saturation and the application of more expensive technical solutions.

For example, when $K_w = 1.6$, it is necessary to provide an increase in the productivity of root-picking machines by 1.8-2.2 times and when $K_w = 3.0$ by 3-3.5 times, etc. A rational strategy can be represented in the form of stagewise increase of K_w, K_{sp} and E_{sp}.

An analytical estimation of the possible nature of the plotted integrated curves and the modification of the initial phase of their trajectories are obtained in accordance with experimental investigations of new types of wide-claw root-picking machines (KS-8 and RKS-6 (USSR), Parma (USA) and others) as follows:

$$K_w(K_{sp}) = 8.4\left[1 - \frac{20.245}{K_{sp} - 30} + \frac{919.147}{(K_{sp} - 30)^2}\right]$$
$$\times \exp\left[-11.21 \times 10^3 (K_{sp} - 30)^{-2.93} + 1;\right.$$

and

$$K_w(E_{sp}) = 7.8\left[1 - \frac{2.310}{E_{sp} - 8} + \frac{11.548}{(E_{sp} - 8)^2}\right]$$
$$\times \exp\left[-196.54\,(E_{sp} - 8)^{-3.044}\right] + 1.$$

These confirm the fact that the maximum effectiveness is achieved at $\partial K_w/\partial K_{sp} \geqslant 0.03$, $K_{sp} \geqslant 95$ rubles/ha and $\partial K_w/\partial E_{sp} \geqslant 0.861$, $E_{sp} \geqslant 17.5$ hp. This gives the basis to consider that at the first stage of development in the introduction and planning of machine complex production intended for separate beet harvesting, itis worth while to preferably use a positive economic effect for power-saturation at 10 hp/t ($K_{sp} = 50\text{-}70$ rubles/ha). The machine complex on the RKS-6 model answers these techno-economic conditions to a large extent.

At the second stage, it is worthwhile to expand the application of machines of a higher power-saturation of the KS-6 type ($K_{sp} = 80 - 120$ rubles/ha). In the next decade, a further increase in effectiveness can be provided from the predicted increase of power-saturation of harvesting aggregates up to 18-22 hp/t and the increase of capital investment above 130-180 rubles/ha. This is possible in view of the development of distinct universal machine complexes with a variable structure of technological capability. Here, it is possible to accomplish a complete development of the integrated curves under the prediction for multistage machine complexes, which can realize separate two-phase harvesting.

Estimation of the correct correlation between the productivity criteria and the cost of unit work has an important significance for the solution of estimation problems and the prediction of machine parameters subjected to tests. This is shown by the analysis of the development of new types of these machines.

For optimization and prediction of machine parameters under development, it is necessary to check the provision of a correct relationship between their productivity and capital investments during tests.

A discrepancy in the prices of machine manufacture appears due to the fact that in the upper limit of the price P_n [58] named by the testers, the normative coefficient of economic effectiveness inadequately conforms to the actual level of machine productivity because often insufficiently, confirmed design normatives are used in calculations:

$$P_n = P_0 + \frac{E_{op}}{E_n} = P_0 + \frac{U_n - U_{op}}{E_n},$$

where P_0 is the price of the old (reference) machine, E_{op}—the economy on operational expenses.

Similar economic calculations decrease the possibility of scientific prediction and distort the economic principles that control the development of complicated systems (to obtain an ever increasing output reduced labor expenditure).

Test experiences and the introduction of new agricultural technological processes show that in individual cases, monetary resources are not economized, and also that direct or deduced costs are not reduced, although the labor is economized.

In these conditions, decision-making about the optimality of the machine parameter seems to be difficult. Therefore, in such cases, for the economic estimation of new agricultural practices and technology, it is worthwhile to use more common indicators of effectiveness and, in particular, the index characterizing the entropy change of the economic system (industrial process). Its effectiveness must be determined while considering the capital investments involved on acquiring and introducing new techniques and technology, expenditure reduction and labor consumption.

In each specific case, it is important to establish the accurate value of the earlier indicated limitation on capital-investments ($K \leqslant D$). The limiting value of specific capital investments, which can be used on acquiring new techniques, is determined by the condition (Fig. 85).

$$D < -\frac{\ln \dfrac{\overset{*}{z}_2}{\overset{*}{z}_1}}{\ln \dfrac{P_2}{P_1}} - 1.$$

It is assumed that as a result of the introduction of new techniques or technology in this process, the quantum of produce (productivity) increases from P_1 to P_2 with a simultaneous decrease in its cost to $\overset{*}{z}_2$.

Simultaneously, it must be kept in view that in due course, the relative effectiveness of the modified design of the new technology is lowered in view of its objective economic patterns, whereas the capital expenditure increases because of the increase and improvement of the machine leveler associated with the given design.

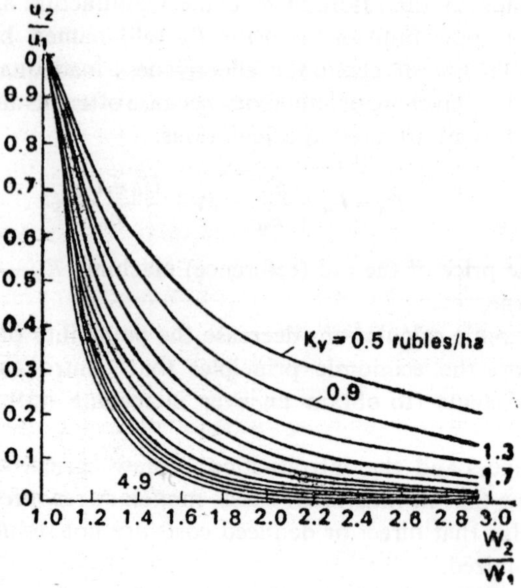

Fig. 85. Curves describing limiting correlations between the degree of expenditure reduction and the increase of productivity at various values of specific capital investments.

Broadly speaking, this is the prediction technique with the use of integrated curves of effectiveness. Comparing indexes of existing and planned machines from the respective prediction of integrated curves, through tests it is possible to reliably establish the directions of their parameter development to estimate the limitable achievable effectiveness of established mechanical-technological principles and also the possible ways and technical means for drawing at each given stage of time their parameters near to the optimum values.

To solve optimization problems with random values of factors, various methods of statistical programming the modeling of operations on the pattern of the Markoff processes, the Monte Carlo method, etc. have been developed and are being used [5, 13, 59, 62].

Graphic methods of optimization should be considered among the most acceptable since the calculation of the aim function's value takes up a lot of machine time for a single integration. The flow charts of the algorithm, for solving such problems, are shown in Figs. 18 and 19.

During testing, main practical significance should be given to experimental methods for solving optimization problems with the use of highly effective methods of an optimum plan for extremum experiments. The method of movement along the gradient, realized simultaneously on the computer and in the actual object, forms the basis of locating the extremum.

An analysis of the results of parameter optimization of some machine aggregates [58] from the criteria of maximum productivity or from minimum deduced costs showed that there is no principal difference in the conclusions obtained by using either of these two criteria. This follows from the important fact that the basic parameter optimum and the working regimes of many (including beet-picking) machines generally valid on the boundaries of limitation on working speed, carrying capacity, the power consumption of the engine and the design parameters (mass, overall dimensions, etc.). This considerably facilitates estimation of the degree of machine parameter optimality in test as well as their prediction for future planning.

The systems method for the estimation of machines and complexes, which is known under the name—the superposition of machines on a model economy—proves to be of great significance along with the study of the machine aggregate-external medium system. Until recently, widespread practical introduction of this valuable method was withheld due to the lack of an explicit systems presentation, insufficient informational objectivity regarding zonal conditions and an imperfect method for obtaining the solution with the help of a computer.

After accumulation, systematization and the provision of problems with objective information, certain programs were approved in tests for the systems analysis of machines complexes. However, in estimating complexes of field-crop cultivation and stock breeding machines, it is also necessary to carry out investigations on the modification of programs and the selection of the necessary information.

The peculiarity of solving a problem using the systems approach, lies in the fact that agricultural aggregates under test are considered to be elements of a complex system, namely, the agricultural production industry. Hence, the effectiveness of the application of new technological processes is determined not only by the data of the machine under estimation, but by the complicated interrelationships existing within these systems.

CHAPTER 12

Estimation of Working Conditions of Machine Operators*

37. AIMS AND METHODS OF ESTIMATING WORKING CONDITIONS OF MACHINE OPERATORS

Although a machine may offer favorable economic effectiveness or it may make use of the most advanced technology, it cannot be recommended for production unless it satisfies the requirements for safety standards and hygienics.

Moreover, machines not conforming to safety requirements are eliminated from testing and are not allowed even for laboratory tests.

Selection of a program for the estimation of dangerous and harmful factors of working conditions in each specific case is made depending on the application conditions of the machine (whether mobile or stationary) and its design peculiarities (presence, or absence, of a working place). In this, the standardized nomenclature of ergonomic indexes of produce quality is followed.

Normative values of basic indexes for safety and labor hygienics for machine operators are also regulated by industry and state standards and other normative-technical documents.

General methods of determination and estimation of working conditions indexes are also stipulated.

For evaluating working conditions, the indexes of experimental samples are compared with the normative as well as with indexes of the best machines of domestic and foreign production which accomplish similar work. The compared machines are tested under strictly comparable conditions: on one field, in similar meteorological conditions and in agrotechnical periods established for the testing zone. Fields, sections and material under processing (on which tests are conducted) are so selected that they characterize the testing zone in turns of relief and the condition of the cultivated crop, soil

*B.I. Ryabtsev has assisted in the preparation of this chapter.

and other conditions which affect the indexes of working conditions of the machine operator.

While accepting the machine for test, the presence and reliability of shields against dangerous units and mechanisms, the presence of signalling devices and instructions regarding safety techniques, the methods of and safety in conducting technical and technological servicing and operation, the meeting of requirements for fire preservation safety and the provision for and correctness of the constructive performance of auxiliary devices and equipment are determined by inspection, on-the-spot measurements and the testing of the machine in operation.

While testing in real operative conditions, the following hygienic indexes are determined: microclimatic conditions, dustiness in the breathing stratum of the machine operator, gas content, noise, high-frequency and low-frequency vibrations, illumination of the working zone, a comfortable working posture for the machine operator, ease and safety of technical and technological servicing, etc.

Microclimatic conditions include temperature, relative humidity and air speed, which are measured in tractor cabins, and the self-propelled chassis and also in the operating cabins of the stationary machines at three points: at the level of the head, chest and legs of the operator. The temperature and relative humidity are measured by thermometers and psychrometers, while air speed is calculated by anemometers. During winters, measurement is made from a closed cabin provided with its own heating and engine warming system. The dust content of the air is determined by the weight method with the use of analytic filters (a cartridge with a filter). Concentration of dust is calculated by the formula:

$$K_d = \frac{(m_2 - m_1)1000}{vt},$$

where K_d is the dust content in air, mg/m^3, m_1, m_2—the filter mass before and after the experiment, mg, v—the bulk recovery rate of the sample, 1/min and t—the duration of selection of the sample in min.

Air pollution by waste gases is determined from the quantity of carbonic oxide with the help of indicator gas analyzers (preliminary estimation) and titrometric gas analyzers of type TG-5.

On machines meant for transportation and the preparation and conveying of chemical weed and pest killers, gas concentration is determined at places where the machine operator is most likely to be present for fulfilling the technological process.

On machines meant for operation with ammonia and other highly volatile toxics, their concentration in the breathing zone of the machine operator is determined at the highest air temperature during the working period of the machine in the MTS zone. If the machine is meant for working with several toxics, then evaluation is made for the most poisonous chemicals. The number of selected and analyzed samples is not less than five.

The mean and maximum values of the contents of harmful chemicals in closed cabin, obtained in the most unfavorable conditions of machine operation, are taken as the basic criteria for the contents of harmful gases, pesticides and herbicides at the working place. These are reported in the machine's instruction manual.

In order to estimate noise, the general sound level on a standard frequency characteristic (scale of noise gage) and also the acoustic pressure level in the octave band with mean geometric frequencies of 63, 125, 250, 500, 1000, 2000, 4000, 8000 Hz are measured. The place of measurement should be protected from outside noise. Measurements are made by the noise gage with noise filters or a measuring guide with a microphone, an amplifier with noise filters, a recorder and device for registering the sound level.

Vibrations are measured on the seat, platforms, footboards, steering wheel, levers and control pedals in three mutually perpendicular directions: vertical, transverse and longitudinal. Accelerometers are rigidly clamped on the measured part. The vibration measuring apparatus should ensure measurement of vibrations within $\pm 3g$ in the frequencies range 0-250 Hz. For estimating vibration, the value of the vibration rate or its level is determined.

Forces on the control parts (steering wheel, levers, pedals) are measured when the machine accomplishes the most complicated technological operation under the maximum possible pressure on the control wheels. Measurements are carried out by tensiometric and dynamometric devices with a permissible error of not more than 4 percent.

Illumination is determined in horizontal and vertical planes directly near the machine, on the working parts that require constant observation and at a distance of 10 m from the machine.

The most comfortable working posture of the machine operator is estimated by the following characteristics: the design and location of the seat and control parts of the machine, the possibility of a free natural position of machine operator on the seat, permitting posture change during the process of working; a convenience in viewing the path of the moving aggregate, the section field under work and the operating part of the aggregate.

In stand testing, the following indexes are measured and estimated: the quality of vision and brakes and the longitudinal and transverse stability and strength of the cabins and frames.

The quality of vision is determined on a specially prepared horizontal platform 20×15 m, which is divided into squares with sides of 1 m. Estimation is carried out by light and shade and by panoramic methods and photography. The position of the eyes of the machine operator on the seat is considered as the center of vision. From this point, the area of vision and dark silhouettes on the ground are determined. In accordance with this, a scanning map is prepared on which the area of the visible part is located, and then the coefficient of vision is calculated.

The quality of brakes is estimated by the effectiveness of the main brake, the possibility of the 'gathering up' of the tractor complex and the effectiveness of the standing brakes.

The longitudinal and transverse static stabilities of the machines are characterized by the value of the limiting angles of stability. They are determined by inclining the machine along a longitudinal and transverse direction on a special rig up to the limiting angles, which are later measured. Besides this, angles of longitudinal and transverse stability can be determined by the calculation of the coordinate values of the center of gravity, on the longitudinal base and on the machine track.

Estimation of tractor cabins and frames for adequate safety protection of the driver from injuries in case the tractor topples over is determined for the hilly version of wheel and caterpillar tractors. A test is carried out by striking once on the cabin and by vertical pressure in the following sequence: stroke from the back, from the front, from the side, at the angle joint and testing on crushing. A pendulum in the form of a cube of mass of 2000 kg strikes on that part of the cabin which first touches the earth if the tractor topples over in the direction of the impact.

The center of gravity of the load and the line of striking should be in one horizontal plane, while the tractor is clamped to its initial position by wooden pieces or by tension ropes. The force of the strokes is selected according to the mass of the tractor from an earlier mapped nomogram. Crushing is achieved by static forces. The force pressure, applied on the roof of the cabin, should be uniformly distributed on all supports. This force is equal to twice the operational mass of the tractor. After running the machine, an estimation of the reliability and durability of the units and parts, ensuring safety for the working of the machine operators, is conducted in operational conditions.

A methodological estimation level of the working conditions during a test should conform to the stipulation and recommendations of the USSR State Standards on the estimation of the quality level of products from an ergonomical viewpoint. Estimation of harmful factors is conducted on a standardized product quality nomenclature of ergonomical indexes. Besides that, the system of standards pertaining to labor safety (SSBT) and to man-machine (SChM) have been introduced in testing.

The inclusion of safety indexes in the basic indexes regarding machine quality, the development of principles for the selection of a program for estimating the safety of various groups of machines, the development of modern methods of processing and estimating test results and a unified system of rigs and equipment for laboratories intended for estimating labor safety have played a vital role in setting up a service for safety estimation and for raising the test quality and authenticity of conclusions drawn from results on machine testing.

A unified system of devices and equipment for the estimation of safety foresees the use of standard industrial samples of domestic and foreign manufacture as well as the development and manufacture of special purpose devices meant for specific test conditions. In KubNIITiM, the rig for determining static stability the indicator for braking distance and the tensiometric devices for the measurement of forces on control parts have been developed. The development of devices for the evaluation of working conditions is also conducted at Baltic, Western, Central and Kirghiz machine testing station. Some original devices for the estimation of the physiochemical composition of air (Figs. 86, 87), and devices for the assessment of working place and work strain on machine operators have been developed in VNIIMOZh.

Fig. 86. Device for collecting dust samples during a shift.

Fig. 87. Device for collecting samples of toxic chemicals.

The schematic diagram of the device (USSR Patent No. 475531) for selecting dust during a shift is shown in Fig. 88. The device has a split housing 1, which has a changeable filtering element 13 between base 2 and cover 14 in the form of an annular cone with a hole in the smaller base 12 for the introduction of pipe 10. The free end of this pipe is placed in the zone where the dust content is studied. Filtering element 13 rests on a net 8 and the circular supporting pad 15 with its large base 11. The hollow of base 2 of the housing is separated from the filtering element by an hermetically sealed partition 3 to which the outlet pipe is connected with rotameter 6. The exit of the rotameter is connected to the evacuator or suction fan 4 with a microelectromotor 5. The device has a meter for adding the time of the sample collection in the form of a multivibrator 7 and an electromechanical pulse counter 9, whose circuits are interlocked with the supply circuit of electromotor 5.

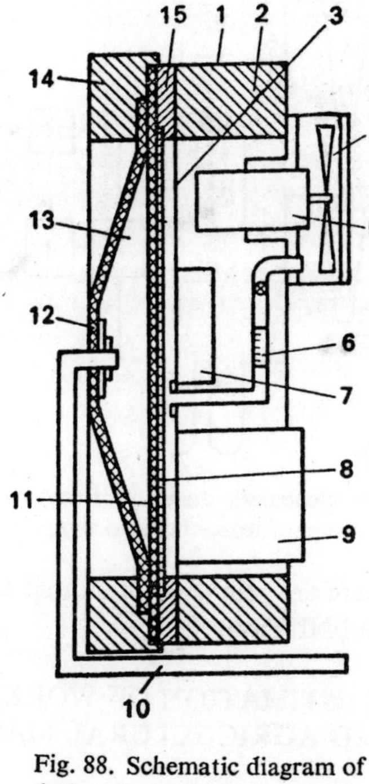

Fig. 88. Schematic diagram of device for selection of toxic chemicals during shift.

With this device, it is possible to obtain the integrated estimation of dust content during the shift which is very important for the estimation of machine operation in field conditions, where the dust content varies over a wide range depending on external conditions.

In testing seed disinfectants, sprayers and dust raisers, the basic index of working conditions is the content of the toxic chemical in the active zone. For estimation of this factor, a high accuracy device (USSR Patent No. 522448) has been developed, which permits the simultaneous selection of two samples in field conditions on stationary as well as on mobile objects.

The device consists of a discharge booster 1 (Fig. 89), a common exhaust chamber 2 for stabilizing the flow and eliminating the effect on the working regime of the switch-off device for aspiration for either of the channels; two parallel branches for aspiration lead into the chamber 2. Each of the indicated branches contain a protective net 3, a fluxmeter 4, a regulating valve 5 and an absorbing device 6 (two Polezhaev absorbers), connected in sequence in the path of aspiration. To account for the duration of selection of samples the device also contains the time counter 7 whose switch is connected with a circuit switch for supply of booster 1.

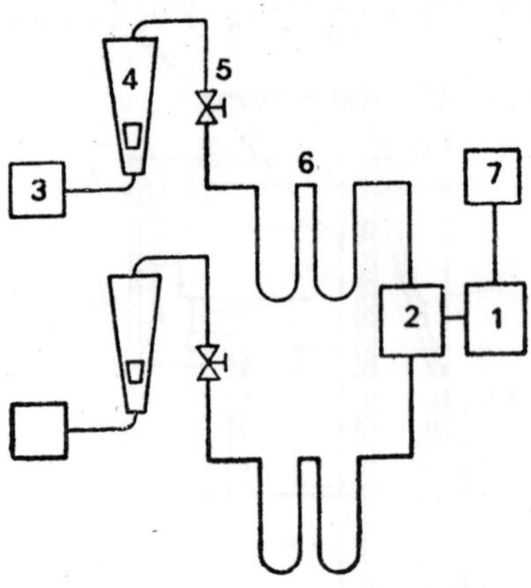

Fig. 89. Schematic diagram of device for selection of toxic-chemical samples.

The devices described here are being manufactured at the Central Machine-Testing Station and at KubNIITiM.

38. COMPREHENSIVE ESTIMATION OF WORKING CONDITIONS ON TRACTORS AND AGRICULTURAL MACHINES

An estimation of the quality level of the agricultural technological process from the hygienic viewpoint with respect to a large number of parameters and a range of contradictory requirements appears to be a complicated problem because the measured indexes do not provide sufficient information for decision-making.

For decision-making on the basis of each quality index, it is necessary to compare its level with the normative level; to estimate the degree of potential risk (weightage of the index); to take into account the degree of non-linearity in the influence of the index on the system as well as the existing domestic and foreign technoeconomic level.

The detailed analysis of a large number of diverse indexes is complicated and requires highly trained specialists. However, in such a case an incorrect decision is not excluded. In order to better justify the controlling decisions while estimating the quality of a product, it is necessary to evolve methods of formalizing decision-making. For example, there is the complex estimation method, which includes the following stages: selection of the nomenclature of the ergonomical indexes of quality, the base indexes $B_1, B_2, \ldots, B_n$, the veto level $C_1, C_2, \ldots, C_n$, determination of the coefficients of the weightage of quality indexes $a_1, a_2, \ldots, a_n$, the relative quality indexes $K_1, K_2, \ldots, K_n$ and the generalized quality index of produce $K_{gen} = K_1 + K_2 + \ldots + K_n$.

In the nomenclature of ergonomical indexes for complex estimation, only those indexes are included which appear in connection with the working of the said machine (Table 26). Risk factors are not included in the nomenclature because their presence in the machine eliminates the possibility of machine operation and index measurement.

Table 26

Estimated index	Mobile machines: equipped with a working place	Mobile machines: non-equipped with a working place	Stationary machines	Complexes
Air temperature	+	−	−	+
Noise	+	−	+	+
Vibration	+	−	−	−
Dust content	+	−	+	+
Toxic chemicals	+	−	+	+
Forces on control and regulating elements	+	+	−	−
Seat design	+	−	−	−
Control element design	+	−	+	+
Illumination	−	−	−	+
Convenience of servicing:				
Technical	+	+	+	+
Technological	−	+	+	+
Condition for control of technological process (worker's posture)	+	+	+	+
Controllability of aggregate (frequency of worker's movement)	+	+	+	+
Convenience in hinging and attaching machine	−	+	−	−
Device for control and warning devices	−	−	−	+
Physical loads	−	−	+	+
Specific index	+	+	+	+

The maximum permissible values are accepted as the basic value of the ergonomical indexes.

The veto represents the level of indexes beyond which the complex estimate of quality becomes meaningless. Selection of veto level depends on the purpose of the estimate, i.e. on the character of the decisions which are to be taken on the basis of the results of the given estimate. For example, the veto levels should be most rigid in the comprehensive estimate meant for decision-making in respect of the release of the machine for production. In these cases, the veto level for the basic indexes can differ from the base level only in the magnitude of the instrumental error in measurement. For example, if the noise base level is 85 dB, then the veto level can be at the level 86-87 dB. In methods of quality estimation of series production, the veto level can be fixed at the level achieved in international practices.

The most complicated stage in working out methods of comprehensive estimation is the determination of the weight factor of the quality index from the extent, the unfavourable impact of each of these indexes has on the operator.

The weight factor of indexes is determined with the help of active experimentation, using planning methods and the subsequent factor analysis of the obtained results [57].

Taking into account the experimental difficulties, it is not at all necessary to achieve high accuracy in the determination of weight factor of indexes because inevitable errors do cause inaccuracies of the second order in determining the generalized index [83]. The weight factor can also be determined by calculations and expert methods [3]. Comparison of the weight factor of ergonomical indexes, measured by experimental analytical and expert methods, has shown satisfactory convergence of the obtained results (Table 27).

Table 27

Index	Method of measuring weight factors		
	experimental	visual	analytic
Air temperature	21	16	14
Noise	13	18	17
Vibration	16	14	14
Dust	—	5	5

In order to obtain a generalized quality index, it is necessary to reduce the index levels $d_1, d_2, \ldots, d_n$, measured on the machine, to nondimensional, relative indexes with the help of analytical transformations. However, the degree of nonlinearity of the action of indexes on the operator should be accounted for.

For example, if it is permissible to double the load on certain control elements of the machine, then the same increase in the temperature or noise level is not permissible.

The relative quality indexes of vibration levels, noise, temperature, the frequency of working motions, dust concentration, toxic chemicals, etc., are determined by the following relation:

$$K_i = a_i\left(1 - \frac{d_i - B_i}{C_i - B_i}\right).$$

Calculation of the relative noise index is carried out on the general level of absolute decibels because it has been corrected for harmful effects. For vibration, the index is calculated relative to the maximum level for human resonance octave bands with a mean-geometric frequency of 4 or 8 Hz.

If the machine or its part is characterized by a set of equally significant parameters (for example, the dimensional characteristics of the seat), then the relative index is calculated as the ratio of the magnitude normative characteristics B_i to the magnitude of characteristics, not corresponding to normative X_i, by the formula $K_i = a_i(1 - X_i/B_i)$.

While calculating the relative index of the forces on the control elements, the element's frequency in using the control is considered to be the ratio of the number of working movements for controlling the machine, M_i, to the number of working movements of the given control element, l_i, in a minute in accordance with the following analytical expression:

$$K_i = a_i\left[1 - \sum_{i=1}^{M} \frac{l_i(d_i - B_i)}{M_i(C_i - B_i)}\right];$$

$$K_i = a_i\left[1 - \sum_{i=1}^{N} \frac{l_i X_i}{M_i B_i}\right].$$

For parameters that do not have a meteorological base (convenience of technical and technological servicing and working posture), estimation scale is assumed. For example, the scale of quantitative estimation of technical and technological servicing convenience takes the following form:

Scale of index	K_i
Convenient	a_i
Marginally inconvenient	$0.78a_i$
Inconvenient	0
Difficult	$0.40a_i$

So as to eliminate the possibility of the leveling of relative indexes, it is necessary to assume that for any value d_i, the magnitude of K_i will not be more than a_i. Hence for the sum of weights $\sum_{i=1}^{N} a_i = 1$, for the machine when the level of parameters satisfies the normative requirements, the generalized working condition index will be equal to one or zero if the indexes are at the veto level. In the case of partial fulfillment of requirements, K_{tot} happens to be less than one. This increases the clarity and information value of the generalized quality index because it practically shows the degree of correspondence to the normatives.

On a similar principle, the models of the comprehensive estimation of working conditions on mobile machines, both equipped and nonequipped with a working place, and stationary machines for control and experimental tests have been worked out. An example of the calculation of a partially generalized index of a working condition on the T-150 tractor for control tests is given in Table 28. In the case when the T-150 tractor is equipped with an air cooler, the index of working condition approaches the normative, which is proved by the fairly high level of the partial generalized index.

The development of methods of comprehensive estimation even permits a more accurate characterization of the differential quality of machine, the simplification of the process of decision-making and the increase of their validity on the basis of the qualitative characteristic of the achievement level of the optimum (base) values of the indexes considering their weightage (weight indexes).

The modern agricultural technological process, in spite of the operational, technological and ergonomical requirements should also conform to certain esthetic requirements, because perfection of the machine in the modern concept is not limited to utilitarian qualities alone. Only in those cases when the product is designed and manufactured with consideration of the human factor, i.e. considering the esthetic aspect, there appears the corresponding esthetic or psychological attitude of the operator toward the machine, and conditions are created for obtaining the maximum economic effect by the careful attitude of the operator toward the machine, which promotes the increase of its durability and reliability in operation.

The shape of the machine has the maximum effect on the esthetic qualities of the machine because it alone initiates the visual perception of workin tools in the human being.

The machine is considered esthetically perfect if its operational quality and shape are organically blended.

Considering the large variety of agricultural technological process is in practice, various combinations of functional and esthetic requirements are possible. However, the most perfect and suitable form can correspond only to the most useful and worthwhile machine products. The perfect form of the product not only acts as a measure of the value of the machine, but also acts as a characteristic of its suitability and usefulness only in the case when it reflects the value of the product to society.

Expert estimation of the esthetic level of the machine is carried out by a

Table 28

Index	Weightage, a_i	Base level index for B_{ij}	Veto level, C_{ij}	Value of index on test data, d_{ij}	Relative quality Index, K_{ij}	Value of decrease of partial generalized index, %	Partial quality index, K_{2i}
Air temperature in cabin	0.16	28	31	27.6	1	0	0.16
Noise	0.14	87	87	86	1	0	0.14
Vertical vibration	0.13	0.1	0.2	0.14	0.96	1	0.12
Frequency of working movement	0.10	25	50	26	0.96	1	0.09
Working posture	0.09	1	4	1	1	0	0.09
Convenience of technical movements	0.10	1	4	2	0.78	2	0.08
Load on control elements	0.08	—	—	Bi=9	0.91	1	0.07
Design of seal	0.08	—	—	Xi=2	0.77	2	0.06
Carbon monoxide	0.04	21	21	13.7	1	0	0.04
Dust content	0.04	15	15	8.3	1	0	0.04
Design control elements	0.04	—	—	Bi=9 Xi=2	0.77	1	0.03

Partial generalized index $K_{2j}=0.92$.

comparison of the sample with the stereotype and by explaining the correspondence of the product form to its contents and function. The composition as an esthetic characteristic of the organizational system of the elements of form and the contents and product compilation are estimated by an expert method. The volume space structure, tectonics, proportionality, rhythm, color of the product and other indexes may then be estimated.

For the solution of these problems, the methods of quantitative estimation of the esthetic level of machines were worked out and introduced at VNIITE', which permits formalization of the process of decision making of the machine under test.

Analysis of the evolved theoretical assumptions, the partial methods of standards and also the modern theory of complicated systems leads to the following main directions of introducing methods of systems investigation in the testing practice:

1. Evolving general formalized principles of the analysis of multipurpose machine ergodic complexes, the methods of formalizing the functioning processes of the system elements and their interaction in the process of real production;

2. Evolving universal mathematical models of working operation and the standardization of machine behavior description and complexes as comprehensive systems; the formation and use of data banks on zonal conditions and on typical system models;

3. Wide introduction of and improvement in the methods of statistical (numerical) modeling of working processes, the development of a complex of machine algorithms of structural analysis of multipurpose machine systems under investigation, the estimation of their effectiveness by modeling methods on the computer in the zonal aspect; and

4. A further development and adoption of scientific prediction methods of various machine and complex characteristics as well as the directions of the development of designs and machine complexes, etc.

Bibliography

1. Adler, Yu.P. 1969. Vvedenie v planirovanie e'ksperimenta (Introduction to Design of Experiment). Metallurgizdat, Moscow.
2. Adler, Yu.P., E.V. Markova and Yu.V. Granovskii. 1976. Planirovanie e'ksperimenta pri poiske optimal'nykh uslovii (Design of Experiment for Optimum Conditions). Nauka, Moscow.
3. Azgal'dov, G.G. and E'.P. Raikhman. 1973. O kvalimetrii (Qualimetry). Izd-vo standartov, Moscow.
4. Akoff, R.A. and F. E'mmeri. 1974. O tseleustremlennykh sistemakh (Meaningful Systems). Sovetskoe Radio, Moscow.
5. Bellman, R. and O. Dreifus. 1965. Prikladnye zadachi dinamicheskogo programmirovaniya (Problems Relating to Dynamic Programing). Mir, Moscow.
6. Bendat, J. and A. Pirsol. 1974. Izmerenie i analiz sluchainykh protsessov (Measurement and Analysis of Random Processes). Mir, Moscow.
7. Biryukov, B.V. 1974. Kibernetika i metodologiya nauki (Cybernetics and Methodology of Science). Nauka, Moscow.
8. Box, J. and G. Jenkins. 1974. Analiz vremennykh ryadov, prognoz i upravlenie (Analysis of Time Series, Prediction and Control). Mir, Moscow.
9. Borodachev, N.A. 1960. Osnovnye voprosy teorii tochnosti proizvodstva (Fundamentals of the Theory of Production Accuracy). Izd-vo AN SSSR, Moscow-Leningrad.
10. Beri, V.V. 1972. Issledovanie i razrabótka mekhanizirovannogo protsessa izvlecheniya iz pochvy kornei sakharnoi svekly (Investigation and Development of Mechanized Processes of Extraction of Beet Roots from Soil). Avtorefrat na soiskanie uchenoi stepeni kandidata tekhnicheskikh nauk. USKhA, Kiev.
11. Beri, V.V., A.Ya. Dorogovtsev and L.V. Pogorelyi. 1973. Opredelenie kharakteristiki odnogo nestatsionarnogo protsessa (Determination of Characteristics of an Unsteady Process). In *Vychislitel'naya i Prikladnaya Matematika*, No. 19, Izd-vo KGU, Kiev.

12. Budko, A.I. 1976. Analiz metodov ispytanii sel'skokhozyaistvennoi tekhniki (Analysis of Test Methods of Agricultural Technological Processes). Obzornaya informatsiya TsNIITE'I V/O Soyuzsel'khoztekhnika, Moscow.
13. Buslenko, N.P. 1978. Modelirovanie slozhnykh sistem (Modeling of Complicated Systems). Nauka, Moscow.
14. Val'd, D. 1960. Posledovatel'nyi analiz (Sequential Analysis). Fizmatgiz, Moscow.
15. Vasilenko, P.M. 1975. Postroenie matematicheskikh modelei mashinnykh agregatov (Construction of Mathematical Models of Machine Aggregates). Mekh. i E'lektr. Sots. Sel. Khoz-va, No. 11.
16. Wentzel, E.S. 1964. Teoriya veroyatnostei (Theory of Probability). Nauka, Moscow.
17. Wiener, N. 1968. Kibernetika (Cybernetics). Sovetskoe Radio, Moscow.
18. Glushkov, V.M. 1974. Vvedenie v ASU (Introduction to Automated Control Systems). Kiev.
19. Glushkov, V.M. 1964. Vvedenie v kibernetiku (Introduction to Cybernetics). Kiev.
20. Gmoshinskii, V.G. and G.I. Fliorent. 1973. Teoreticheskie osnovy inzhenernogo prognozirovaniya (Theoretical Principles of Engineering Prediction). Nauka, Moscow.
21. Gozman, G.I., A.Ya. Dorogovtsev and V.V. Beri. 1974. Planirovanie e'ksperimenta pri ogranichennom vremeni nablyudeniya (Planning of Experiment with Limited Observation Time). Mekh. i E'lektr. Sots. Sel. Khoz-va.
22. Golubentsev, A.N. 1967. Termodinamika proizvodstvennogo protsessa (Thermodynamics of Industrial Process). Tekhnika, Kiev.
23. Goncharenko, F.T. and E'.A. Sidorshin. 1960. Tekhnologicheskie predposylki avtomatizatsii protsessov v polevodstve (Technological Assumptions for Automation of Processes in Field-cultivation). Vestnik sel'skokhozyaistvennoi nauki, No. 12.
24. Dronova, N.F. and D.M. Kaun. 1968. K issledovaniyu statisticheskikh svoistv mikroprofilei polei, tipichnykh dlya raboty gusenichnykh traktorov (Investigation on Statistical Properties of Microprofiles of Fields, Typical for the Working of Caterpillar Tractors). In *Voprosy Mekhanizatsii i E'lektrifikatsii Sel'skokhozyaistvennogo Proizvodstva/ VNIIME'SKh*, No. 11. Izd-vo Rostovskogo Universiteta.
25. Zheligovskii, V.A. 1960. E'lementy teorii sel'skokhozyaistvennykh mashin i mekhanicheskoi tekhnologii sel'skokhozyaistvennykh materialov (Elements of the Theory of Agricultural Machines and Mechanical Technology of Agricultural Materials). SKHI. Tbilisi.
26. Zade, L. 1966. Ponyatie sostoyaniya v teorii sistem (Idea of State in Systems Theory). In *Obshchaya Teoriya Sistem*. Mir, Moscow.

27. Zaidel', A.N. 1968. E'lementarnye otsenki oshibok izmerenii (Preliminary Estimation of Errors in Measurement). Nauka, Leningrad.
28. Ivanova, V.S. and V.F. Terent'ev. 1975. Priroda ustalosti metollov (Nature of Fatigue in Metals). Mashinostroenie, Moscow.
29. Ivanov, V.V. 1975. Metody algoritmizatsii nepreryvnykh proizvodstvennykh protsessov (Methods of Algorithmization of Continuous Industrial Processes). Nauka, Moscow.
30. Iofinov, S.A. 1974. Kompleksnaya otsenka rezul'tatov ispytaniya mashin (Complex Estimation of Results of Tests on Machines). Mekh. i E'lektr. Sots. Sel. Khoz-va, No. 7.
31. Iofinov, S.A. and O.V. Kul'kov. 1968. Ob opredelenii e'kspluatatsionnykh pokazatelei raboty MTA po kharakteristikam dvigatelya (Determination of Operational Indexes of Working of MTA with Respect to Engine Characteristics). In *Mekhanizatsiya Sel'skokhozyaistvennogo Proizvodstva*, vol. 119, No. 3, Leningrad.
32. Kardashevskii, S.V. 1977. Puti uskoreniya sozdaniya i vnedreniya novoi sel'skokhozyaistvennoi tekhniki (Ways of Rapid Development and Introduction of New Agricultural Technological Processes). In *Puti Uskoreniya i Vnedreniya Novoi Tekhniki*. VISKhOM, Moscow.
33. Kardashevskii, S.V. 1972. Printsipy razrabotki i vnedreniya sistemy upravleniya kachestvom produktsii v otrasli (Principles of Development and Introduction of Systems for Quality Control of Product in Industry). Traktory i sel'khozmashiny, No. 12.
34. Kardashevskii, S.V. 1977. Ispytaniya v sisteme upravleniya kachestvom mashin (Testing in the System of Quality Control of Machines). Mekh. i E'lektr. Sots. Sel. Khoz-va, No. 11.
35. Kardashevskii, S.V. 1973. Vysevayushchie Ustroistva posevnykh mashin (Sowing Devices of Seeding Machines). Mashinostroenie, Moscow.
36. Kardashevskii, S.V. and L.I. Komarov. 1972. Ispytaniya sel'skokhozyaistvennoi tekhniki za rubezhom (Testing of Agricultural Technological Processes in Foreign Countries). Obzor TsNIITE'I V/O Soyuzsel'khoztekhnika, Moscow.
37. Kassandrova, O.N. and V.V. Lebedev. 1970. Obrabotka rezul'tatov izmerenii (Processing the Results of Measurements). Nauka, Moscow.
38. Katsygin, V.V. 1964. Osnovy teorii metodiki vybora optimal'nykh rabochikh skorostei i zakhvatov sel'skokhozyaistvennykh mashin (Theory of the Methodology for Selecting Optimum Working Speeds and Claws for Agricultural Machines). Voprosy sel'skokhozyaistvennoi tekhniki, vol. 13. Minsk.
39. Kirtbaya, Yu.K., L.V. Pogorelyi and V.P. Maksimchuk. 1970. Veroyatnostno-statisticheskie predposylki model-irovaniya proizvodstvennykh protsessov (Probabilistic and Statistical Hypotheses for Modeling of Industrial Processes). Vestnik sel'skokhozyaistvennoi nauki, No. 10.

40. Kliland, D. and V. King. 1974. Sistemnyi analiz i tselevoe upravlenie (Systems Analysis and Aim Control). Sovetskoe Radio, Moscow.
41. Kukta, G.M. 1964. Ispytaniya sel'skokhozyaistvennykh mashin (Testing of Agricultural Machines). Mashinostroenie, Moscow.
42. Korbut, L.A. and A.T. Korobeihikov. 1976. Zadachi MIS v uskorenii nauchno-tekhnicheskogo progressa v mekhanizatsii rabot v zhivotnovodstve i kormoproizvodstve (Role of MTS in Accelerating Scientific-technical Progress in Mechanization of Stock-breeding and Fodder Production). In *Novoe v Metodakh Ispytaniya Tracktorov i Sel'skokhozyaistvennykh* Moscow, No. 19.
43. Lebeg, A. 1960. Ob izmerenii velichin (On Measuring Parameters). Uchpedgiz, Moscow.
44. Livshits, N.A. and V.N. Pugachev. 1963. Veroyatnostnyi analiz sistem avtomaticheskogo upravleniya (Probability Analysis of Automated Control Systems). Sovetskoe Radio, Moscow.
45. Lisichkin, V.A. 1972. Teoriya is praktika prognostiki (Theory and Practice of Prediction). Nauka, Moscow.
46. Lur'e, A.B. 1970. Statisticheskaya dinamika sel'skokhozyaistvennykh agregatov (Statistical Dynamics of Agricultural Aggregates). Kolos, Leningrad.
47. Makovetskii, O.A. 1975. Vliyanie gustoty i ravnomernosti razmeshcheniya rastenii na produktivnost' sakharnoi svekly pri malykh normakh vyseva semyan v usloviyakh pravoberezhnoi lesostepi UkrSSR (Effect of Density and Uniformity of Spacing of Plants on Productivity of Beet Root for Small Sowing Norms with Littoral Forest Steppe in the Ukraine). Avtoreferat na soiskanie Uchenoi stepeni kandidata sel'skhozyaistvennykh nauk. VNIS, Kiev.
48. Markova, E.V. and A.N. Lisenkov. 1973. Planirovanie e'ksperimenta v usloviyakh neodnorodnostei (Design of Experiment in Inhomogeneous Conditions). Nauka, Moscow.
49. Mesarovich, M., D. Mako and I. Takahara. 1973. Teoriya ierarkhicheskikh mnogourovnevykh sistem (Theory of Multilevel Hierarchical Systems). Mir, Moscow.
50. Mel'nikov, S.V., V.P. Aleshkin and P.M. Roshchin. 1972. Planirovanie e'ksperimenta v issledovaniyakh sel'skokhozyaistvennykh protsessov (Design of Experiment during Investigations on Agricultural Processes). Kolos, Moscow.
51. Mitkov, A. and S.V. Kardashevskii. 1978. Statisticheskie metody v sel'skokhozyaistvennom mashinostroenii (Statistical Methods in Agricultural Machine Building). Mashinostroenie, Moscow.
52. Mitropol'skii, A.K. 1971. Tekhnika statisticheskikh vychislenii (Methods of Statistical Calculations). Nauka, Moscow.
53. Nalimov, V.V. 1971. Teoriya e'ksperimenta (Theory of Experiment). Nauka, Moscow.

54. Novikov, O.A. and S.I. Petukhov. 1969. Prikladnye voprosy teorii massovogo obsluzhivaniya (Applied Problems of Queuing Theory). Sovetskoe Radio, Moscow.
55. Novoe v metodakh ispytanii sel'skokhozyaistvennoi tekhniki (New Approaches in the Test Methods for Agricultural Technological Process). Nos. I-10. TsNIITE'I, Moscow, 1963-1974.
56. Okun', Ya. 1974. Faktornyi analiz (Factor Analysis). Statistika, Moscow.
57. Plokhinskii, N.A. 1970. Biometriya (Biometrics). Izd-vo MGU, Moscow.
58. Pogorelyi, L.V. 1974. Tekhnologicheskie i tekhnicheskie osnovy sovershenstvovaniya mekhanizirovannykh protsessov uborki sakharnoi svekly (Technological and Technical Developments in Mechanized Beet Harvesting Processes). Avtoreferat Dissertatsii.
59. Pogorelyi, L.V. and V.V. Brei. 1976. Primenenie metodov sistemnogo analiza pri ispytaniyakh sel'skokhozyaistvennoi tekhniki (Application of Methods of Systems Analysis for Testing of Agricultural Technological Process). Obzornaya informatsiya TsNIITE'I V/O Soyuzsel'khoztekhnika, Moscow.
60. Pogorelyi, L.V. and V.P. Maksimchuk. 1968. Veroyatnostnyi metod opredeleniya optimal'nykh parametrov agregatov (Probability Method of Determination of Optimum Aggregate Parameters). Mekh. i E'lektr. Sots. Sel. Khoz-va, No. 4.
61. Pogorelyi, L.V. and M.I. Yadrenko. 1972. Nekotorye modeli protsessov, opisyvayushchikh funktsionirovanie mobil'nykh agregatov (Some Models Describing the Functioning of Mobile Aggregates). In *Vychislitel'naya i Prikladnaya Matematika*, No. 18, Izd-vo KGU, Kiev.
62. Pogorelyi, L.V. and V.V. Brei. 1976. Metody prognozirovaniya e'ffektivnosti sel'skokhozyaistvennoi tekhniki na stadii ispytanii (Methods of Predicting Effectiveness of Agricultural Technological Process at the Test Stage). Obzornaya informatsiya TsNIITE'I V/O Soyuzsel'khoztekhnika, Moscow.
63. Rokitskii, P.F. 1973. Biologicheskaya statistika (Biological Statistics). Vysheishaya shkola. Minsk.
64. Roslyakov, V.P. 1962. Opredelenie statisticheskikh kharakteristik usiliya na kryuke sel'skokhozyaistvennogo traktora (Determination of Statistical Characteristics of Load on the Drawbar of an Agricultural Tractor). Doklady TSKhA, No. 73. Moscow.
65. Rukovodstvo po primeniyu latinskikh planov pri planirovanii e'ksperimenta s kachestvennymi faktorami (Handbook on the Use of Latin Plans for Design of Experiment with Qualitative Factors). Pod red. E.V. Markova. Chelyabinsk, 1971.
66. Rytov, S.M. 1966. Vvedenie v statisticheskuyu radiofiziku (Introduction to Statistical Radiophysics). Nauka, Moscow.

67. Sveshnikov, A.A. 1968. Prikladnye metody teorii sluchainykh funktsii (Applied Methods of the Theory of Random Functions). Nauka, Moscow.
68. Sveshnikov, A.A. 1972. Osnovy teorii oshibok (Principles of Theory of Errors). Izd-vo LGU, Leningrad.
69. Smirnov, N.V. and I.V. Dudin-Barkovskii. 1969. Kurs teorii veroyatsnostei i matematicheskoi statistiki (Course on the Theory of Probability and Mathematical Statistics). Nauka, Moscow.
70. Sturis, A.I. 1971. Ispytaniya sel'skokhozyaistvennoi tekhniki na ustalostnuyu prochnost' (Testing of Fatigue Strength of Agricultural Technological Process). Obzornaya informatsiya TsNIITE'I V/O Soyuzsel'khoztekhnika, Moscow.
71. Tarakanov, K.V., A.A. Ovcharov and A.N. Taryshkin. 1974. Analiticheskie metody issledovaniya sistem (Analytical Methods of Investigation of Systems). Sovetskoe Radio, Moscow.
72. Terekhov, A.P. 1971. Tsifrovoe modelirovanie mekhanizirovannykh protsessov sel'skokhozyaistvennogo proizvodstva (Digital Modeling of Mechanized Processes of Agricultural Production). Mashinostroenie, Moscow.
73. Trapeznikov, V.A. 1966. Avtomaticheskoe upravlenie i e'konomika (Automatic Control and Economics). Avtomatika i telemekhanika, No. 1.
74. Pedorov, V.V. 1971. Teoriya optimal'nogo e'ksperimenta (Theory of Optimization of Experiment). Nauka, Moscow.
75. Feller, V. 1967. Vvedenie v teoriyu veroyatnostei i ee prilozheniya (Introduction to Theory of Probability and its Application). Vols. 1 and 2. Mir, Moscow.
76. Fini, D. 1970. Vvedenie v teoriyu planirovaniya e'ksperimentov (Introduction to the Theory of Design of Experiments). Nauka, Moscow.
77. Kharman, G. 1972. Sovremennyi faktornyi analiz (Modern Factor Analysis). Statistika, Moscow.
78. Khedi and Dillon. 1965. Proizvodstvennye funktsii v sel'skom khozyaistve (Industrial Functions in Agriculture). Nauka, Moscow.
79. Hicks, C. 1967. Osnovnye printsipy planirovaniya e'ksperimenta (Principles of Design of Experiment). Mir, Moscow.
80. Khinchin, A.Ya. 1963. Raboty po matematicheskoi teorii massovogo obsluzhivaniya (Works on Mathematical Queuing Theory). Fizmatgiz, Moscow.
81. Khorafas, D.N. 1967. Sistemy i modelirovanie (Systems and Modeling). Mir, Moscow.
82. Kh'yutson, A. 1971. Dispersionnyi analiz (Dispersion Analysis). Statistika, Moscow.
83. Khrunov, E.V., L.S. Khachatur'yants, V.A. Popov and A. Ivanov.

1974. Chelovek-operator v kosmicheskom polete (Man as an Operator in Space Flights). Mashinostroenie, Moscow.

84. Chetyrkin, E.M. 1975. Statisticheskie metody prognozirovaniya (Statistical Methods of Prediction). Statistika, Moscow.

85. Chuev, Yu.V., Yu.B. Mikhailov and V.I. Kuz'min. 1975. Prognozirovanie kolichestvennykh kharakteristics protsessov (Prediction of Quantitative Characteristics of Processes). Sovetskoe Radio, Moscow.

86. Shabranskii, V.A. 1975. Otsenka tochnosti vyseva i razmeshcheniya vskhodov svekly po obobshchennym pokazatelyam (Estimation of Accuracy of Sowing and Placing of Beet Seedlings of Generalized Indexes). Mekh. i E'lektr. Sots. Sel. Khoz-va, No. 4.

87. Shatunovskii, G.M. 1971. Obshchie printsipy prognozirovaniya i otsenki urovnya mekhanizatsii sel'skokhozyaistvennykh rabot (General Principles of Prediction of Estimation of Level of Mechanization of Agricultural Operations). In *Analiz i Otsenka E'ffektivnosti Konstruktsii Sel'skokhozyaistvennykh Mashin*. Rostov.-Don.

88. Shepovalov, V.D. 1975. Avtomaticheskaya optimizatsiya rezhimov raboty agregatov (Automatic Optimization of Working Regimes of Aggregates). Mekh. i E'lektr. Sots. Sel. Khoz-va, No. 1.

89. Shenk, Kh. 1972. Teoriya inzhenernogo E'ksperimenta (Theory of Engineering Experiment). Mir, Moscow.

90. Sheffe, G. 1963. Dispersionnyi analiz (Dispersion Analysis). Fizmatgiz, Moscow.

91. E'ikoff, P. 1975. Osnovy identifikatsii sistem upravleniya (Principles of Identification of Control Systems). Mir, Moscow.

92. Yang, E'. 1974. Prognozirovanie nauchno-tekhnicheskogo progressa (Prediction of Scientific-technical Progress). Progress, Moscow.

93. Yankovskii, I.E. 1978. Sistemnyi printsip ispytaniya mashino-traktornykh agregatov (Systems Principle for Testing of Machine-tractor Aggregates). Mekh. i E'lektr. Sots. Sel. Khoz-va, No. 3.